Broadband Access Networks

TELECOMMUNICATIONS TECHNOLOGY AND APPLICATIONS SERIES

Series editor:
Stuart Sharrock, Consultant
The Barn, Sugworth Lane, Radley, Abingdon,
Oxon, OX14 2HX, UK

This series covers research into and the development and application of a wide range of techniques and methods used in telecommunications. The industry is undergoing fundamental change under the combined impact ot new technologies, deregulation and liberalization, and the shift towards a service oriented philosophy. The field of communications continues to converge, encompassing all of the associated technologies of computing, networking, software, broadcasting and consumer electronics. The series presents this material in a practical and applications-based manner which equips the reader with the knowledge and tools essential for an engineer working in the industry.

Titles available

1. **Coherent Lightwave Communications Technology**
   Edited by S. Shimada

2. **Network Management**
   Concepts and tools
   Edited by ARPEGE Group

3. **The Informatics Handbook**
   A guide to multimedia communications and broadcasting
   S. Fist

4. **Mobile Communications Safety**
   Edited by N. Kuster, Q. Balzano and J.C. Lin

5. **The ISDN Subscriber Loop**
   N. Burd

6. **High Speed Optical Communications**
   R. Sabella and P. Lugli

7. **Handbook of Data Communications and Networks**
   W. Buchanan

8. **Data Mining Techniques in Speech Synthesis**
   R. Damper

9. **Intelligence and Power in Global Information Systems**
   W. Whyte

10. **Broadband Access Networks**
    L.A. Ims

Broadband Access Networks

Introduction Strategies and Techno-economic Evaluation

Edited by

Leif Aarthun Ims

Telenor Research and Development,
Kjeller,
Norway

SPRINGER-SCIENCE+BUSINESS MEDIA, B.V.

First edition 1998

Thomson Science is a division of International Thomson Publishing

ISBN 978-0-412-82820-1 ISBN 978-1-4615-5795-1 (eBook)
DOI 10.1007/978-1-4615-5795-1

A catalogue record for this book is available from the British Library

Contents

Preface

The access network is expected to be one of the major battlegrounds of telecommunications network operators, since upgrades of the existing narrowband access network will be the critical factor in supplying multimedia broadband services in a competitive market. The future broadband access network architecture needs to be flexible enough to efficiently support the provision of a full set of broadband and narrowband services with a wide range of capacity demands. A wide range of broadband access technologies are available. Furthermore, the key issues in the upgrading of the very cost sensitive access network are financial as well as technological, both for incumbent and new entrant operators. Thus, in order to identify minimum-risk introductory strategies the economic viability of access network broadband upgrades needs to be carefully assessed.

However, despite the definite need for techno-economic evaluations, very few books have been published in this field. One of the reasons might be that broadband access network upgrading only very recently gained wide recognition as a key challenge for broadband delivery. Secondly, this kind of strategic work and these studies tend to be considered rather sensitive by operators, and thus both results and methodologies are not usually readily available. Thirdly, the work reported in this book in many respects was a major pioneering effort, which quite ambitiously aimed at modelling the whole life-cycle costs and revenue streams of access network upgrades, as opposed to several other efforts, which often are limited to pure investment cost comparisons.

The material presented in the book is mainly the result of work performed during the period from 1992-1996 by the project RACE (research for advanced communications in Europe) 2087/TITAN (tool for introduction strategies and techno-economic evaluation of access network). TITAN was carried out within the second framework programme for telecommunications research in the European Union, RACE II

One of the main objectives of the TITAN methodology and the accompanying application of the tool is to establish guidelines for introduction of advanced communication networks in a competitive multiservice environment. The work and the guidelines here presented will hopefully generate valuable information and background material for development of broadband access networks. In that respect the intention has been to provide material relevant both to telecommunications network operators, equipment manufacturers business consultants and telecommunications regulatory bodies, including professionals like managers, engineers, equipment developers, research scientists and strategic network planners.

The authors gratefully acknowledge the support of our employers and the European Commission, who have provided the financial support for this work.

The authors owe a debt of gratitude to a number of our colleagues and friends who have supported the work through advice and numerous discussions. In particular, we would like to express our sincere debt of gratitude to the members of the TITAN project, some of whose contributions during the project proved crucial for the results accomplished: Ade Ajibulu (Analysys), L. Filipe Dias (University of Aveiro), Jean-Luis Dufau-Richet (SAT), Raimondo Diaz de la Iglesia (Telefonica), Jan van Hoecke (Jan Van Hoecke International), Dennis Joram (CNET-France Telecom), Matti Kalervo (Nokia), Natalie Kerteux (MET), Jens Kraushaar (Ericsson), Guido Mossotto (CSELT), Eddie Murphy (Analysys), Mario Jorge Oliveira (University of Aveiro), Carlos Rabadão (University of Aveiro), Ingunn Schjetne (Telenor), Sigurd Sivertsen (Telenor), Andre Socard (SAT) and Torge Øverli (Telenor).

Some of the material in the book is based on results achieved in the EURESCOM (European institute for research and strategic studies in telecommunications) projects P306 (Access network evolution and preparation for implementation) and P413 (Optical networking). The contributions from the participants in these projects are gratefully acknowledged.

Special acknowledgements are given to Erling Hansen for his work on the questionnaires in the Delphi survey.

We would like to thank Åse Aardal and Britt Kjus, whose artistic effort in creating several of the line drawings is appreciated.

We are indeed indebted to Judith Gordon, who carefully read the whole manuscript, and through numerous suggestions improved the consistency and the english language in the book. The book has been co-authored by some 20 authors, of which none has english as their mother tongue.

In addition, we would like to express or sincere thanks to the editor of Chapman & Hall, Dave Hatter, for his continuous encouragement.

Finally, I would like to mention my wife Anne Berit and our children Julie and Andreas, who continuously encouraged the demanding work and showed patience throughout the editorial process of this book.

Leif Aarthun Ims
Oslo

Contributors

Stefano Balzaretti (2,4)
CSELT,
Turin, Italy.

Marco De Bortoli (2,4)
CSELT,
Turin, Italy.

Eduardo de Castro (7)
University of Aveiro,
Aveiro, Portugal.

Marc Drieskens (5,6,11)
Jan Van Hoecke International,
Brussels, Belgium.

Manuel de Oliveira Duarte (7)
University of Aveiro,
Aveiro, Portugal.

Bjørn Erik Eskedal (12)
Telenor Mobil AS,
Oslo, Norway.

Umberto Ferrero (2,4,13)
CSELT,
Turin, Italy.

Nikolaus Gieschen (5,11)
Deutsche Telekom AG,
Technologiezentrum Darmstadt,
Berlin, Germany.

Rolf B. Haugen (12)
Telenor Research and Development,
Kjeller, Norway.

Frank Fleuren (5,6)
KPN Research,
Leidschendam, The Netherlands.

Leif Aarthun Ims
(1,4,11,13,14,16,18,19,20)
Telenor Research and Development,
Kjeller, Norway.

Markku Lähteenoja (2,5,13,17)
Telecom Finland Research Centre,
Helsinki, Finland.

Sotirios Markatos (1,5,11)
EUROCOM Expertise S.A.,
Athens, Greece.

Jari Mononen (2,5,13,17)
Telecom Finland Research Centre,
Helsinki, Finland.

Dagfinn Myhre (13,14,16,18,20)
Telenor Research and Development,
Kjeller, Norway.

Borgar Tørre Olsen
(1,4,8,11,12,13,14,16,18,19,20)
Telenor Research and Development,
Kjeller, Norway.

Mauro Ravera (2,4)
CSELT,
Turin, Italy

Kjell Stordahl (1,3,4,9,11,14,16,19)
Telenor Network AS,
Oslo, Norway.

Markku Tahkokorpi (5,6,10,11,15)
Nokia Research,
Helsinki, Finland.

Ilari Welling (5,6,15)
Nokia Research,
Helsinki, Finland.

Alcibiade Zaganiaris (1,11,13,14)
CNET-France Telecom,
Lannion, France.

Numbers in parentheses indicate the chapters which the author has contributed to.

Introduction 1

Alcibiade Zaganiaris, Kjell Stordahl,
Borgar Tørre Olsen, Sotirios Markatos, Leif Aarthun Ims

The telecommunications sector has experienced dramatic technology changes during the last two decades. New technologies, like the introduction of optical fibres, the explosion in mobile communications, and the development of satellite personal services are revolutionising the whole field of telecommunication services. The new services and technologies coupled with the extensive market deregulation, for new and even the traditional voice services, are already shaking up the market. New service providers entering into the marketplace is challenging the established service providers and threatening even the core business of traditional public network operators (PNOs). The deregulation wind, which started to blow in the late eighties, and nowadays global competition, as well as international co-operation play together a prominent role in gradually changing the environment, in which meeting customers' needs is considered the key to success and survival.

On the other hand, telecommunication networks and services constitute the backbone in the emerging information society. Indeed, almost all recent technological and organisational innovations are dependant on and sensitised by easy access to services provided by telecommunication networks. The *information superhighways,* like the motorways in the fifties, electricity at the beginning of the century, railways in the nineteenth century, are becoming the vital and prerequisite instruments for economic growth in modern society.

In order to cope with these new developments and trends, the PNOs are forced to establish new strategies for upgrading their network technologies to facilitate the emerging new services and meet the real information and communication needs of the customers. If they fail to meet these needs, new actors in the marketplace will satisfy them, threatening the dominant role of the PNOs over the recent years. In contrast, the new players in the deregulated telecommunication market also need to develop specific strategies if they are going to capture a profitable share of the new market, through development of new applications and services, more flexible tariff structures, and/or through co-operation and strategic alliances. Thus, one thing is certain; new investments will be needed by all operators, and strategic decisions have to be taken on how and where the investments should be placed in order to satisfy customer demand.

The telecommunications network operators of today are faced by the challenge of how to develop the existing narrowband copper based network infrastructure into a broadband access network. Cost effective, future proof broadband networks accommodating a wide range of demographic diversity for a set of bearer services with different bandwidth requirement will be required. Strategies for developing the traditional network, along a cost effective path, flexible enough to serve a complex set of customer demands are crucial for the operators, service providers and equipment manufacturers. The access network is the most cost sensitive part of the telecommunications network, and in addition the one most closely related to service demand.

Around 1990 most of the experts expected an evolution toward fibre to the home (FTTH) connections in the access network. However, economic analyses indicated that the overall project values of network architectures for the residential market are reduced significantly when the fibre is installed close to the customer premises. FTTH solutions have proved very expensive, due to the high cost of civil works and the low customer share of optics and electronics equipment. In general, the cost of network elements decreases during time, whilst there is no reduction in civil works costs. The recent development of both cable modems and digital subscriber line (DSL) modems has changed the situation completely. New technology development enables the utilisation of the existing twisted pair and coaxial cable base for high capacity transmission to the customer as alternatives to the FTTH solutions.

The accelerated development of broadband applications is the most important factor for the evolution of a broadband market. Some years ago there was a strong belief in the need for one killer application, and in particular video on demand (VOD) was focused as a potential killer application. However, the residential broadband market has proved to be more sophisticated than earlier assumed. It is obviously difficult to predict the killer application(s). Moreover, there are reasons to believe that in the future the demand for residential broadband connections will be created the by a wide range of applications rather than a single one. The explosive evolution of Internet with a variety of applications during the last years has a great impact on the telecommunications market. Both the traffic and the number of subscribers are now increasing exponentially. In addition, the number of applications on Internet are increasing, and the functionality and content of the applications are under continuous development. The cost of cable modems and DSL modems is expected to decrease significantly over the next years due to mass production. This will impact the subscriber tariffs, which accordingly will fall to a level close to the current narrowband connection tariffs and give the opportunity to extend the present narrowband applications on Internet to broadband applications.

In a competitive environment, network infrastructure changes are forced to be viable in the usual economic sense, i.e. enable a reduction of the annual charges. Furthermore, the investments needed for any network upgrade must generate positive returns, consistent with typically short term expectations of investors. The decision to invest in new technology is subject to the risks and uncertainties inherent in the competitive marketplace. In the former monopoly environment, the network deployment was implemented by a single regulated entity, namely the

PNO. Product life time as well as number of users over time were then predictable and the risks associated with the investment were readily known and manageable. The introduction of deregulation and competition increasingly focuses the attention of the operators on satisfying new customer needs, as the only way to expand their revenue base. Since plain old telephony service (POTS) is growing at an average of only 3 % in Europe, the long-term growth and financial strength of the operators imply the emergence of additional services. Furthermore, the investment in state-of-the-art technologies, like for example residential fibre systems, is not likely to be recouped by telephony services alone. With increasing pressure to minimise costs and to maximise revenues, a large variety of access network architectures must be rigorously examined in order to determine the most appropriate ones for the different area types and service demand profiles.

Techno-economic evaluation in telecommunication network projects is required in order to derive suitable introduction strategies. This task is usually performed by a variety of proprietary software tools in addition to commercially available software tools. Some of these tools are very specialised and calculate only the installed first costs for established technology with very sophisticated models (for e.g. network optimisation). Optimisation tools are most suited to find the optimal solution for well known and established technologies. Special tools are developed for a detailed study of life cycle costs with particular attention to operation and maintenance of network elements. Usually they need very detailed statistical data on component performance. Other tools are very general in nature and not optimised for detailed network architecture evaluation, because cost data, cost predictions, demand data, and geometric models must be included. This book presents a methodology and tool for techno-economic assessment developed by the project RACE 2087/TITAN (tool for introduction strategies and techno-economic evaluation of access network) within the second framework programme for telecommunications research in the European Union (EU), RACE II (research for advanced communications in Europe). The major contribution of the TITAN methodology and tool is the ability to combine low level, detailed network parameters of significant strategic relevance (e.g. variations in civil works costs, network termination cards, splitting ratio of passive optical networks), with high level, overall strategic parameters such as density of subscribers, tariff elasticity and broadband take rate.

1.1 THE TITAN METHODOLOGY AND TOOL

The work presented in this book relates to concepts developed under the first framework programme of the EU for telecommunications research, RACE I. The evolution prospects and frameworks (EPF) group of the RACE project R1044 "Integrated broadband communications development and implementation strategies" developed a first techno-economic approach and the SYNTHESYS tool. This study was based on a simple geometric model, for quantifying the duct and cable lengths. The basis was a series of assumptions regarding e.g. the flexibility point locations and the distribution of users. The cost database included component costs for the starting year only. The SYNTHESYS tool enabled calculations of the installed first cost as well as sensitivity analyses with respect to the subscriber density, hub size and splitting ratio. At that stage, the cost of innovating optical networks appeared to be prohibitive, since the calculations were based on the initial component costs, which at that time were very high due to low production volume.

The work continued in 1992 through the RACE II framework programme. The project R2087 TITAN focused in more detail on the likely steps which must be taken by operators to develop broadband networks and services from the basis of the existing public switched telephone network (PSTN). The project developed a methodology for the assessment of optical access networks for the residential and small business user. The objective of TITAN is the calculation of the overall financial budget of any kind of access system, and hence the estimation of the economic viability of different kinds of access network evolutions. The TITAN methodology was developed to include:

- all kinds of network architectures,
- any kind of transmission media: fibre, copper or radio,
- various feeder lengths,
- cost data over a study period, calculated from the initial component cost, learning curves and volume demand parameters,
- civil works differentiating the various transmission media and based on different technologies yielding quite different unit prices,
- operation, administration and maintenance costs,
- life cycle costs,
- demand assessment for a wide range of services, using an extensive European Delphi survey,
- cash balance based on the above demand assessment and on a first approach of the tariff and revenue aspects,
- a risk assessment regarding the investment in the access network in different environments with a variable degree of regulation, existing infrastructure or competition,
- introduction scenarios for selected case studies, themselves inspired by the European diversity.

This enables the comparison of various optical or hybrid architectures through a global system assessment and contributes to the identification of minimum-risk introduction strategies. The methodology and the tool developed within this project have been validated by comparisons with operating European networks and field trials.

1.2 ORGANISATION OF THE BOOK

The book is split in three main sections, of which the first is devoted to the evolution of the access network towards broadband in general. The second chapter introduces the concepts of evolutionary paths and scenarios, and points out a number of key evolutionary factors present in the telecommunications environment today. The framework described provides guidelines for quantitative analyses of access network strategies. Chapter 3 addresses the main driver for broadband network deployment, namely the demand for new services. A comprehensive Delphi survey on emerging services and applications performed by the TITAN project is presented. This includes a description of both the methodology and the forecast results. Chapter 4 gives a brief introduction to the access network technologies and architectures that enable broadband service delivery, and charts evolution paths from the starting points of both incumbent telecommunications and cable operators towards a full service network (FSN).

Section II presents the methodology and tool developed in the TITAN project in order to assess future introduction scenarios for broadband services and networks. Chapter 5 describes the general modelling framework and the general inputs to the model. The approach adopted by TITAN for the calculation of annual operation, administration and maintenance (OAM) costs is presented. OAM costs are calculated as a percentage of capital expenditure. They include preventive and corrective maintenance, network management, use of installed plant and repair of defective equipment. In addition, the chapter includes a brief introduction to the network component database that has been developed by TITAN. This database contains data gathered from many European sources, with cost elements of different technological maturity and commercial availability. Costs of material, installation and civil works are included in the database for a given reference year. Finally, the key financial outputs used for the evaluation of the overall financial budget of various architectures and upgrade projects are described. These outputs include annual cash flows, project cash balance, net present value and internal rate of return. The following chapter 6 introduces the concept of the geometric model for calculation of the overall cable length of the network, including descriptions of several different models. In chapter 7 a geometric model for mixed demographic structures is presented. The model is applied to different urban and rural structures, reflecting varied patterns of settlement typical of different European regions. Chapter 8 includes a detailed description of the calculations of the network costs, in which the evolution of component prices is taken into account. The cost trends of the various network elements are derived from initial cost, appropriate learning curve coefficients and network penetration assessment over the study period considered. Chapter 9 addresses the issue of sensitivity analysis and risk assessment of upgrade technologies and strategies, focusing in particular on the implementation in TITAN. The development of the risk models was based on probability distributions of critical parameters, including the impact of competition. Section II of the book concludes with a brief description of the general structure of the TITAN tool and how the methodology has been implemented in software. The tool is based on Excel® spreadsheet software.

Section III of the book contains work which illustrates applications of the methodology and the tool. Results from selected extensive access network studies are presented in this section. Even though some of these studies have study periods which start a couple of years back in time (1995,1996), the issues addressed, the services and technologies examined, the analysis approach used and the majority of the results remain very relevant. The section starts with a study of technology alternatives for narrowband services (chapter 11). Optical access and radio in the local loop (RLL) have been assessed and compared to conventional copper layout in a greenfield environment. Chapter 12 illustrates how the tool has been applied to assess radio access network evolutions. The study presented in this chapter deals with various technologies for RLL.

Access network upgrade strategies for emerging new broadband services have been evaluated in several studies, with technological options ranging from enhanced copper to hybrid fibre coax and broadband passive optical networks. Chapter 13 describes an extensive study on different broadband access network upgrade alternatives for PNOs, in which the diversity of conditions met today by European countries was mapped. Chapter 14 presents a study in which the effect of competition in the access network was modelled assuming two main operators sharing the market. Broadband upgrade alternatives were evaluated for a new cable operator, competing with the dominant PNO. Chapter 19 describes an in-depth risk analysis of these two actors in the same competitive market. Video on demand and Internet access upgrades have been analysed, as shown in chapter 15. This study examined PNO and cable operator multimedia upgrades for the services video on demand and Internet access at various access capacities.

A set of studies has been carried out in which upgrade alternatives and strategies for the mixed residential and business market were assessed. Chapter 16 compares and contrast migration paths in two significantly different network areas, one suburban area with a dominating residential market and one downtown area with a majority of business customers. Chapter 17 presents a case study in which the residential market segment dominates, and hence is assumed to govern the network evolution. An analysis of the opposite situation - a business dominated area - is described in chapter 18.

Finally, chapter 20 summarises the main findings from some of the studies presented in the previous chapters, and draws general conclusions in the context of the studies presented in previous chapters.

1.3 HOW TO OBTAIN THE TITAN SOFTWARE

The TITAN 4.0 software may be downloaded free of charge either from Chapman & Hall at *http://www.chapmanhall.com/chaphall/eesoft.htm or from* the RACE 2087/TITAN homepage at *http://www.fou.telenor.no/optimum/titan.* TITAN has been developed as a research tool, and thus may not comply with the standard of quality usually set by manufacturers of fully commercialised software products.

The software package includes the TITAN 4.0 program core, a limited cost component and service component database and an architecture template sheet. The TITAN 4.0 programs runs under Excel© 4.0. Macro errors may occur if the program is run under Excel© 5.0 or higher versions of Excel©. The TITAN 4.0 user manual is included in appendix C of this book.

The Excel© add-in program Crystal Ball©, is used in combination with TITAN for risk analysis. Crystal Ball© is not delivered with TITAN, but may be purchased from Decisioneering Inc., 1515 Arapahoe Street, Suite 1311, Denver, CO 80202 USA. http://www.decisioneering.com, telephone: +1-303-534-1515, telefax: +1-303-534-4818.

Section 1

Access network evolution

Evolutionary paths and scenarios 2

Mauro Ravera, Stefano Balzaretti, Umberto Ferrero, Marco De Bortoli, Markku Lähteenoja, Jari Mononen

2.1 INTRODUCTION

The present situation in the telecommunications environment is characterised by the presence of a number of evolutionary factors, which concurrently contribute to make less predictable the future scenarios for the service market and the network implementation. Among these factors the liberalisation process taking place in most European countries, the technical innovation and the introduction of new services are of the utmost importance. Particularly for the access network a number of changes can be identified, starting from a "natural monopoly" situation based on the deployment of a limited number of techniques and the provisioning of few basic services, and leading to a competition in the service and infrastructure provision with a high diversification of architectures and services.

This evolution is now leading to a major change in the telecommunication company strategies: the market liberalisation and the open network provision (ONP) rules raise competition in the area until recently controlled by the public network operator (PNO). Hence, other network operators like community antenna television (CATV) operators and public facilities providers are now able to provide alternative access infrastructures and handle the telecommunication services on their own. As a consequence the strategy of the telecommunication operator will be to preserve its current market space from the entry of competitors, mainly by providing the transport service to the service providers at a competitive price and by upgrading its network to enhance its performances and transport capabilities, at least for a set of privileged customers. On the other hand the telecom operator will require powerful and effective technical solutions to reach new customers outside its traditional market. Hence, there is a strong requirement for technologies which are capable of:

- providing a large number of existing and new services, so-called "open architectures",
- guaranteeing a high profitability, i.e. high revenues compared to the required investments.

This chapter contains a framework for the description of access network evolutionary paths and related scenarios, taking into account market considerations, services and technical aspects [1]. This is intended to provide a guideline for quantitative analyses, such as the techno-economic evaluations carried out in the RACE (research for advanced communications in Europe) 2087/TITAN (tool for introduction strategies and techno-economic evaluation of access network) project. For this purpose the perspectives of different types of operators like current telecom operators, CATV operators and alternative infrastructure providers have been considered. Particular attention has been paid to the provisioning of wideband and broadband services on different types of wireline access network architectures. The scenarios proposed are mainly focused on residential and small business customers. Neither distributive services, interactive services via satellites, nor mobile services, have been considered to be within the scope of this work.

2.2 GENERAL ASPECTS

A brief introduction to the terminology adopted in the work is required [2]. The basic terms used in the chapter are "*scenario*", "*evolutionary path*", "*target*" and "*strategy*". Fig 2.1 illustrates the relationship between them.

The *target* represents the telecommunication company objectives both with respect to economic and technology aspects of interest of the company. The strategy represents the means by which these objectives are reached, starting from the present situation.

A *strategy* is defined as the set of actions and initiatives that the telecommunication company plans to take in order to reach a predefined target. The definition of a strategy implies a choice among different evolutionary paths.

An *evolutionary path* is a time-ordered sequence of network architectures and the set of transitions between pairs of consecutive architectures. There can be alternative transitions, and thus alternative evolutionary paths, from a starting point to the selected target.

A *scenario* is the description of a network environment, including one or several operators, in which there is provided a set of services to a number of users within a certain area and timeframe. It includes both a starting situation and a target for the network and services evolution. The scenario is thus described by attributes for both its initial situation and its time evolution. Due to the complexity of elaboration of overall scenarios including all these aspects, a scenario is proposed to be decomposed in a number of "specific" scenarios such as:

1. regulatory scenario
2. environmental scenario
3. service scenario
4. technology scenario.

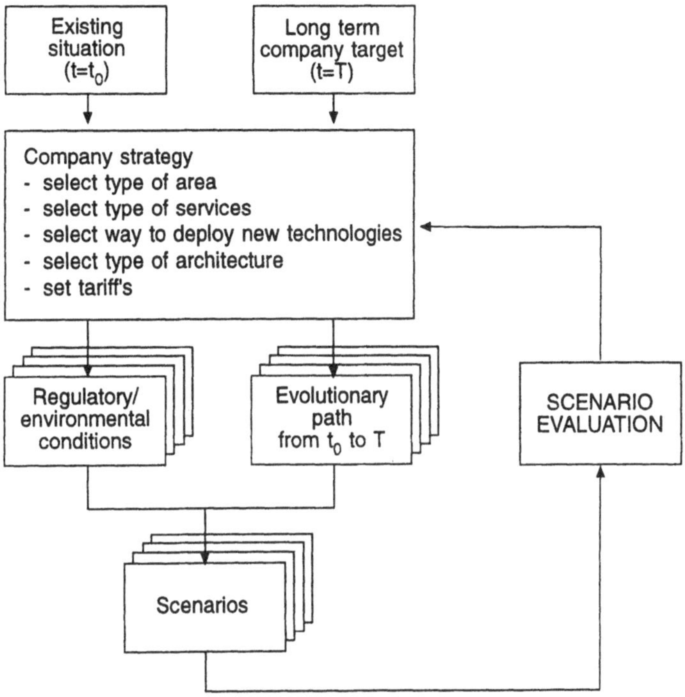

Fig. 2.1 The relationship between the terms "target", "strategy", "evolutionary path" and "scenario".

2.3 DEFINITION OF SCENARIO ATTRIBUTES

A scenario can be characterised by a number of attributes related to the initial situation. Such attributes define the existing network structure, the type of area under consideration, the service type and penetration and the number and type of companies competing for the telecommunication services.

In the scenario description the evolution opportunities from the initial situation should also be described, in order to link them to other subsequent situations. The transitions are mainly based on the introduction of new architectures, on the evolution of the service demand, on the development of the existing network and on the entrance of new competitors or development of existing ones. In the following a list of the attributes is provided: for each of them it has been indicated whether the attributes needs both an initial and a trend value.

2.3.1 Regulatory scenario

The *regulatory scenario* describes the sharing of the potential market (for end-user services and for transportation) between the competitors, the tariff structures and the revenues of the telecommunication operators. An example of the relationship between service providers and network operators is depicted in Fig 2.2. The attributes of the regulatory scenario are:

- number of competitors in the service market
- number of competitors in the access network provisioning
- percentage share of the service market between the competitors
- percentage share of the initial and new access network provisioning between the competitors.

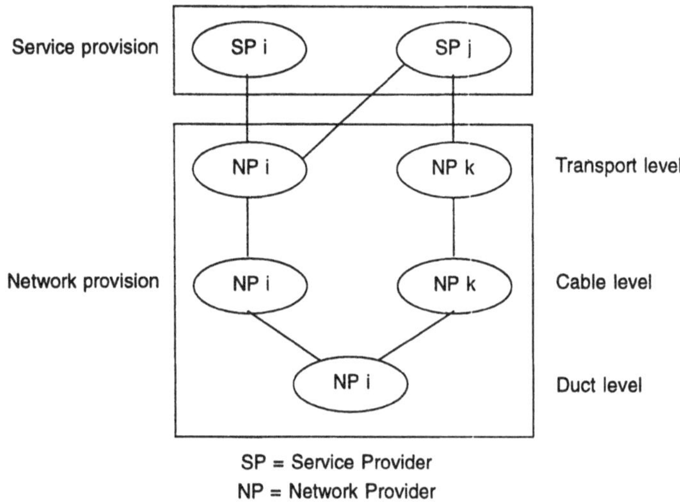

Fig 2.2 Example of competition between different types of operators according to the ONP rules.

2.3.2 Environmental scenario

The *environmental scenario* characterises the existing and upgraded network infrastructure in addition to the geographic and demographic situation. This enables the mapping of technology and market assumptions onto a specific network area. The attributes of the environmental scenario are related to area type and regulation and the existing network infrastructure:

1. *Type of area*
- customer density (initial, forecast)
- type of living units
- customer expenditures on telecommunication services (initial, forecast)
- cost of civil works
- mean loop length.

2. *Existing network infrastructure*
- initial duct availability
- percentage of network infrastructure which needs substitution (initial, forecast)
- initial network occupancy.

The distinction between urban, suburban and rural areas is implicit in the attributes. Hence, it has not been highlighted with a separate attribute. For the existing network infrastructure only the free resources (ducts, transmission capacity) and the resources disposal policy are considered. The customer expenditures on telecommunication services, along with the service tariffs, in principle allows the estimation of the service penetration in the selected geographic area. However, the service penetration will be introduced explicitly in the service scenario.

The amount of required material in the network is influenced by the mean loop length and by the number of independent living units. Given a certain area size, the mean loop length is nearly constant and the number of living units is given by the subscriber density and the number of households per living unit. If the new architecture provides the existing services (i.e. it is integrated), the replacement of existing infrastructures is done with the new systems. Otherwise replacements are performed using the previous type of systems. No forced replacement of existing systems is foreseen in this environment.

2.3.3 Service scenario

The *service scenario* describes the services and applications provided by the service operators. It identifies the packages of services to be supported by the selected architectures, and provides the description of their penetration over the period under study. In the Delphi survey carried out by TITAN the service penetration was correlated to the level of tariffs and the willingness to pay of the customers [3]. The attributes of the service scenario are:

1. *Service type*
2. *Penetration of services (initial, forecast)*
3. *Subscription tariffs.*

Service penetration is the percentage of customers which would subscribe to the service if provided at a reasonable price. 100 % penetration means that one service subscription is delivered to all the households.

2.3.4 Technology scenario

The *technology scenario* contains the description of relevant technologies, systems and architectures to provide the services in the form of selected target and intermediate access network architectures. The attributes of the technology scenario are related to network architectures, network technologies, network equipment and installation and operation, administration and maintenance (OAM) procedures:

1. *Network architectures and technologies*
   - existing network architecture
   - intermediate network architecture
   - final network architecture.
2. *Cost of network equipment and installation*
3. *Cost of operation, administration and maintenance procedures.*

In addition to the proposed attributes, other aspects which are external to the proposed scenario might be taken into consideration, e.g. the discount factor and its variation. However, in order to limit the complexity of the analysis to a manageable level, it may prove beneficial to assume some parameters to be invariant during a certain period of time. These may include the tariffs for the services, the cost of OAM procedures, the customer density and the mean loop length. The qualitative relationship between the different attributes of a scenario is illustrated in Fig 2.3. The figure highlights which attributes are influenced by the value assigned to another attribute.

The impact of the scenario on the company strategy may then be assessed by an economic evaluation of the project on the basis of the information provided by the scenario attributes.

The values of the scenario attributes should be integrated with the degree of uncertainty in each value. The main sources of uncertainties are the competition between operators, the varying costs of network components caused by the unpredictable service market, the lack of knowledge about costs of operating the new architectures and the errors in the prediction of service demand. All these items will influence the capability of the operator to achieve its strategy, because they change the values of the scenario attributes. The selection of the statistical attributes and the identification of the relationship between such attributes and the results of the techno-economic tool application is a matter of risk analysis, as described in more detail in chapter 9.

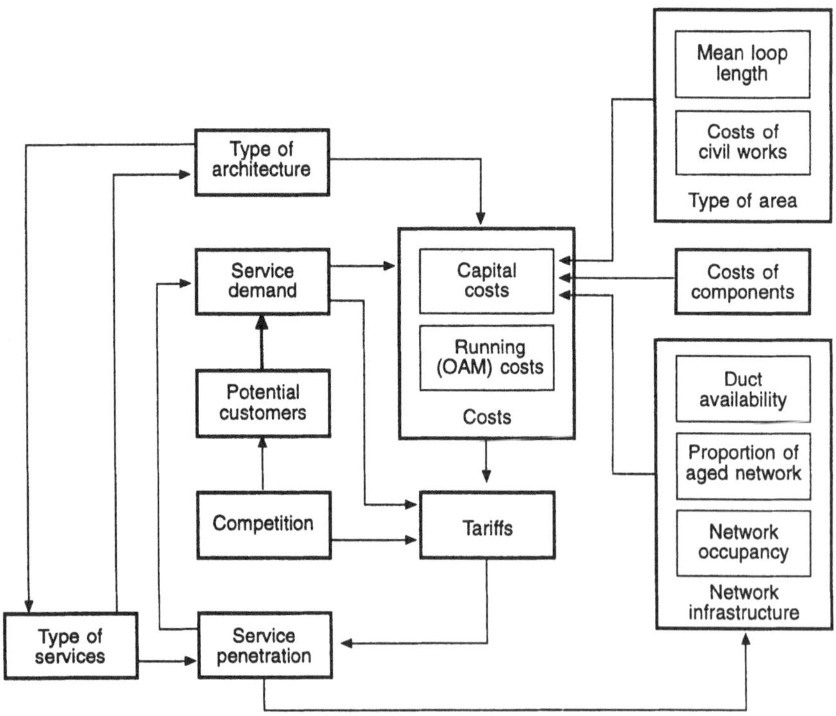

Fig 2.3 The relationship between scenario attributes.

2.4 REGULATORY ASPECTS

In this section the main issues regarding the regulatory aspects of the provisioning of telecommunication and entertainment services are described, with an emphasis on the situation in the European countries. The purpose is to highlight the possible bottlenecks in the evolution of the network and service market towards a fully liberalised environment and to provide basic information for the description of the future environment from a market point of view.

2.4.1 Provision of video services by telecommunications operators

The regulations concerning the provision of video services by telecommunications operators (TOs) varies throughout the European Union (EU). In Germany virtually all CATV services are carried over networks provided by Deutsche Telekom. In the United Kingdom (UK) British Telecom (BT) as the dominant TO is explicitly forbidden from offering broadcast entertainment services. The UK position will be reviewed in 1998. In most other European countries the CATV service is provided by private organisations, or at least organisations other than the TO. Until now most TOs have been satisfied with this situation. However, recent developments which indicate that broadband services may generate significant revenues, combined with developments in technology which will allow broadband services to be delivered over existing telephony networks (at least as an intermediate solution), have made TOs very eager to enter into the business of broadcast entertainment.

2.4.2 Provision of telecom services by CATV operators

In the Netherlands CATV operators are allowed to carry telecommunication services with the exception of voice telephony. In France, telecommunications between members of a closed user group (CUG) are possible over CATV networks. In Germany the TO *is* the CATV operator. However, it is in the UK that CATV operators are currently allowed to enter the telecommunications market at whatever level they wish, at least for fixed links in their own franchise areas. In other countries in the EU, CATV operators are not allowed to carry telecommunication services. It is likely that this situation will change with time. Where TOs are gaining access to the market for entertainment, it is reasonable to expect that they will be forced to allow CATV operators to offer telecommunications services.

2.4.3 Open network provision

The objectives of open network provision (ONP) as defined by the European Commission are to strengthen the European telecommunication and information technology industries by opening the market to competition and to provide users with access to an increased range of equipment and services across the EU. One of the corner stones of ONP is that tariffs should reflect costs. In recent years TOs have tended to use profits from trunk transmission to subsidise the cost of local loop provision. One recommendation is that TOs should be obliged to publish a tariff for use of local loop access by telephony service providers, which will also apply to lines used by the TO's own telephony service. This will have the effect of forcing the TO to subsidise its competitors if it wishes to subsidise its own local loop provision. Until now the Commission has been pushing TOs to rebalance their tariffs between local access and core network.

In most EU member states a TO earns less revenue from line rental charges than the cost of maintenance of the local loop. This discrepancy is known as the access deficit. Where a TO is prevented from raising its tariffs to properly cover this cost, a service provider using the local loop should have to pay a proportion of the deficit.

Another recommendation being considered by the Commission is that entertainment services utilising "*video or audio signals individually selected by the user*" should be added to the definition of telecommunications services. This would ensure that video on demand services would be subject to ONP regulation. If co-location of equipment is a technical prerequisite in order to offer a service which is open to competition, then the TO will be obliged to offer such access on a non-discriminatory cost oriented basis.

2.4.4 Universal service

The issue of universal service is important when looking at telecommunications regulation. The laudable desires to make tariffs related to costs and to avoid discrimination against remote parts of the EU are in conflict. It is clear that providing advanced services in the centre of London or Paris can be done much more economically than in western Ireland or on a Greek island. This implies that there most likely will be defined a basket of services which must be provided everywhere, in order to prohibit undue discrimination of remote regions.

2.5 EVOLUTIONARY FACTORS

A number of external factors can drive the evolution of the networks and services considered in this book. Among those factors the following were considered of relevance, taking the scope of this book and the TITAN methodology and tool into account:

- change of demographic environment
- new technologies become available at a reasonable price
- decrease in the price for existing technologies
- network saturation/installation of new infrastructures
- requirement for new services, or services with a better quality of service
- entrance to the service market by competitors, due to change of regulation and/or capability to provide a new service
- change of tariffs for existing services.

The *change of the demographic environment* in the considered area can lead to a variation in the mix of user types and a variation of user density due to migration of population, typically found at the vicinity of the metropolitan areas.

Such changes are relatively slow and lead to a change in the service demand and service mix for the considered area.

The *technology innovation* leads to cheaper network components. The aspects of this are two-fold: the reduction of the price for new technologies enables the change of network architectures when the costs of new architectures become less expensive than the costs of a continuing installation of the existing ones. On the other hand the costs of the existing architecture tend to decrease during time.

The *saturation of the existing systems and infrastructures* is another factor which should be taken into consideration during network planning. The need for new capacity (for new services or new customers) leads to the installation of additional equipment and cables. The new technologies like passive optical networks (PONs) and synchronous digital hierarchy (SDH) should allow a high flexibility in the provisioning and reconfiguration of logical connections, thus avoiding the need for continuous upgrading of the physical network. On the other hand their introduction usually requires a certain amount of civil works in order to lay down the required new ducts.

The need to introduce *new services* is closely related to the service demand from the customers and to the presence of *competitors*. The evolution in service demand generally depends on the demographic environment and on the applied *tariffs*. On the other hand the presence of competitors makes the market more dynamic and unstable, thus usually lowering the tariffs and enabling the introduction of new architectures and services.

2.5.1 General targets for telecommunications companies

The definition of the strategy of a telecommunications company requires a definition of the target. The target can vary, depending on the type of company (service provider, network operator), on the existing regulation and on the type of area being considered. In this work, the following targets have been considered for determining the strategies of the telecommunications operators:

- guarantee universal non-discriminating service provision
- minimise risk of losses
- exploit revenue opportunities
- limit time for return on investments
- minimise life-cycle costs
- minimise installed first costs.

In a monopolistic situation, the telecommunications operator is generally required to provide a small number of basic services to all customers in every type of service area. The immediate consequence of this is the need to invest part of the revenues in the provisioning of services to remote customers with low profits or no profit at all. Additionally, in absence of huge uncertainties in the service market, the network evolution is mainly technology driven, i.e. the introduction of new technologies is linked to their availability and maturity rather than to the real business opportunities. This leads to a policy of general containment of costs, rather than to the search for new sources of revenues.

With the entrance of new competitors the target of the telecommunications operators will most likely be more diversified. There will be stringent requirements to exploit any type of opportunity to provide profitable services even on a small scale, especially before a competitor can seize a large market share. Such target requires the operator not only to *retain revenues* and to minimise the risk of losses, but also to be prepared to invest significant amounts of money in profitable activities in the short term in order to generate new revenues, and hence maximise the overall revenues. The short term profitability can be achieved by investments limited to small areas and scheduled over a rather long period of time before the coverage of the overall territory is achieved. This quick return on investments will be an advantage for an operator with respect to his competitors because it limits the risks of his strategy and allows the service provision to fund itself without the need for additional subsides. The aspects related to the risk of loosing opportunities or wasting investments are dealt with in chapter 9.

2.5.2 Characterisation of operator strategies

Once the target for the telecommunications operator has been set, it has to define a strategy, i.e. the appropriate set of actions enabling him to achieve his target. The strategy of the operator may be described by the decisions listed below. The options are indicated in brackets:

- selection of customers (all, customers with high expenditure capability)
- selection of services (keep existing ones, introduce new services)
- schedule of investments (concentrated at the beginning, distributed in time)
- tariffs for new services (high, low)
- selection of type of architecture
- exploitation of the existing network (reduction, stability, growth)
- strategy for the deployment of the new technology (overlay network, introduction in new areas, replacement of the existing network)
- allocation of new services (on overlaid network, integrated in the existing network).

The *selection of customers* is strongly related to the amount of investments the company is able to make and to the need for short term revenues. Provisioning of services to all customers is a typical requirement of dominant operators: even for such operators new services will be initially provided to a small number of customers to limit both risks and volume of investments and to maximise returns on investments, so-called "cream skimming". If in the long term the network architecture is expected to provide the existing services in addition to the new ones, high profitability can only be guaranteed by using the architecture to serve as many customers as possible and hence achieving an adequate economy of scale.

The *selection of services* is critical because it mediates between user requirements, the operator capability to invest and his revenue expectation. The provision of new services by the dominant telecommunication operator is generally pushed by the shrinkage of revenues. This may be caused by the entrance of new competitors, the consequent loss of customers for basic services and the risk of being excluded from highly profitable markets.

Investments can be scheduled differently, depending on the market situation. Investments may be carried out at the beginning of the upgrade period, implying the operator having to risk a large negative cash flow, if services are required to be provided in advance. Distribution of investments over time may be enabled by a market which is somewhat protected against competing entrants. The choice of investment timing may be strongly influenced by the financial situation of the company and the financial forecast, i.e. the capability to obtain external funding, the need to support investments only by its revenues and the variations in inflation and discount rate.

Tariffs are set to comply with the need for revenues from the investments and to avoid overwhelming the willingness of the customer to pay. There are also requirements originating from the regulatory bodies. These requirements on the one hand try to prevent monopolists exploiting their position and on the other hand aim to guarantee their profits in case they serve an unprofitable market.

If a telecommunications company has considerable capital invested in an existing network, it could be necessary to exploit it as much as possible before replacing it with a new network. The *exploitation of the existing network* can be achieved by enhancing its capabilities, either by providing new services on separated networks when enhancement is not feasible or by integrating them with the existing services only in case of growth or greenfield situations. New technologies can be introduced in the old part of the network only where refurbishment is needed. In this case the benefits of depreciation are fully achieved, whilst in the opposite case the sale of the equipment usually does not guarantee adequate revenues.

Replacement of the existing network has to be justified by a strong need for a new architecture or type of equipment to provide new services integrated with the previous ones. In this case it may be necessary to consider the loss coming from the missed depreciation of existing assets and tax reduction. The provision of both existing and new services on the same network will allow the operator to achieve the economy of scale required to fund a technical innovation introduced on a small scale.

Two main approaches can be followed regarding the economies that can be achieved by the network operator, namely to achieve economies of scale or to achieve economies of scope [4]. The two approaches are closely related to the market situation and degree of competition in the area being considered, as is shown in Table 2.1.

Table 2.1 Relationship between economies in the network provisioning and market structure [3].

	Weak economies of scale	*Strong economies of scale*
Weak economies of scope	Single service networks, several network operators	Single service networks, few network operators
Strong economies of scope	Multiservice networks, several network operators	Multiservice networks, few network operators

The first result of the application of the selected strategy is the identification of the environmental conditions and of a number of evolutionary paths. Their definition follows a set of *criteria* that are common to most strategies:

- do not remove equipment recently installed,
- removal of equipment usable in other parts of the network is allowed,
- move the optical network unit (ONU) towards the customer when justified by sharing of optical technologies, e.g. in high density areas,
- exploit coaxial cable resources when available,
- use overlay solutions for distributive services if capacity requirements or other technical characteristics of the signal may cause problems in transport on existing architectures,
- use active star solutions based on point-to-point systems when requirements for capacity or flexibility are stringent,
- share fibre systems when no drawback is foreseen,
- use radio systems in rapidly changing areas and low demand areas or to reach customers dispersed on a wide area,
- use enhanced copper systems when willingness to pay for new services is high and no ducts are available in the short term.

2.5.3 Evaluation criteria

The identification of a number of evolutionary paths and the selection of scenarios has to be matched with the needs of the operator to select the optimal strategy to achieve his targets. The evaluation criteria listed below can be introduced and qualitatively applied to the selected scenarios and strategies. The TITAN methodology described in this book illustrates one approach for performing the quantitative evaluations of scenarios and strategies.

- capability to reuse the existing network and infrastructures
- capital and running costs
- potential revenues
- exploitation of market opportunities or risk of missing them
- upgradeability to more advanced solutions
- compliance to ONP criteria
- need for disposal of installed network
- network residual value.

The network residual value should be estimated at the end of the study period since it contributes to the total amount of resources available to the operator. The residual value of the capital investments can still be significant after the upgrade period, and may increase the value of the project when it is integrated with the cumulative cash balance estimate.

2.6 SUMMARY

Guidelines for quantitative analyses, such as the techno-economic evaluations carried out in the TITAN project, have been proposed through the description of a framework for access network evolutionary paths and related scenarios. The framework comprises both market considerations, services, technical aspects and characteristics of operator strategies. The proposed analysis of the strategies of the operators and market, service and network scenarios highlights a number of crucial issues for the evolution of the telecommunications sector in the short and medium term.

REFERENCES

[1] Initial Scenarios and residual value analysis. Editor: Ravera, M. Brussels, Belgium, 1994. (RACE 2087/TITAN Deliverable 4.1).

[2] Matthews, D, Darabi, F. The Local Loop - Market, Technical and Regulatory Strategies. OVUM Ltd, 1994.

[3] Stordahl, K, Murphy, E. Methods for forecasting long term demand for wide and broadband services in the residential market. *IEEE Communications Magazine*, (33), (2), 1995, 44-49.

[4] Reed, D P. Residential Fiber Optic Networks - An Engineering and Economic Analysis. Artech House, Norwood, MA, USA, 1992.

Demand for new services

3

Kjell Stordahl

3.1 INTRODUCTION

The total western European market demand for fixed voice telephony, plain old telephony services (POTS) and narrowband integrated services digital network (N-ISDN), is expected to increase only marginally over the next few years. However, the demand in the developing European countries is expected to increase significantly. The two major factors affecting the new demand for POTS are the economic growth and demographic changes. Commercial ISDN facilities are now offered in an increasing number of European countries.

Despite recent field trials and market surveys the new broadband residential and small business customer applications are still not clearly envisaged. Nevertheless, some trends can be recognised: 1) The integration of telecommunications and information processing is escalating. 2) The use of interactive video services, such as videoconferences, is growing. 3) Entertainment services are increasingly important. Several methods for forecasting the penetration of new applications and bearer services have been reported, including market surveys and extrapolations from current household spending patterns [1]. The RACE II (research for advanced communications in Europe) 2087/TITAN (tool for introduction strategies and techno-economic evaluation of access network) project performed a comprehensive Delphi survey among European telecoms experts in ten European countries in order to forecast the penetration of future services and applications [2,3]. This chapter describes the results from this Delphi survey, which forms the basis for the forecasts of the total market of broadband bearer services and ISDN as used in this book and presented in several international publications. The Delphi survey results include demand forecasts for wideband and broadband bearer services in the residential market, as a function both of service price and of disposable income. The leading applications, are forecasted, and forecasts are presented of application demand as a function of both price and demand. The terms "application" and "bearer service", or "service", are not unambiguously defined in the telecommunication environment. We assume that the application refers to the use of the telecommunication service made by the customer and that the bearer service or service is the product sold to the customer by the service provider.

3.2 THE DELPHI SURVEY

A Delphi survey is a method by which the opinions of experts are canvassed, in order to achieve consensus on a particular issue. The methodology involves asking a set of questions, analysing the results and resubmitting the questions to the experts, together with a summary of first-round results. The experts then resubmit their opinions, which may have changed following consideration of results from the previous round. This procedure can be repeated a number of times, and usually leads to a reduction in the variance of the answers received. Between the middle of 1993 and the middle of 1994 the TITAN project carried out a comprehensive two-round Delphi survey among experts in ten European countries.

3.2.1 The questionnaire

The applications covered by the questionnaire were video on demand (VOD), multimedia telegames, videotelephony, telecommunity (telemedicine), teleshopping, advertising and marketing, interactive TV, electronic newspapers, home office (simple and advanced), and remote education (home and studio). Fig 3.1 lists the applications in the Delphi survey.

Teleinteraction (Entertainment) 1	Teleshopping (Electronic market) 2	Thematic channels (Information) 3	Teleworking 4
Video on Demand 1A	Home ordering with electronic bill payment 2A	Interactive-TV, and specialised-channels 3A	Home office (simple) 4A
Multimedia telegames (Video and virtual-reality games) 1B	Advertising and marking 2B	Electronic newspaper 3B	Home office (advanced) 4B
Videophony 1C	2C	3C	Remote-education (home) 4C
Telecommunity (Telemedicine) 1D	2D	3D	Remote-education (studio) 4D

Fig 3.1 The applications in the Delphi survey.

For each application detailed questions were asked concerning likely demand (in terms of penetration of households) for services which deliver that application. The general description and the technical assumptions of the VOD application as included in the questionnaire are shown in Fig 3.2.

General description:
This is an application where a video-library can be reached by a teleservice,and programs may be ordered and transmitted to the home.

Technical assumptions:
The user may either use an advanced telephone or a PC to communicate with the video-library over a teleservice. The PC- or telephone-keyboard is used to select videos from the library. The transmission of the video may either be done via a Cable-TV network, or a telecommunication network. We may here assume that the household must subscribe to either a *wideband* or a *broadband* teleservice to support this application. The broadband-service will give faster transmission and better image qualities, but both the user-equipment and the teleservice will be more expensive.

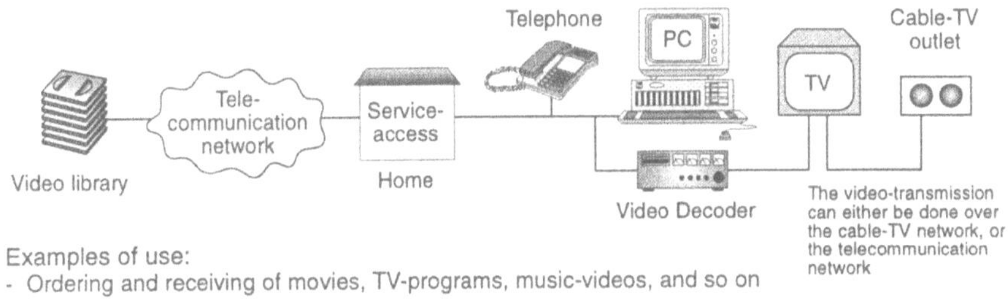

Fig 3.2 The general description and the technical assumptions of the VOD application as presented in the questionnaire.

The questions asked covered the expected level of demand:

- over a range of different prices
- at different times in the future (1995, 2000, 2005 and 2010)
- for various levels of household disposable income.

In addition, qualitative information was sought regarding typical users of each application. Questions of a similar nature were asked concerning three bearer services: basic rate ISDN (up to 144 kbit/s), wideband (up to 2 Mbit/s), and broadband (up to 8 Mbit/s). Respondents were also asked to indicate the number of applications a customer would subscribe to over each bearer service, and which three applications are likely to be the most popular. In total, 398 questions were asked in the first round of the survey. This was reduced to 365 for the second round as some of the qualitative questions were not resubmitted to the experts.

3.2.2 Execution of the survey

The Delphi survey was carried out by TITAN partners across Europe. It was decided to ask ten experts in each of ten countries to complete the questionnaire. The countries chosen were Belgium, Finland, France, Germany, Greece, Italy, the Netherlands, Norway, Portugal, and the United Kingdom. Where possible, five of the ten experts approached in each country were from the dominant telecommunication operator. The other five were chosen from companies within the information technology industry, universities and other companies and institutions concerned with consumer demand.

It proved more difficult than anticipated to persuade the experts to complete the questionnaire. By early December 1993, 58 questionnaires had been returned, and in order to keep to the agreed schedule it was decided to begin analysis of the replies. Second-round questionnaires were sent out to the same experts in February 1994, incorporating details of the results obtained from the first round. By the end of May 1994, 50 second-round questionnaires had been returned, and it was decided to proceed with the analysis of the second-round results.

3.2.3 Presentation of the results

The results in the following sections represent European means and are based on data from the Delphi survey. The results are presented by use of medians and not mean values, which normally are used in statistics. The median is the middle observation or the 50 % fractile when all observations (answers from the experts) are ordered in an increasing sequence. The median is commonly used in Delphi surveys since it is a more robust estimator than the mean value, i.e. it is not affected by extreme answers to the same extent as the mean value. For most services, the results from the first and second rounds, in terms of medians, are very similar. However, there are significant changes if we look at the spread in the observations,. After the second round the observations are concentrated in a smaller interval. This means that the experts after the second round have come to a more common opinion.

3.2.4 Services and typical customers

A set of quantitative questions was asked about future application, including the characteristics of typical customers for each application. Fig 3.3 summarises the information gathered from this part of the survey.

Service	Sex	Age	Educational level of users	Purpose
Video on Demand	Male	All	All	Entertainment
Telegames	Male	Young	All	Entertainment
Videotelephony	Both	All	High & Medium	Communication
Telecommunity	Both	Old	All	Communication
Advertising & marketing	Female	All	High & Medium	Information
Interactive TV/ specialised channels	Male	All	All	Entertainment & education
Electronic newspaper	Male	All	High	Information for pers-onal & business use
Home office	Both	All	High & Medium	For work
Remote education	Both	All	High & Medium	For study and entertainment

Fig 3.3 The characteristics of typical users for each application.

3.2.5 Ranking of the applications

A household will usually demand more than one application to justify a wideband or a broadband connection. Since the terminal equipment is not very different for some services, the household gets more value for a subscription to a connection with more than one service. In the Delphi survey the experts were asked about the average number of applications provided on wideband and broadband connections. The average number of applications per wideband and broadband connection were respectively 3.2 and 3.8.

The experts in the Delphi survey were asked to point out the three most important applications. Fig 3.4 arranges the applications in order for wideband and broadband connections, based on the responses from the experts.

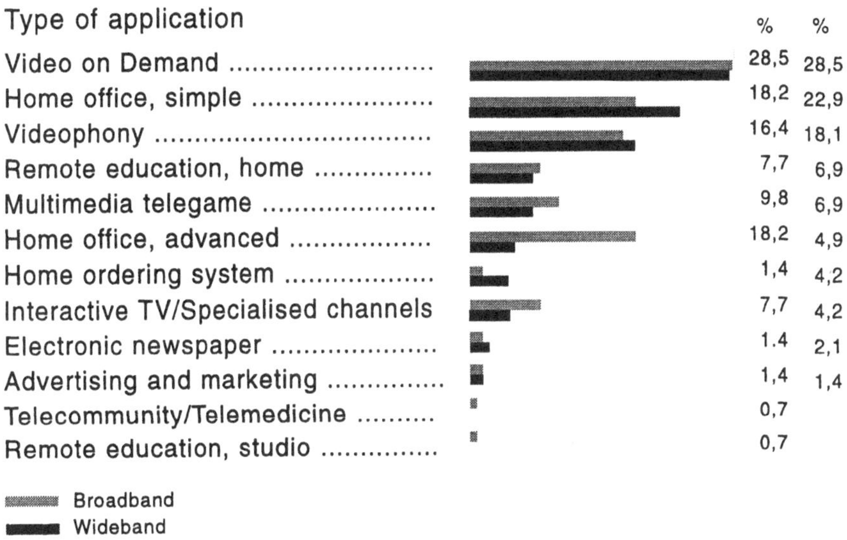

Fig 3.4 Ranking of the wideband and broadband connection applications, based on the responses from the experts.

The results in Fig 3.4 are given based on the data from the second round. It is interesting to note that videophony was the highest ranked service in the first round and in the second round it was ranked as service number three. VOD has got a significantly higher percentage ranking in round two compared with the first round. The changes can be explained by the distance in time (eight to nine months) between round one and round two. During this time the technology for VOD especially has developed and a number of field trials were underway. Fig. 3.4 shows that VOD, home office (simple) and videophony are the most popular services for wideband connections, while advanced home office together with the mentioned services are most popular for broadband connections.

3.3 DEMAND, FORECASTS AND SATURATION OF APPLICATIONS

3.3.1 Demand

For every application the experts were asked to indicate the demand for a given set of annual tariffs. Hence, a demand curve can be constructed for each application based on medians from the survey. The demand for a set of annual costs of the application provided on a wideband connection is given in Fig 3.5 and Fig 3.6 for comparison. 1 ECU ≈ 1.25 $US.

Fig 3.5 and Fig 3.6 show that VOD and videophony are very attractive applications. The quality of the videophone applications ciurrently being offered on ISDN is limited, but it is expected to improve through advanced coding techniques. The price for the applications provided on ISDN will certainly be much lower than the price for a 2 Mbit/s connection. In the future the combination of high quality and decreasing price for a 2 Mbit/s connection may compete with the lower quality and lower price on ISDN. An intermediate solution may be a 384 kbit/s service on ISDN.

Fig 3.5 and Fig 3.6 indicate sizeable demand for multimedia teleshopping, home office (simple and advanced), interactive TV, multimedia telegame, and electronic newspaper. It is noteworthy that the demand for the home office application is inelastic compared with the other applications. This means that the demand for this application is less dependent on price. Companies and public organisations may subsidise employees to work at home by providing communication equipment and expenses.

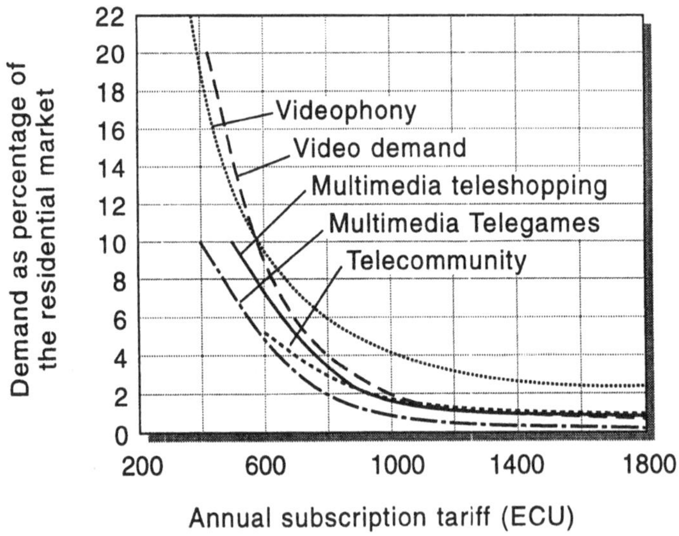

Fig 3.5 Demand curves for a set of applications, I. 1 ECU ≈ 1.25 $US.

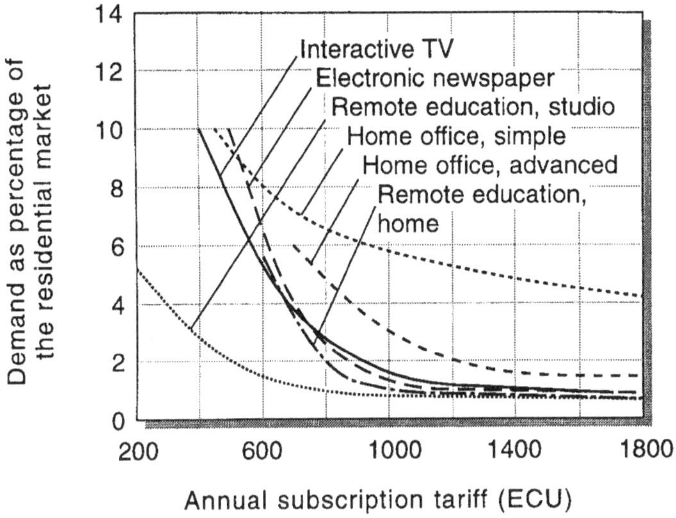

Fig 3.6 Demand curves for a set of applications, II. 1 ECU ≈ 1.25 $US.

3.3.2 Long term forecasts and saturation of applications

The experts were asked to indicate long term forecasts for each application. The results from the survey for wide and broadband services indicate that the forecasts of home office (simple), multimedia teleshopping, multimedia telegame, and interactive TV/specialised channels increases in the same way as for wideband connections (see Fig 3.9). For VOD and videophony the forecasts are significantly higher, while the forecasts are lower for the other applications. Since wideband (2 Mbit/s) is the bearer service, it is not consistent to have higher forecasts for specific applications. However, VOD needs a wideband connection. The inconsistency may come from an increased interest for VOD during the last months of the survey not yet reflected in the forecasts for wideband connections. Videophony will most probably be transmitted on a lower bandwidth (ISDN).

Fig 3.7 shows the expected saturation for each wideband application both in round one and two. There are only minor changes from round one to round two, except for videophony and VOD. The estimated saturation has decreased from 50 % to 41 % for videophony and from 35 % to 30 % for VOD. The services with highest saturation are: Videophony, VOD, multimedia teleshopping, multimedia telegame and home office (simple).

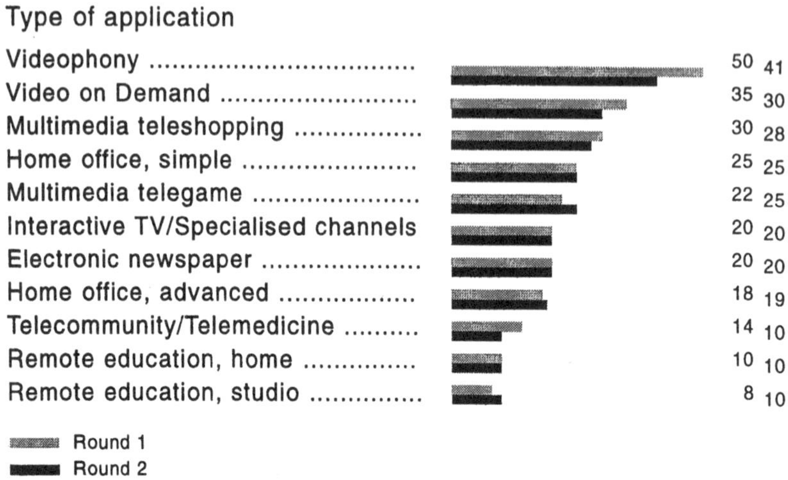

Fig 3.7 The expected saturation for each wideband application in rounds one and two (in percentage of the residential market).

3.3.3 Willingness to pay for the application

Fig 3.8 gives an indication of how much a household with an average disposable annual income is willing to pay per year for specific wideband application. The answers from the experts are based on the assumption that the household demands the application. As pointed out before a household will order only a limited number of applications.

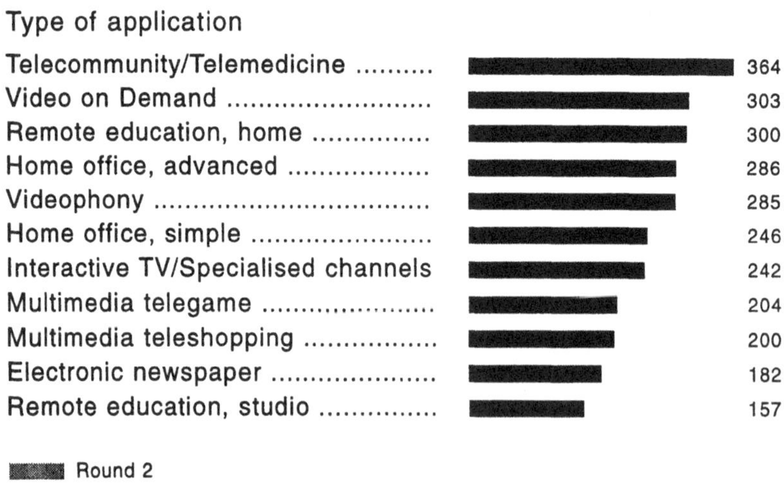

Fig 3.8 The willingness to pay of a household per year (in ECU) per application, given an average disposable income of 20,000 ECU.

A household is willing to pay more for the telecommunity/telemedicine application compared to other applications. We have previously seen that the forecast/demand for the particular application is not that high. However, the household who really wants the application is willing to pay for it! In addition, telecommunity/telemedicine, VOD, remote education, home and advanced home office are the applications with the highest willingness to pay.

3.4 DEMAND AND FORECASTS FOR BEARER SERVICES

The experts were asked to indicate the percentage penetration in the residential market for the services carrying the above described applications, the so-called bearer services. The bearer services considered were ISDN, 2 Mbit/s (wideband) and 8 Mbit/s (broadband), and market penetration was given for the years 1995, 2000, 2005, 2010. In addition, the experts were asked to indicate the saturation penetration.

3.4.1 Demand and forecasts for bearer services

Fig 3.9 shows the Delphi survey results in terms of medians. The figure shows low demand for broadband connections for the residential market. The estimated penetration in 2010 is 5 %. There is a significantly higher figure for wideband connections; in 2010 the penetration is estimated to 15 %. The development of the demand for ISDN is faster - within a ten year period from 1995 to 2005, the forecasts increase to 20 %.

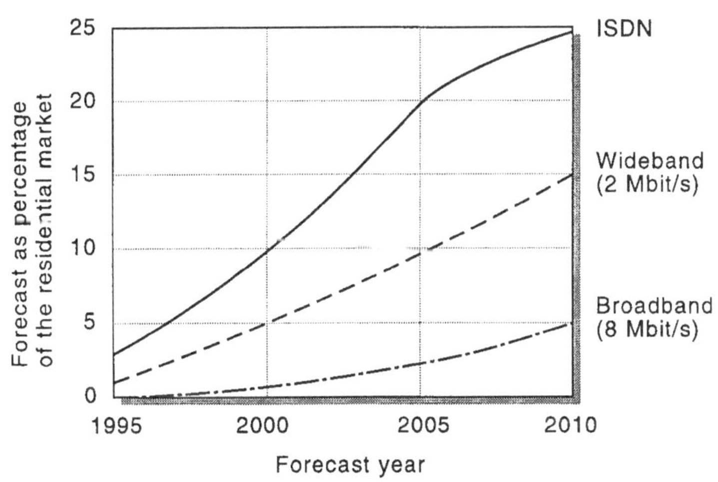

Fig 3.9 Forecasts for ISDN, wideband and broadband services.

Fig 3.10 shows the variation in the opinion of the experts regarding forecasts for wideband connections. The standard deviation of the observations increases from 1.8 % (1995) to 14.1 % (2010) while the difference between the 75 % fractile and the 25 % fractile increases from 2 % (1995) to 10 % (2010). Thus, the variation in the opinion of the experts in the Delphi survey must be taken into account. The distribution of the observations may be used as an input to risk analysis in order to reflect the uncertainty in critical variables. Such risk analyses with distributions of observations as inputs were performed in the TITAN project.

Forecast year	Round 1				Round 2			
	Standard deviation	75% fractile	25% fractile	Differ-ence	Standard deviation	75% fractile	25% fractile	Differ-ence
1000 ECU	20.4	30	5	25	6.5	12	5	7
1400 ECU	17.3	20	2	18	4.9	10	3	7
1800 ECU	13.9	15	0.5	14.5	3.3	5	2	3
2300 ECU	11.4	10	0	10	2.1	3	0	3
4600 ECU	7.4	5	0	5	1.0	1	0	1

Fig 3.10 Standard deviation, 75 % fractile, 25 % fractile and difference between 75 % fractile and 25 % fractile of the forecasts of the two rounds.

3.4.2 Demand for bearer services as a function of tariff

Fig 3.11 shows the estimated demand curves for ISDN, wide- and broadband connections. The curves are based on the medians for each cost alternative in the Delphi questionnaire. The difference between the demand curves is very small and indicates that residential users are not willing to pay much more for a high capacity connection despite better quality.

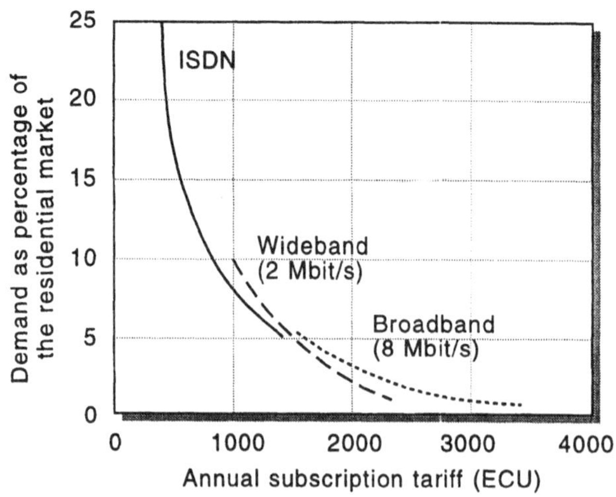

Fig 3.11 Demand curves for ISDN, wideband and broadband connections as a function of annual subscription tariff.

The demand curves are also used to make forecasts. Fig 3.11 shows a static demand curve where the demand only is dependent on the users annual tariff. In chapter 8 a methodology for forecasting network component costs is presented. The component costs decrease as a function of time due to mass production. In the TITAN project the cost of developing various network architectures is calculated. These results are used as an input to the demand curve for forecasting.

3.4.3 Demand versus disposable income

Based on answers from the experts, willingness to pay as a function of disposable family income is estimated for ISDN, wideband and broadband connections. Disposable income is the household income after tax, the part of the income which is available for savings and purchasing goods and services. Fig 3.12 shows that the demand is equal for ISDN, wideband and broadband when the disposable family income is in the interval 8,000 - 12,000 ECU. These households cannot afford to pay more for a high capacity connection than for ISDN. For higher family income there is a slight difference in willingness to pay between ISDN and wide- and broadband connections. However, Fig 3.12 shows that the incremental willingness to pay for wideband and broadband connections, is very small, even for wealthy families.

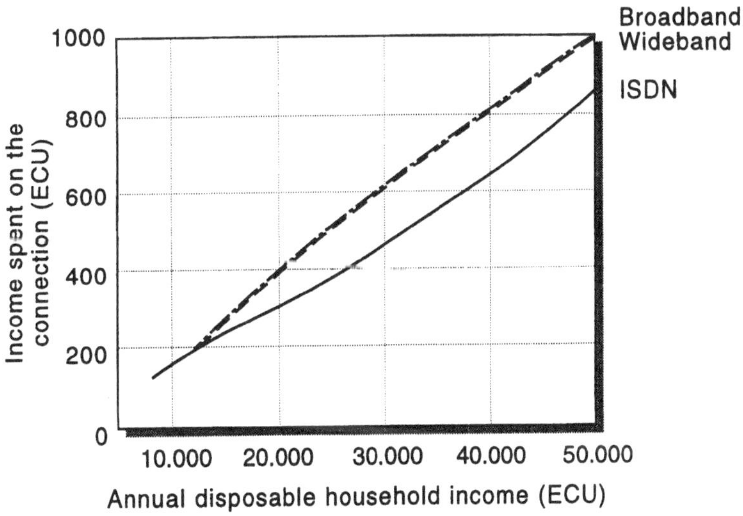

Fig 3.12 Willingness to pay for ISDN, wideband and broadband connections.

The results from the survey, illustrated by Fig 3.11 and Fig 3.12, indicate that residential users are not willing to pay much more for higher capacity and better quality when ISDN is the reference connection.

3.5 ESTIMATION OF PRICE ELASTICITY

The demand for a service will often be affected by a change in its price. The concept of price elasticity is used to measure the response to a price change. The Delphi survey results were used to estimate the price elasticity of ISDN, wideband and broadband connections.

3.5.1 Demand model

The demand for wideband and broadband connections were estimated based on the questionnaire in the Delphi survey. A set of different functions have been examined. The best model turned out to be:

$$y = \alpha\, e^{\beta p} \qquad (3.1)$$

where

y: The demand in percentage of the residential market,
p: Annual subscription cost,
α,β: Parameters in the model.

Statistical estimation procedures are used to find the value of the parameters which fits best. Regression analysis is used. However, a logarithmic transformation must be performed in order to get a linear expression for the parameters in the model. The new model is:

$$\ln y = \ln \alpha + \beta p \qquad (3.2)$$

Now, the regression procedure called "ordinary least squares regression" (OLS) finds the optimal values of $\ln \alpha$ and β.

3.5.2 Estimation of demand curves

The model fitting is based on the median results from the questionnaire in the Delphi survey, in which the experts were asked to estimate the proportion of subscribers for a set of annual subscription tariffs. The estimated parameters from the regression analysis are:

Wideband: $\alpha = 58.65$, $\beta = -0.001735$
Broadband: $\alpha = 33.25$, $\beta = -0.001289$

The estimated significance level for $\ln \alpha$ and β in the wideband model are 0.003 and 0.006 respectively, while the significance levels for the same parameters in the broadband model are 0.003 and 0.004 respectively. In the test applied a parameter will not be included if the significance level is greater than 0.05.

The multiple correlation coefficient tells how good the fit is. The size of the coefficient varies between - 100 % and 100 %. In the linear transformed model the estimated multiple correlation coefficients are 99.44 % and 99.64 % respectively! The fit is very good. The reason is that the number of observations is limited since the total number of responses from the experts for each question is reduced to one observation, the median.

The analytical expression of the demand curve for wideband connection is given by:

$$y = 58.45\, e^{-0.001735p} \tag{3.3}$$

The analytical expression of the demand curve for broadband connection is given by:

$$y = 33.25\, e^{-0.001289p} \tag{3.4}$$

In analysis of consumer demand there is always a question about demand when the price is 0. In some situations the demand will be 100 %, but not necessarily. When we are looking at the development of the access network, we also have to bear in mind that the subscribers have to pay for the terminal equipment and for some services as well as for the traffic produced. Hence, it is not obvious that the demand will be 100 % if the annual tariff is 0. Work has also been carried out to model the demand curve based on the assumption that zero tariff gives 100% demand. Then the restriction implies that $\alpha = 100$. The estimated parameters from the regression analysis are:

Wideband: ($\alpha = 100$), $\beta = -\,0.002037$
Broadband: ($\alpha = 100$), $\beta = -\,0.001703$

The estimated significance level for β in the wideband model is 0.0001, while the significance level for the same parameter in the broadband model is 0.0002.

In the linear transformed model the estimated multiple correlation coefficients are 99.87 % and 99.73 % respectively. The fit is as for the previous models, very good. The analytical expression of the demand curve for wideband connection is given by:

$$y = 100\, e^{-0.002037p} \tag{3.5}$$

The analytical expression of the demand curve for broadband connection is given by:

$$y = 100\, e^{-0.001703p} \tag{3.6}$$

When the annual tariffs are not close to zero, it is recommended to use equation (3.3) and (3.4).

3.5.3 Price elasticities

The price elasticity, *e*, is derived from the derivative of equation (3.1) and dividing the expression for the derivative by the factor *y*/*p*.

$$e = \frac{\frac{\partial y}{y}}{\frac{\partial p}{p}} = \frac{\partial y}{\partial p} \cdot \frac{p}{y} = \beta p \tag{3.7}$$

The price elasticities show the relative change in demand when the price changes. For wideband and broadband connections, assuming the price is not close to zero, the price elasticity functions are:

$$\text{Wideband: } e = -0.001735p \quad (3.8)$$
$$\text{Broadband: } e = -0.001289p \quad (3.9)$$

The formulas reflect the relative change in demand in percentage, when the tariff changes with one ECU as a function of the tariff.

3.6 UNCERTAINITY IN THE RESULTS

In the first round of the Delphi survey 58 questionnaires were returned - in the second round 50. The opinion among the experts varied, but the variation decreased significantly from round one to round two.

It is very difficult to derive long term forecasts for new services. Hence, it is of great importance not only to present the most probable values, represented by the median and the mean value in the survey, but in addition to use some estimates of the uncertainty. The uncertainty in the results described by the fluctuation in the answers from the experts are calculated by the use of standard deviation and by the difference between the 75 % and 25 % fractiles.

These two estimates of the uncertainty have been calculated for all questions both in round one and two. The uncertainty measure in round one has been divided by the same measure in round two. For every question we get a ratio indicating the uncertainty reduction from round one to round two. We have used the median and not mean values for the results, since some of the answers give very high ratios due to lower uncertainty in the second round. Based on all the answers in the two rounds the medians are calculated to:

Ratio of standard deviations = 3.1.
Ratio of the differences between 75 % and 25 % fractiles = 3.0.

Fig 3.10 and Fig 3.13 illustrate how the answers vary between the two rounds. In Fig 3.13 for the alternative of 1,000 ECU annual tariff, the standard deviations are 20.4 and 6.5 in the two rounds respectively, yielding a ratio of 3.1. The calculated ratios are increasing with increasing annual tariffs. For an annual tariff of 4,600 ECU the standard deviations are 7.4 and 1.0 respectively and the calculated ratio 7.4. The ratio between 75 % fractile and 25 % fractile in the two rounds fluctuates around 3.

Fig 3.10 shows that the ratio between the standard deviations in the two rounds decreases from 3.8 in 1995 to 1.4 for maximum demand. The proportions between the fractile difference in the two rounds are between 2.5 and 3. The conclusion is that the variation in the answers from the experts has been reduced significantly from round one to round two. Hence, the experts have approached a common opinion which is one of the objectives of a Delphi survey.

Forecast year	Round 1				Round 2			
	Standard deviation	75% fractile	25% fractile	Differ-ence	Standard deviation	75% fractile	25% fractile	Differ-ence
1000 ECU	20.4	30	5	25	6.5	12	5	7
1400 ECU	17.3	20	2	18	4.9	10	3	7
1800 ECU	13.9	15	0.5	14.5	3.3	5	2	3
2300 ECU	11.4	10	0	10	2.1	3	0	3
4600 ECU	7.4	5	0	5	1.0	1	0	1

Fig 3.13 Standard deviation and difference between 75 % and 25 % fractile of the forecasts for the two rounds.

The uncertainty measures presented give an estimate of the uncertainty to be used in the risk analysis in order to evaluate the effect of the different results. The set of answers from the experts on various questions constitutes a probability distribution, which may be input to simulations for risk analysis. The application and implementation of risk analysis in TITAN is described in chapters 9, 10 and 19.

3.7 SUMMARY

Where an acceptable level of quality is provided by wideband bearer services, the Delphi survey indicates that customers will be unwilling to pay more for broadband bearer services. Low forecasts were made for broadband connections in the year 2010 (5 % of the residential market). The TITAN Delphi survey suggests that the services with the most potential are video on demand, videotelephony and the simple home office. There is expected to be widespread take-up of video on demand. Demand for the simple home office is expected to be less dependent on price than other residential services, as employers use different criteria from private individuals when deciding whether to purchase a service or not. Experts consulted for the TITAN project believe that each household with a wideband or a broadband connection will have between three and four service subscriptions.

In a situation when there is no historical data of the demand, it is of course very difficult to make a good forecast. The traditional forecasting methods are not applicable. The results from this study are based on a Delphi survey. The way the experts iterated throughout the questions to a sort of common opinion shows the strength of this methodology.

REFERENCES

[1] Luck, D P. Broadband To The Home: Evolution Scenarios for Australia. *In: Proc. Thirteenth European Conference on Fiberoptics and Networks (EFOC&N '95), Brighton, UK, June 27-30, 1995, 1, 173-176.*

[2] Stordahl, K, Murphy, E. Methods for forecasting long term demand for wide and broadband services in the residential market. *IEEE Communications Magazine*, 33, (2), 1995, 44-49.

[3] Stordahl, K, Olsen, B T, Murphy, E, Zaganiaris, A. de Castro, E, Ravera, M. Delphi Survey: Forecasting Demand for Wide- and broadband services. From RACE 2087/TITAN. *In: Proc 6th International IEEE workshop on Optical Access Networks (OAN '94), Kyoto, Japan, Oct. 16-19, 1994, 7.3-1 - 7.3-10.*

Access network technologies

4

Leif Aarthun Ims, Mauro Ravera, Stefano Balzaretti, Umberto Ferrero, Marco De Bortoli, Kjell Stordahl, Borgar Tørre Olsen

4.1 INTRODUCTION

The selection of transmission medium or sets of transmission media and corresponding network topologies, node configurations and system technologies are among the key strategic issues in the development of the access network. Capacity demand, service types, flexibility and costs are the main differentiators in the evaluation of the various access network architectures. In the following the term architecture denotes (a) specific combination(s) of transmission medium, network topology, node configuration and system technology.

This chapter describes the main access network technologies and architectures commercially available today, some of the emerging technologies and some alternatives which are expected to reach the market within the next years. The intention is neither to provide a complete overview of access network technologies and architectures, nor to give a thorough description of the respective technology and architecture options. The main objective is to provide a brief introduction to this aspect, at the level of detail considered to be sufficient for performing techno-economic analyses in the context of the TITAN (tool for introduction strategies and techno-economic evaluation of access network) methodology and tool and similar assessment methods.

4.2 SERVICE CHARACTERISTICS AND NETWORK ARCHITECTURES

The selection of the network architecture is strongly related to the services and applications to be provided over the network [1,2]. Hence, the service and application attributes are of great importance in the selection of architectures. As mentioned in chapter 3, the terms "application" and "service" are not unambiguously defined in the telecommunication environment. In the context of this book the application refers to the use of the telecommunication service made by the customer and that the service is the product sold to the customer by the service provider. Additionally, the telecommunications services can be divided into information services and communications services. Information services permit the access to a service centre for access to data bases, movies on demand, etc. Communication services allow the interconnection of remote users and the dialogue between them for videotelephony, file transfer, etc. The impact of the service selection on the choice of the access network architecture may be based on the values of the following attributes:

- modulation technique
- bit rate for digital services
- bandwidth required
- symmetry
- communication configuration
- communication establishment
- mobility.

The *bandwidth* and bit rate required to support the service are a fundamental attribute of services with respect to the selection of the architecture, since some types of physical media and transport systems may not provide the required capacity. The bandwidth required is not an independent parameter as it is determined by the type of modulation and coding adopted for the information transfer.

The *symmetry* of the service differentiates between architectures which have been conceived for asymmetric services, like asymmetric digital subscriber line (ADSL), traditional coaxial cable systems some radio systems, and symmetric architectures like high speed digital subscriber line (HDSL) and optical fibre point-to-point systems.

The point-to-multipoint, or multicast and broadcast *communication configuration* are most readily implemented in shared medium architectures as passive optical networks (PONs), tree-and-branch coaxial cables and point-to-multipoint radio systems. In other types of architectures, including synchronous digital hierarchy (SDH) rings, the point-to-multipoint configurations require some additional functionality in the active equipment. However, point-to-point, or narrowcast configurations require additional capabilities like filtering, blocking, and encryption in shared medium architectures.

The *communication establishment* impacts the mix of capacity permanently allocated in the access network and capacity allocated on demand. Video distribution services such as television broadcasting, pay per channel (PPC), pay per view (PPV) and near video on demand (NVOD) require a fixed amount of bandwidth per customer. Additionally the establishment of connections on demand and the control of information flow and format requires the presence of a signalling channel up to the nearest network element capable of realising such functions.

Mobility implies that some category of services can be implemented either on a fixed architecture or on a wireless architecture only, at least in the distribution part. Services which allow mobility are necessarily implemented on wireless architectures, but other features for control of the position of the subscribers and call handling are required.

The relationship between the applications and the proposed bearer services is addressed in more detail in chapter 3.

4.3 THE PRESENT ACCESS NETWORK INFRASTRUCTURE

The dominant telecommunication access network infrastructure is still based on twisted pair copper cables for telephony services. In addition many business customers are now connected to the access network with multiples of 2 Mbit/s provided on optical fibres. The latter is mainly due to the digitalisation and the high demand for primary rate integrated services digital network (ISDN), leased lines and data communication services. Coaxial cable networks and direct to the home (DTH) satellites support the residential market with distributive broadband applications.

In urban areas today fibres are deployed not only in the junction networks but also in parts of the access network. Fibre loops with service access points (SAPs), transmission equipment and remote subscriber units are established in order to increase reliability and flexibility. The SAP refers to the localisation of the concentrator and/or add and drop multiplexers in the network, as commonly used in European countries with advanced access network infrastructures. The development of the new architecture is advancing by the introduction of SDH technology and add and drop multiplexers in the local loop. Shown in Fig 4.1 is a typical public switched telephone network (PSTN) access network infrastructure.

The cable operators typically have a coaxial cable distribution network for broadcast television services. The signals are fed from a headend, which passes some tens of thousands of homes. The headend serves several hubs, which typically passes some thousands of homes. Optical fibres are usually used for the transmission of signals from the head end to the hubs, whilst the network between the hub and the subscriber is a coaxial cable network, with bandwidths of 300 MHz, 450 MHz, 606 MHz, 750 MHz or 860 MHz. Fig 4.2 shows a coaxial cable network for distribution of television services.

In summary, the main telecommunications network platforms applied today are [3]:

- PSTN/ISDN network
- NMT (Nordic mobile telephone system)/GSM (global system for mobile communications) network
- PSDN/CSDN (packet and circuit switched datacommunications network)
- FrameRelay
- ATM (asynchronous transfer mode)
- Internet
- leased lines
- CATV (community antenna television) network
- satellite systems

The fast expansion of Internet has created a new market and increased the demand for a set of applications. Further development is expected to create demand for broadband applications. The dominating network operator has already integrated the traffic from the different platforms in the transport network at long distance and junction level and also in parts of the access network. The technology used today is mainly fibre and plesiochronous digital hierarchy (PDH) transmission systems and in addition radio links in rural areas. During the next years a dramatic change in technology is expected with respect to price and functionality. The network operators will start to implement SDH transmission systems mainly for network capacity expansion, and during the next few years also replace the existing PDH equipment due to reduced life cycle costs of SDH technology. In parallel the introduction of ATM switching technology has started.

However, the difference in existing access networks for some of the competitors calls for quite different upgrade strategies. This advocates an in-depth study of the starting point for upgrades to broadband, namely the existing competing infrastructures, and technology alternatives available.

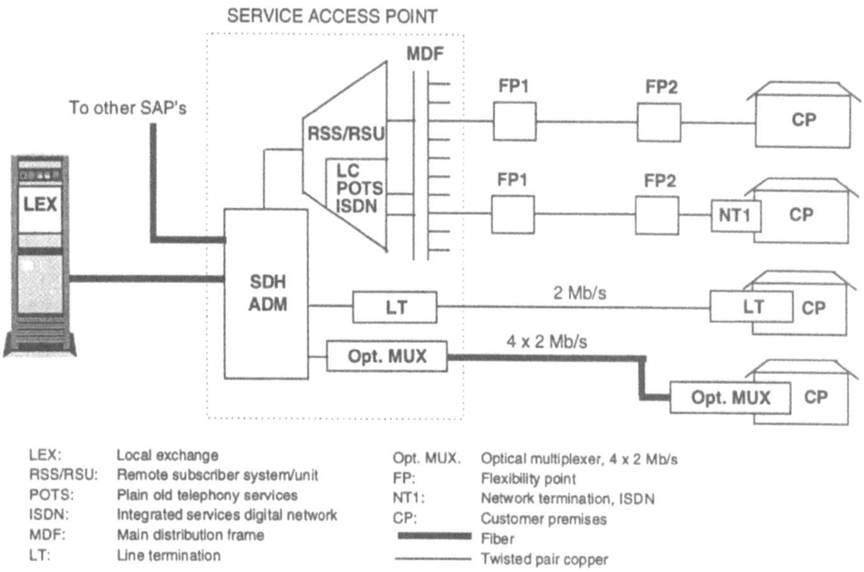

Fig 4.1 A typical PSTN access network infrastructure.

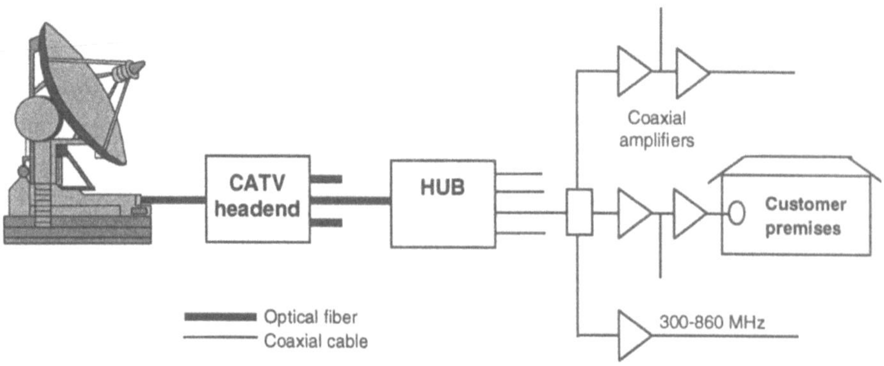

Fig 4.2 A typical coaxial cable network for distribution of television services.

4.4 EMERGING WIRELINE ACCESS NETWORK ARCHITECTURES

Today there is a wide range of alternatives for upgrading the physical layer of the access network to broadband [4,5]. The different available transmission media are twisted pair copper cable, coaxial cable, fibre cable, radio and satellite. For each of the media there are several different options with respect to network topology, node configuration and system technology.

The physical network topology refers to the physical network links. The network topology can be rather complex, depending on the number of levels of the network itself. However, the main topology options are point-to-point networks and point- or mulitipoint-to-multipoint networks. In the former each physical link is dedicated to connect two (active star, double active star) network terminations. The latter connects several network terminations (single/double passive star, bus, tree).

The node configuration refers to the configuration of the network terminations of one particular transmission medium, i.e. the network nodes at which the terminal equipment is located. For fibre architectures, several node configurations have been proposed: fibre to the cabinet (FTTCab), fibre to the node (FTTN), fibre to the curb (FTTC), fibre to the building (FTTB) and fibre to the home (FTTH) are commonly used, depending on the local network area conditions.

The system technology attributes are the multiplexing technique, the access protocol for point-to-multipoint configurations and the modulation techniques (basically classified in analogue and digital and digital techniques). Typical multiplexing techniques are frequency division multiplexing (FDM), time division multiplexing (TDM), subcarrier multiplexing (SCM) wavelength division multiplexing (WDM), high density wavelength division multiplexing (HDWDM) and optical frequency division multiplexing (OFDM). Commonly used access protocols are frequency division multiple access (FDMA) time division multiple access (TDMA), code division multiple access (CDMA), subcarrier division multiple access (SCMA) and wavelength division multiple access (WDMA).

A set of different architectures expected to be relevant for the access network towards the turn of the century is listed below.

- digital subscriber line (DSL) technologies, like ADSL (asymmetric digital subscriber line), HDSL (high bit rate digital subscriber line), VDSL (very high speed digital subscriber line), and SDSL (symmetric digital subscriber line)
- FTTH using TPONs (telecommunication over passive optical networks) and ATM PONs (ATM over broadband passive optical networks)
- FTTC/FTTB using TPONs and ATM PONs
- FTTN using TPONs and ATM PONs
- FTTN using point to point fibre
- FTTC/FTTB using hybrid fibre coax (HFC)
- FTTN using HFC
- fixed radio access systems, point to point
- fixed radio access systems, point to multipoint
- mobile radio systems, cellular: NMT/GSM/DCS (digital cellular system)1800
- mobile radio systems, cordless: DECT (digital European cordless telecommunication)/CT2 (cordless telephony, second generation)
- DTH satellite

- communication satellites: Iridium, ICO Global Communications, Globalstar, Teledesic.

4.4.1 Digital subscriber line technologies

Until recently twisted pair has been considered to be a significant bottleneck with respect to capacity. Recent developments of complex modulation schemes have enabled the extension of the line capacity by order of magnitudes. In the short term the main advantage of copper is a variable cost option, alleviating the need for high and risky upfront investments. In addition the installed base of 600 million lines world-wide constitutes an immediate advantage [6]. The enhanced copper or digital subscriber line (DSL) technologies differ with respect to transmission capacity, transmission distance and the number of twisted pairs used. In general for the DSL options, there is a trade off between distance and capacity. The DSL system technologies include ADSL, VDSL (very high-speed digital subscriber line), HDSL (high bit-rate digital subscriber line) and SDSL (symmetric digital subscriber line) [7]. At present only ADSL and HDSL are commercially available, VDSL and SDSL are expected to become available towards the turn of the century.

Fig 4.3 shows a detailed sketch of an enhanced copper architecture. An architecture with an SDH ring between the local exchange (LEX) and the SAP is depicted. The infrastructure for plain old telephone system (POTS) and N-ISDN are included for clarity.

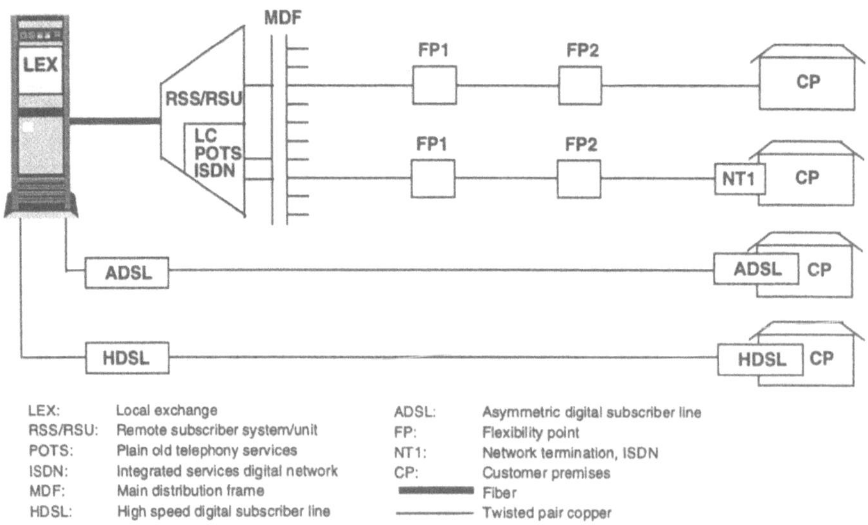

Fig 4.3 The enhanced copper upgrade option with ADSL for 2 Mbit/s asymmetric demand and HDSL for 2 Mbit/s symmetric demand.

ADSL uses one twisted copper pair for transmission of between 2 Mbit/s (4 km) and 8 Mbit/s (2 km) downstream and 176 kbit/s and 1 Mbit/s upstream. Analogue telephony is transmitted on the same twisted pair.

HDSL requires two or three pairs for symmetric 2 Mbit/s transmission over 2.5 km. Analogue telephony may not be transmitted simultaneously on the same twisted pair. The one-pair version of HDSL, the so-called SDSL, may include the option of transmission of analogue telephony on the same twisted pair.

VDSL provides both asymmetric and symmetric transmission, with asymmetric capacities as high as 26 Mbit/s (1 km). The 13 Mbit/s capacity VDSL standard modem has a 1.5 km transmission distance. For both downstream capacity options the upstream capacity is at least 2 Mbit/s. The symmetric VDSL versions will have 26 Mbit/s (0.3 km) and 13 Mbit/s (1 km) transmission capacities. Analogue telephony is to be transmitted on the same twisted pair. However, there is an uncertainty with respect to the coexistence of symmetric and asymmetric VDSL systems in the same cable sheath.

4.4.2 Coaxial cable upgrades

The need to support analogue broadcast video in the access network has focused the interest on hybrid fibre coax (HFC) as an attractive solution. The HFC technology for coaxial cable networks allows provision of return capacity in the coaxial cable distribution networks [8].

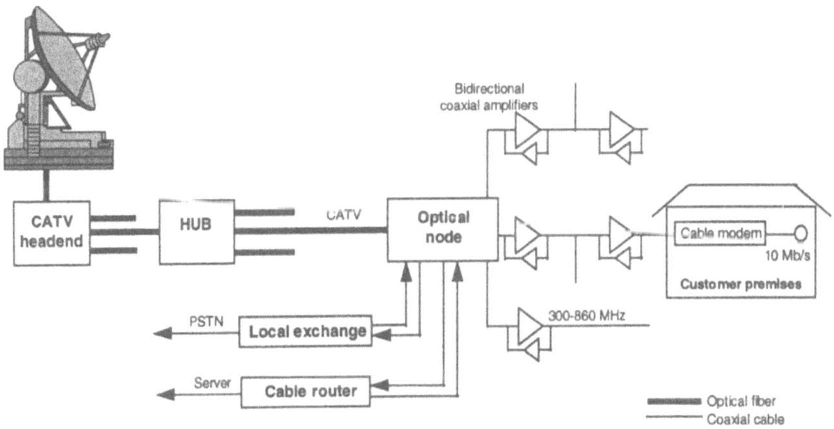

Fig 4.4 The hybrid fibre coax upgrade option with cable modems.

Both dedicated channel systems for asymmetric and symmetric broadband transmission and shared capacity systems (cable modems) have been proposed. An HFC upgrade implies splitting the coaxial network into smaller coaxial segments in order to achieve the required return path capacity. In the case of each subscriber being fitted with a 10 Mbit/s cable modem typically 5-600 subscribers share 50-60 Mbit/s upstream capacity, implemented by an Ethernet protocol. At the hub side a cable router terminates the coaxial cable network segment. Fig 4.4

depicts the HFC upgrade option with cable modems as analysed here. The existing coaxial cable infrastructure between the access node serving 1,000 subscribers and the homes is upgraded to a bi-directional network with return path and cable modems are installed at the customer premises.

4.4.3 Fibre upgrade technologies

Optical fibre is the transmission medium which offers the highest bandwidth, with transmission capacities potentially as high as Tbit/s [9]. However, introduction of fibre optics requires the deployment of fibre cable, which usually is associated with high civil works costs. In addition, at present the fibre optics terminal equipment is rather costly, due to low production volumes. Hence, it is expected that it will still be years before the fibre deployment for broadband in the access network reaches economically justifiable levels in non-greenfield areas.

Telecommunication over a passive optical network (TPON) was the first generation fibre in the loop (FITL) systems. These systems utilise a shared fibre infrastructure to deliver a range of services to a set of customers. TPON is used to support telephony, ISDN basic access (BA) and ISDN primary rate access (PRA) and leased lines.

The second generation FITL systems will provide broadband capacity. The concept of BPON (ATM based broadband passive optical network) is already well-known as systems that typically offer 622 Mbit/s or 155 Mbit/s downstream capacity and 155 Mbit/s upstream capacity [4]. The access is shared between 16 or 32 optical network units (ONUs) with an inherent statistical multiplexing capability. The optical line terminal (OLT) terminates the network at the exchange side. A typical BPON architecture is shown in Fig 4.5.

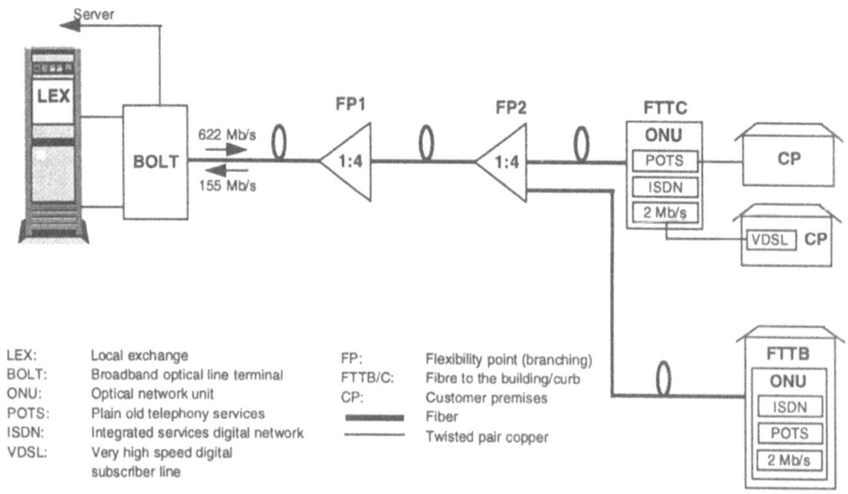

Fig 4.5 A typical BPON architecture.

More advanced optical technologies like wavelength division multiplexing (WDM) may enable future access networks with significantly increased flexibility and capacity. However, the near-term and medium term challenges in the access network will most likely be solved by further utilisation of the above mentioned set of technologies. In the long term, the deployment of WDM may become a cost effective solution also in the access network for very high capacity demand.

The incomplete list of hybrid FITL architectures below illustrates the wide variety of solutions available for the access network operators today:

- SDH ring (FTTR (fibre to the remote node)/FTTN) and twisted pair in the distribution network
- PON (FTTC/FTTB) and twisted pair drop
- Optical active star (FTTSA (fibre to the service area)/FTTN) and tree and branch coaxial cable in the distribution network (traditional CATV network)
- Optical active star (FTTN) and tree and branch coaxial cable in the distribution network (integration of interactive and distributive services)
- Parallel PON/SDH ring (FTTN) in the feeder network and tree and branch coaxial cable in the distribution network
- PON (FTTC/FTTLA (fibre to the last amplifier)/FTTB) and coaxial cable drop (for provision of distributive CATV services)
- Optical active star (FTTC/FTTB) with multiplexer and parallel twisted pair/coaxial cable drops
- PON (FTTC/FTTB) and twisted pair/coaxial cable drops
- SDH ring and PON in the distribution network (FTTC/FTTB) with parallel twisted pair/coaxial cable drop
- SDH ring and point-to-point fibre in the distribution network (FTTC/FTTB) with multiplexer and parallel twisted pair/coaxial cable drop

4.5 WIRELESS ACCESS TECHNOLOGIES AND ARCHITECTURES

The remarkable growth of wireless communications over the past years, including terrestrial radio systems and satellite systems, has led to an increased interest in wireless technologies for broadband access. The present radio and satellite systems mainly offer narrowband interactive services (telephony) or broadband distributive services. Several technology alternatives and concepts for wireless broadband access have been proposed and are currently under research and development. Currently the maturity of the broadband radio access systems lags behind their wireline counterparts, at least if significant broadband take rate capabilities are considered. Nevertheless, the broadband radio access solutions may in due time prove to be a key technology in the emerging broadband market, mainly because they enable potentially very rapid network roll-outs, low capital costs compared to wireline alternatives in scarcely developed areas, and flexibility in planning and deployment [10]. Cell size, capacity and return channel capability are the main differentiating attributes of the wireless access solutions. Wireless architectures for broadband access as such have not been analysed in the TITAN work, but have been included in brief in this section in order to relate the wireline technologies to the wireless alternatives and vice versa in the context of future broadband access networks.

4.5.1 Narrowband radio access systems

Today wireless local loop (WLL) and radio in the local loop (RLL) systems are deployed for the provision of narrowband services. The RLL technologies may either be fixed wireless access systems, intended as a replacement for the conventional copper loop, or provide the additional feature of mobility to the user. In general, the present radio access systems are either fixed radio access systems (point to point or point-to-multipoint), cordless systems (DECT, CT2) or cellular systems (NMT/GSM/DCS1800). Listed below are some of the hybrid RLL architectures for fixed radio access:

- point-to-point radio links and twisted pair drop
- FTTR plus point-to-point radio links and twisted pair drop
- point-to-multipoint radio links and twisted pair drop
- FTTR plus point-to-multipoint radio links and twisted pair drop
- point-to-multipoint radio links and parallel twisted pair/coaxial cable drops. The solution supports video services
- FTTR plus point-to-multipoint radio links and parallel twisted pair/coaxial cable drops. The solution supports video services.

Some of the hybrid cellular technology architectures are listed below:

- active star with twisted pairs between the local exchange and the distribution node, plus wireless systems based on typically GSM or DCS 1800 with mobile or fixed terminals
- active star with twisted pairs between the local exchange and the distribution node, plus wireless systems based on CT2, DECT or DCS 1800 with mobile or fixed terminals
- FTTR plus wireless systems based on GSM or DCS 1800 with mobile or fixed terminals
- FTTN plus wireless systems based on CT2, DECT or DCS 1800 with mobile or fixed terminals
- FTTN optical active star plus coaxial cable and wireless systems based on CT2, DECT or DCS 1800 in the distribution network and mobile terminals.

The radio access technologies are expected to evolve from the current stage with systems offering POTS, N-ISDN and n·64 kbit/s leased lines to systems with up to 2 Mbit/s capacity, e.g. local multipoint distribution service (LMDS) and multipoint multichannel distribution system (MMDS). Finally, the next generation systems are expected to be wireless ATM systems with broadband capacities, such as the European mobile broadband system (MBS) and universal mobile telephony system (UMTS).

4.5.2 Cellular systems for wideband and broadband access

Several broadband radio access alternatives are available or under development. The next generation mobile system is expected to be UMTS, which will have at least a 2 Mbit/s symmetric transmission capacity. MBS is specified for symmetric transmission of 155 Mbit/s over some few hundred meters in the 40-60 GHz frequency band. The multichannel multipoint distribution service (MMDS) system has been in operation since 1989, and it is currently being developed from a 30 analogue channels capacity broadcasting system with digital transmission and significantly higher capacities. MMDS requires line of sight, and has a 50 km reach [11]. The evolution of such systems towards future interactive broadband services is likely to include small cell size systems, with less stringent requirements for line of sight, as for example local multipoint distribution service (LMDS) systems. Analogue LMDS systems are available today, which utilises the 27.5-29.5 GHz frequency band yielding a service area of between 5-10 kilometre radius. This is significantly smaller than the up to 50 kilometre radius cell size of similar MMDS systems [12]. Future digital LMDS systems will have higher capacity than the analogue LMDS systems. However, the interactivity is one challenge of these broadcast tailored systems. The return channel may either be provided by the use of the existing twisted pair cable network, by the use of an overlaid cellular radio technology or through the use of a return channel in the 27.5-29.5 GHz frequency band [11]. The European version of LMDS is called microwave video distribution service (MVDS). The LMDS, MBS, UMTS systems all will probably need another several years in the making before they become commercially available systems.

4.5.3 Satellite systems for wideband and broadband access

Communication satellites at present include fixed service satellites (e.g. for television distribution, distance learning and data communications), DTH broadcast service satellites and mobile service satellites (voice services and digital communications) [13].

Conventional fixed service satellites are expected to be used for the implementation of the Direct PC concept, which offers a downstream capacity of up to 10 Mbit/s, including applications such as Internet access, real-time video and audio and virtual private networks. The DTH systems may be used to provide distribution of broadband services such as pay per view, asymmetric Internet access and near video on demand.

The new communication satellite systems ICO Global Communications (10 satellites), Globalstar (48 satellites) and Iridium (66 satellites) will support mobile telephone and data communication limited to 9.6 kbit/s. ICO Global Communications is a mid-earth orbit (MEO) system, whilst the others are low-earth orbit (LEO) systems. In addition, the satellite system Teledesic with 840 satellites in LEO has been proposed for high bit rate multimedia applications in the access network.

However, all of these satellite systems for wideband and broadband access are definitely better suited for asymmetric, distributive and downloading services than continuous wideband or broadband services with significant requirements for upstream capacity. The lack of symmetric capability, the limited total system capacities compared to the wireline alternatives and the lag in commercial maturity are the main inhibitors for these systems constituting anything else but niche systems, and not full-scale, alternative infrastructures to the wireline systems within the next five to ten years.

4.6 SUMMARY

At present no single network architecture seems to be the obvious choice. In stead, boundary conditions given by the existing infrastructure and customer segment characteristics probably will require simultaneous roll-out of several technology alternatives. The DSL and cable modem upgrades will probably emerge within the next few years, and be the predecessors for the future access network architectures based on optical fibre, radio and satellite. However, the two latter alternatives at the moment lag both with respect to capacity (niche products) and maturity compared to the wireline alternatives.

REFERENCES

[1] Miki, T. Toward the Service-Rich Era. *IEEE Communications Magazine*, 32 (2), 1994, 34-39.

[2] LeBel, P, Oliver, R. Residential Broadband Architectures Strategies and Economics. *In: Proc. The 7th International IEEE workshop on Optical Access Networks (OAN '95), Nuremberg, Germany, September 24-27, 1995, 4.4-1 - 4.4-10.*

[3] Stordahl, K, Olsen B T, Ims, L A. Do we need a Pan-European network and what is the need for new applications? *In: Proc. 22nd European Conference on Optical Communication (ECOC '96), Oslo, September 15-19, 1996, 2, 2-3 - 2-10.*

[4] Pugh, W, Boyer, G. Broadband Access: Comparing Alternatives. *IEEE Communications Magazine*, 33 (8), 34-45, 1995.

[5] Kopf, D. Internet Race - xDSL vs Cable Modems. *America's Network,* August 1, 1996.

[6] Kelly, D. Digital Subscriber Lines (DSLs): expanding the use of the copper loop plant. *In: Proc. Eurescom Workshop on Optical Multiservice Access Networks (WOMAN), Turin, Italy, September 23-24, 1996.*

[7] Maxwell, K. Asymmetric Digital Subscriber Line: Interim Technology for the Next Forty Years. *IEEE Communications Magazine*, 34, (10), 1996, 100-106.

[8] Paff, A. Hybrid Fiber/Coax in the Public Telecommunications Infrastructure. *IEEE Communications magazine*, 13, (4), 1995.

[9] Cook, A, Stern, J. Optical Fiber Access - Perspectives Towards the 21st Century. *IEEE Communications Magazine*, 32 (2), 1994, 78-86.

[10] Garg, V K, Sneed, E L. Digital Wireless Local Loop Systems. *IEEE Communications Magazine*, 34, (10), 1996, 112-115.

[11] de Passoz, G, Clausse, J L, Karam, G. Can Digital Techniques give a new boost to MMDS? *In: Proc. 11th International Symposium on Subscriber Loops and Services (ISSLS '96), Melbourne, Australia, February 4-9, 1996, 134-138.*

[12] Nowak, R J. Wireless: Part of the Broadband Future. *In: Proc. 11th International Symposium on Subscriber Loops and Services (ISSLS '96), Melbourne, Australia, February 4-9, 1996, 122-127.*

[13] Seidman, L P. Satellites for Wideband Access. *IEEE Communications Magazine*, 34, (10), 1996, 108-111.

Section 2

Methodology and tool for techno-economic evaluation of the access network

Modelling in TITAN 5

Markku Lähteenoja, Jari Mononen, Markku Tahkokorpi, Ilari Welling, Frank Fleuren, Marc Drieskens, Nikolaus Gieschen, Sotirios Markatos

5.1 INTRODUCTION

Deriving suitable access network upgrade strategies requires strategic analyses and techno-economic evaluations of network projects. A sufficient access network strategic analysis to some extent relies on the combination of high level parameters and low level parameters of significant strategic relevance, dependant on the access network itself. The former typically are parameters such as density of subscribers, tariff elasticity and broadband take rate, whilst the latter may include network termination cards and splitting ratio of passive optical networks etc.

The RACE (research for advanced communications in Europe) 2087/TITAN (tool for introduction strategies and techno-economic evaluation of access network) methodology and tool is one approach which enables strategic analysis on the basis of this combination of differential level parameters. This chapter describes the framework of the TITAN model [1]. In addition, the approach used for costing operation, administration and maintenance is described, together with the TITAN database and the key outputs from the calculations. Major features of the model, such as the geometric models, the component price evolution over time and sensitivity and risk analysis, are described separately in chapters 6, 8 and 9 respectively. The application of the TITAN model to different urban and rural structures is presented in chapter 7. Chapter 10 very briefly describes the tool implementation

5.2 THE FRAMEWORK

The framework of TITAN comprises the following steps, which are necessary in the techno-economic evaluations of access network solutions:

1. The services to be provided must be specified and the market penetration of these services over some study period has to be defined. The services have associated tariffs, i.e. the part of the tariff that is attributed to the access network. From the combination of annual market penetration and annual tariff information TITAN calculates the revenues for each year for the selected service set.
2. The architecture scenarios to provide the selected service set must be defined next. This needs network planning expertise and is partly outside of the framework of TITAN methodology. However, TITAN includes several geometric models (chapter 6), that will facilitate the network planning by automatically calculating lengths for cables and ducting. These geometric models are optional parts of the methodology and TITAN can be used without them, e.g. for radio access technology evaluation, where no geometric models are necessarily needed. The result of architecture scenario definition is a so called shopping lists. This lists indicates the volumes of all network components (equipment, cables, cabinets, ducting, installation etc.) for each year of the study period and the location of these network components in different flexibility points and link levels.
3. The costs of the network components are calculated using an integrated cost database (chapter 5.4) developed within the TITAN project, containing data gathered from many European sources. Architecture scenarios together with the cost database give investments for each year.
4. TITAN has adopted a proportional approach for the calculation of operation, administration and maintenance (OAM) costs, in which the OAM cost is related to the capital investment (chapter 5.3). Each network component in the cost database is classified by an OAM class. This OAM class defines a percentage of the cumulative investments. This percentage is used to calculate OAM costs for this network component on an annual basis.
5. Investment costs together with the OAM costs give the life cycle costs for the selected architecture scenarios.
6. Finally, by combining service revenues, life cycle costs and general economic inputs (e.g. discount rate, tax rate) TITAN gives cash flows and other economic results, such as net present value (NPV), internal rate of return (IRR) and payback period.

The framework of the TITAN techno-economic evaluations is shown in Fig 5.1.

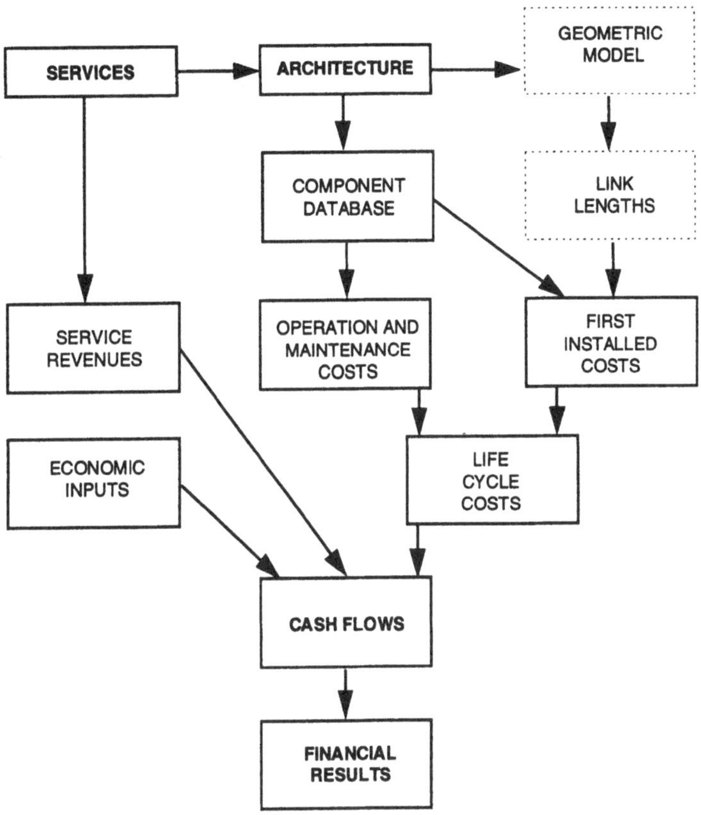

Fig 5.1 The TITAN framework.

5.3 OPERATION, ADMINISTRATION AND MAINTENANCE COSTS

The OAM costs are calculated using inputs from the cost database in order to produce life cycle cost [2,3]. OAM costs include preventive and corrective maintenance, network management and repair of defective equipment.

An interesting quantity is the share of the OAM cost, compared to the life cycle cost. Experience from operating networks suggests a ratio between 20 % and 35 % over ten years. The data gathering process for preventive maintenance has been largely based on periodic routine testing, usually on an out-of-service basis. Both planned and unplanned maintenance-related outages result in customer problems; a recent study performed by a large network operator has reported a 50/50 ratio between those two types of outages in the local exchange area context [4].

The conventional analytical method for calculating OAM costs is a very time-consuming process because a vast amount of very detailed data from many different sources has to be collected. The result of this calculation whenever possible, is accurate only for very specific network in a specific and well defined environment. The application of these results to other network architectures is not possible without significant loss of accuracy. For this reason TITAN adopted a simpler solution, the "proportional" approach, which relates the OAM cost to the capital investment. Each cost item in the cost database is classified by an OAM class. This OAM class defines the percentage of the global investments, i.e. the investments made for this cost item so far. This percentage is used to calculate OAM costs for this cost item on an annual basis [1].

The justification of this approach is simply based on the fact that capital usually can be related in some way to complexity, functionality, size, power and construction of hardware. The effect of equipment ageing on the OAM cost is considered to be compensated by the preventive maintenance, which is included in this percentage. This approach leads to the basic OAM cost and does not include the specific OAM characteristics of the various systems such as savings due to a fully automatic management system and those due to enabling-disabling functions for service provisioning. The total OAM cost may be divided in two parts: 1. Operation and administration, representing 60 % of the total; 2. Maintenance and fault management, accounting for the remaining 40 %. Systems with automatic management, may enjoy a significant saving of the operation and administration part of the OAM cost. This is frequently the case with innovative fibre-based, radio or hybrid architectures. In the tool the default value for savings of the operation and administration part is 50 %, which yields 30 % saving of the total OAM costs.

Table 5.1 Operation, maintenance and administration classifications as used in the tool.

OAM class	*OAM cost*
Civil works	3.0 %
CopperCable	4.2 %
Electronics	7.0 %
Enclosures	5.0 %
FibreCable	4.1 %
OptTransceiversAtLT	1.0 %
OptTransceiversAtNT	2.1 %
PassOptComp	4.2 %

The above approach for calculating OAM costs is based on the simple assumption that OAM costs are directly related to the cumulative investments in the network. This method may be further enhanced by dividing the OAM costs into three separate components, two maintenance parts and one combined operation and administration part.

The first maintenance component, the cost of repair parts is driven by the investments, and hence may be determined by the "proportional approach".

The second maintenance component, the cost of repair work is defined by three input parameters, namely the labour cost per hour, the mean time between repairs, (MTBR) and the mean time to repair (MTTR).

The third OAM component, the operation and administration costs, typically is driven by services, say by the number of customers, or by the number of critical network elements.

5.4 THE TITAN DATABASE

The TITAN database includes a service section and a cost component section. The services database includes service specific information for each service or revenue component, such as the initial tariff, a reference year for the initial tariff, the tariff unit and a learning curve coefficient. The penetration of the service in the initial year and the saturation penetration are also contained in the service database.

The cost database includes costs for a given reference year for components, installation, civil works, operations and maintenance. The data structure of the cost database is depicted in Fig 5.2 below.

The price curve for each cost component can be calculated by means of learning curve data and volume data. Learning curve data and volume data are classified in relevant groups. This simplifies the addition of new components and makes it easier to compare price evolution of various components with different functionality. For each component, a learning curve class and volume class have to be specified in order to predict the price curve. The classes are stored in the database.

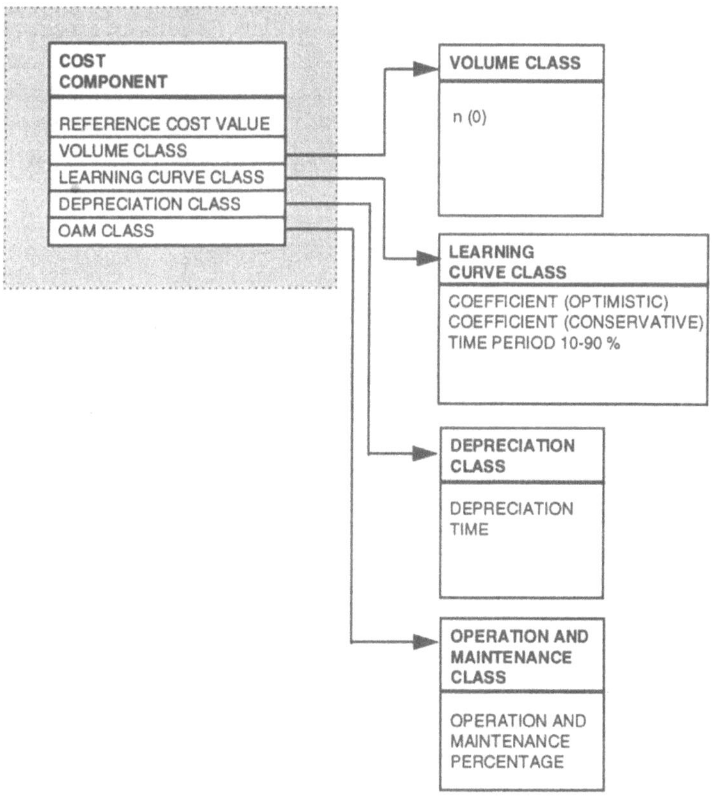

Fig 5.2 The structure of the TITAN cost database.

Another important financial figure is the depreciation. The depreciation time is additionally classified into several classes. The class number refers to the number of years over which a component is completely depreciated.

The final financial figure is the OAM class. This class specifies the percentage of the investment that contributes to the yearly running cost of the network. Other fields in the database are:

1. Cost unit, or the currency of the cost data
2. TypeID, a type identification of the component
3. ConfClass, the confidence class of the cost component which defines the uncertainty and the relative standard deviation in the price
4. Importance, the financial importance of the cost component.

5.5 KEY OUTPUTS

In the evaluation of investment scenarios there are often several aspects of interest and significance. Various commonly used indicators, like payback period, net present value or internal rate of return, can give differing results in comparisons, depending on the complexity of the scenario. Thus, it is often necessary to use several figures of merit for the studies in order to get a thorough understanding of the economic issues related to each scenario. In most of the cases these evaluations have some less known inputs, and it is advisable to apply sensitivity analysis and/or risk assessment using multiple figures of merit as indicators.

5.5.1 Revenues

Usually the calculation includes the revenues from the services. The TITAN methodology handles the revenues simply by assuming a certain annual tariff for each service per connected customer. However, it must be noted that revenues in TITAN refers to the revenues to the access operator, and not to the total amount paid by the customers for their telecommunication services. The revenues are not considered a major output in TITAN, but their inputs have to be correct in order to enable the calculation of economic outputs like profits, retained profits and cash flows.

5.5.2 Investments

The first of the main outputs of the TITAN methodology is the investment cost of the network project. The investments are usually spread over the study period, as the TITAN methodology is applied to scenarios. The future investments are discounted to the start of the study period using a conventional discounting formula, in order to get a single figure of merit for the total investment. The total discounted investment cost is referred to as the installed first cost (IFC).

In the TITAN methodology a network is divided into hierarchical flexibility points and connecting links. All links or flexibility points at the same hierarchy level form a so called network level. The current implementation of the methodology allows cost breakdowns on network level, i.e. the investments may analysed on the basis of location of the cost item in the network. In addition, the investments may be categorised by type of cost component. The available cost component types are:

- civil works
- installation
- materials.

The materials are further sub divided into following groups:

- copper cable
- fibre cable
- passive optical components
- electronics
- optical transceivers
- enclosures.

The investments can also be depicted as a function of time. In total, these three main output possibilities enables a good overview the key cost items in the scenario under study.

5.5.3 Life cycle costs

The life cycle cost is defined as the sum of the cumulative discounted investments and the cumulative discounted running costs. This gives the total costs for constructing and running the network over the study period.

Often the sole reason for a techno-economics study is the evaluation of the cost of the proposed network project. The reason might be a new network or an upgrade of the existing network. The inclusion of service revenues in the study is not always required in such cases, e.g. in comparative studies of competing network technologies.. Thus, the final result of the study can be given either in terms of the investment costs or in terms of the life cycle costs of the project.

5.5.4 Cash balance

The cash balance, or cumulative cash flow time series is a very informative figure for a network and service scenario. In particular for a greenfield case it gives much information in a single chart. An example of a cash balance time series is depicted in Fig 5.3.

A typical cash balance curve for a network scenario is first deeply negative because of the high initial investments. If the scenario is profitable, the cash flow turns positive fairly soon and the cash balance curve starts to rise. The lowest point in the cash balance curve gives the amount of funding required for the project. The point in time when the cash balance turns positive gives the payback period for the project. This is the usual case when studying a single technology scenario in a greenfield situation.

In an investment scenario where a majority of the expenditure occur in the beginning of the study period, the payback period gives a good indication of the efficiency of the investment. If the scenario is more complicated, that is, if there for example are several technology steps in an upgrade situation, it is not possible to define a single payback period. It is still possible to use the cash balance curve as an indicator for the profitability of the scenario. In such cases it is important to study the cash flow trend at the end of the study period.

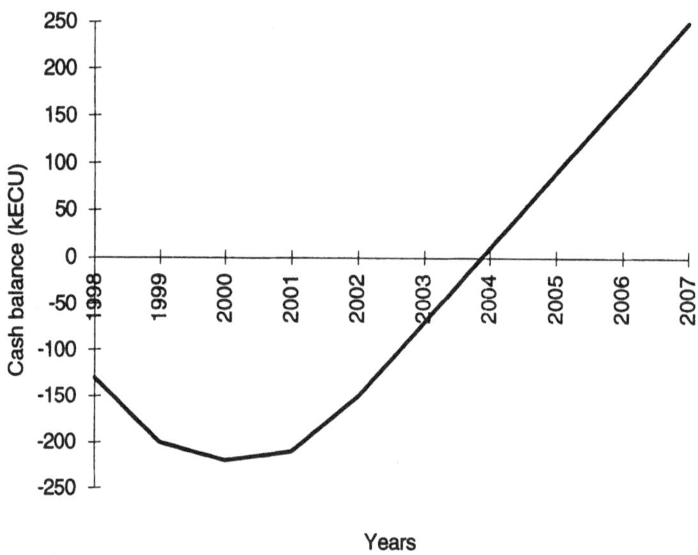

Fig 5.3 Cash balance curve

5.5.5 Net present value

The net present value (NPV) gives a single figure of merit for a project. Its definition is the sum of the discounted cash flows plus the discounted rest value of the project. The NPV gives the monetary value of the whole project in today's money. It is a good indicator for the profitability of the scenario, especially in cases where the payback period is difficult to use due to major investments at several points in time of the project period.

The definition or calculation of the rest value of the network is the weakest point in this figure of merit. There are several ways to try to define the rest value. The approach selected in TITAN uses the bookkeeping value of the network as the rest value, since this is the only figure that can be calculated from the inputs already available. Other possible approaches are simply to assume that the rest value is zero at the end of the study period. This can be true in some cases, for example if the operator has a license only for a fixed time and there is no guarantee that the license will be extended. Another approach is to use the cash flow of the network as an indication of the value of the network. The rest value may then be defined as three times the average cash flow during the three last years of the study period.

5.5.6 Internal rate of return

The internal rate of return (IRR) is the discount rate at which the NPV is zero. If the IRR is higher than the opportunity cost of money (the interest of an average long term investment), the project is generally viable. The net present value is rather difficult to use for comparisons of scenarios which differ significantly, e.g. in the size of networks under study. In these cases internal rate of return gives a good indication of how "good value for the money" these networks give.

5.6 SUMMARY

The schematic diagram in Fig 5.4 illustrates the TITAN methodology with the associated inputs, calculation stages and outputs.

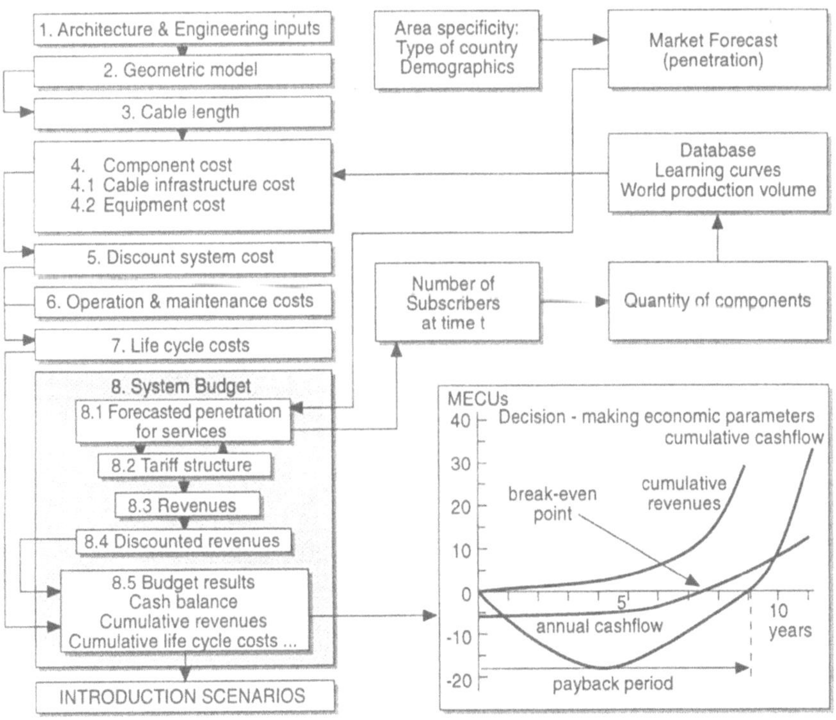

Fig 5.4 The TITAN methodology.

REFERENCES

[1] Olsen, B T, Zaganiaris, A, Stordahl, K, Ims, L A, Myhre, D, Øverli, T, Tahkokorpi, M, Welling, I, Drieskens, M, Kraushaar, J, Mononen, J, Lähteenoja, M, Markatos, S, De Bortoli, M, Ferrero, U, Ravera, M, Balzaretti, S, Fleuren, F, Gieschen, N, De Oliveira Duarte, M, de Castro, E. Techno-economic evaluation of narrowband and broadband access network alternatives and evolution scenario assessment. *IEEE Journal on Selected Areas in Communications*, 14 (6), 1996, 1184-1203.

[2] Zaganiaris, A, Gieschen, N, Olsen, B T, Stordahl, K, Tahkokorpi, M, Drieskens, M, Ajibulu, A. A methodology for achieving life-cycle costs of optical access networks-from RACE 2087/TITAN. *In: Proc. 11th Annual Conference European Fibre Optic Communications and Networks (EFOC '93), the Hague, the Netherlands, June 30-July 2, 1993, 2, 136-141.*

[3] Zaganiaris, A, Tahkokorpi, M, Drieskens, M, Van Hoecke, J, Olsen, B T, Markatos, S, De Bortoli, M. Life-cycle costs and economical budget of optical and hybrid access networks. *In: Proc. 5th IEEE Conference on Optical/Hybrid Access Networks, Montreal, Canada September 7-9, 1993.*

[4] Diaz de la Iglesia, R. Service-affecting optoelectronic failures in FITL systems: downtime, repair actions and maintenance expenses. *IEEE Journal on Selected Areas in Communications*, 12, (2), 1994.

Geometric models 6

Ilari Welling, Markku Tahkokorpi, Frank Fleuren, Marc Drieskens

6.1 INTRODUCTION

In the RACE (research for advanced communications in Europe) 2087/TITAN (tool for introduction strategies and techno-economic evaluation of access network) project geometric models have been developed and used to estimate the amount of cable/fibre and ducting required in the network. At the conceptual level the geometric model is a function that takes several inputs such as subscriber density, network topology (star, ring, bus), average cable over length, duct availability etc., and gives two outputs, which are the total amount of cable/fibre required in the network and the total amount of new duct required. The TITAN methodology has used several different geometric models, which are introduced in the following subsections. Various models have been developed in order to model several different area types.

6.2 THE SYNTHESYS MODEL

The SYNTHESYS geometric model has been developed by the former RACE 1044/SYNTHESYS [1]. It is the basis for most of the geometric models used in the TITAN project. The SYNTHESYS model is a polygon based geometric model for the access network, as shown in Fig 6.1.

The first level links consist of A-B, B-C, C-D and C- E links. Second level links are denoted as D-F links. Third level links are the links between F and the subscriber premises entrance.

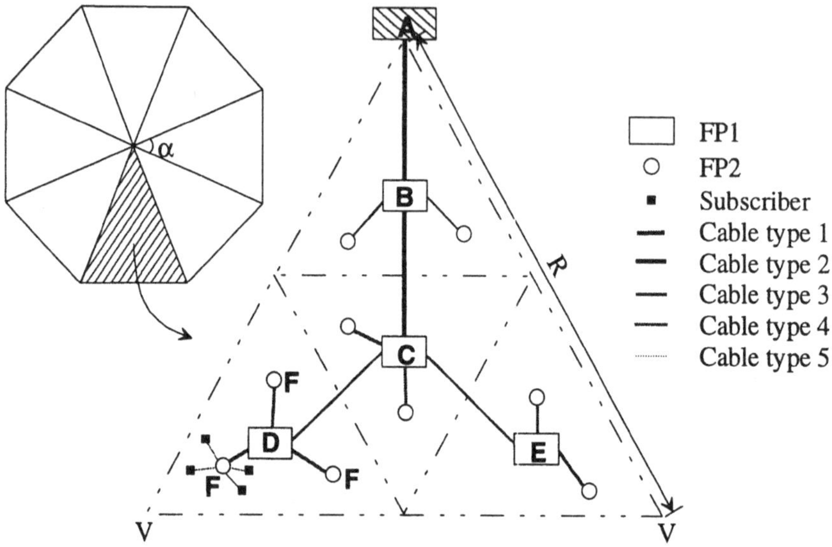

Fig 6.1 The SYNTHESYS model.

A set of equations has been derived in order to calculate cable length estimates at different network levels. In these equations the following variables are used:

> d = number of potential users per km^2
> N = number of potential users per hub
> n = number of fibre cables leaving the hub (n is also the rank of the polygon)
> M = Number of potential users per branching box.

The radius of the polygon is:

$$\left|\overline{AV}\right|^2 = R^2 = \frac{2 \cdot N}{d \cdot n \cdot \sin(\alpha)} \tag{6.1}$$

where α is the peak angle of the triangular sector:

$$\alpha = 360^\circ / n \tag{6.2}$$

Using these variables cable length estimates can be expressed as follows:

$$\left|\overline{AB}\right| = \left|\overline{BC}\right| = \frac{R}{3} \cdot \cos\left(\frac{\alpha}{2}\right) \tag{6.3}$$

$$\left|\overline{CD}\right| = \left|\overline{CE}\right| = \frac{R}{6} \cdot \sqrt{1 + 8 \cdot \sin^2(\alpha/2)} \tag{6.4}$$

$$|\overline{DF}| = R \cdot \left(0.132 + \frac{0.336}{n}\right) \tag{6.5}$$

Equations (6.1) - (6.4) are directly derived from the geometry, whereas (6.5) is obtained by simulation. The area of surface *S* corresponding to a branching box is:

$$S = \frac{M}{d} \tag{6.6}$$

By approximating that this surface is equivalent to a circle, i.e. $\pi \cdot r^2$, the average distance *b* between the branching box and building entrance is:

$$b = \frac{2r}{3} = \frac{2}{3} \cdot \sqrt{\frac{M}{\pi \cdot d}} \tag{6.7}$$

An average distance of 15 meters is assumed between the building entrance and the subscriber premises. Therefore the total branching distance L_{branch} is:

$$L_{branch} = b + 15 = 15 + \frac{2}{3} \cdot \sqrt{\frac{M}{\pi \cdot d}} \tag{6.8}$$

The additional 15 meters is assumed to be within the building for which no ducting is included. The maximum length of the subscriber line, L_{max}, is:

$$L_{max} = 3 \cdot |\overline{AB}| + |\overline{CD}| \tag{6.9}$$

Cable lengths are calculated from the geometric model using the same equations as for the duct lengths. To these values further cable overlength due to field constraints is added. Overlength is expressed in percentages and typically 40 % has been used.

Depending on the situation the estimates for the total cable and duct lengths to be used in economic models are calculated from equations 6.3, 6.4, 6.5 and 6.8. In some cases it is convenient to form a single equation for the average subscriber loop length by combining the above mentioned equations. However, this equation would not be generally applicable, e.g. because some link levels might use different types of cable and hence links could not be combined in modelling.

The SYNTHESYS model as described above, has been successfully applied to e.g., passive optical network (PON) architectures. However, the large diversity in access network architecture types requires a more flexible model.

6.3 THE MODIFIED SYNTHESYS MODEL

The modified SYNTHESYS model is similar to the basic SYNTHESYS model with a feeder part of the network added to it [2]. The feeder part is shown at the top of Fig 6.2. The feeder network includes two flexibility points and link levels. The lengths of the feeder cables are case dependent and therefore input separately, i.e. they are not estimated by means of trigonometry.

The modified SYNTHESYS model is the geometric model implemented in most versions of the TITAN tool.

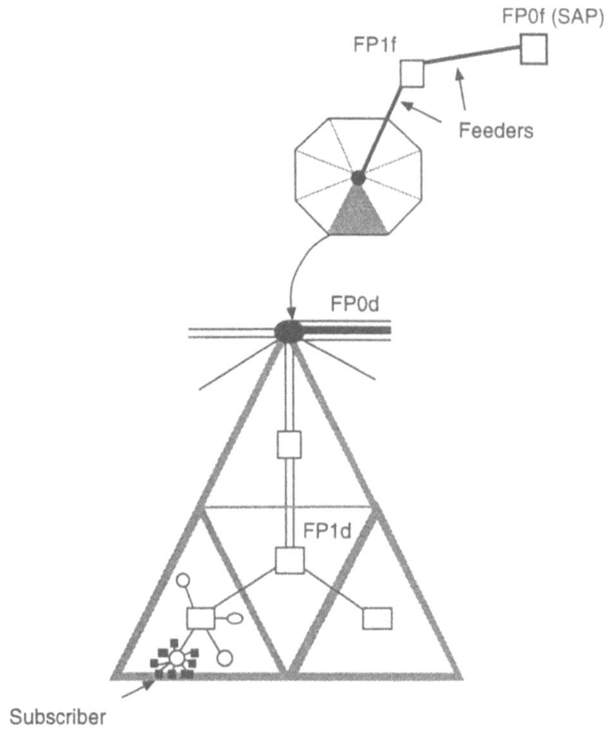

Fig 6.2 The modified SYNTHESYS model. Generalised access network model.

The SYNTHESYS and the modified SYNTHESYS models have been developed to model star topologies. In order to model also other sorts of networks, particularly community antenna television (CATV) networks, the SYNTHESYS model has been enhanced further.

6.4 THE BUS/STAR MODEL

The third variant of the SYNTHESYS model is called the bus/star model [3,4]. In this model higher levels of the network (from FP1d and higher) are equal to the modified SYNTHESYS model.

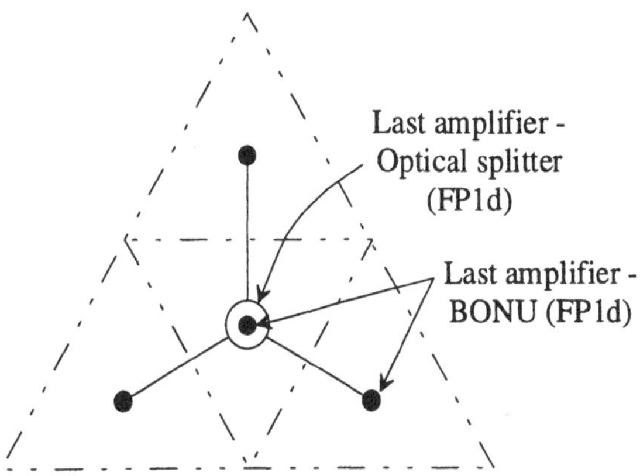

Fig 6.3 The basic SYNTHESYS model used at the highest levels of the bus/star model.

Levels down from FP1d are implemented as bus topologies. Fig 6.4 and Fig 6.5 show the bus structures. One should note that now the splitting ratios do not indicate the number of cables, but the number of new flexibility points.

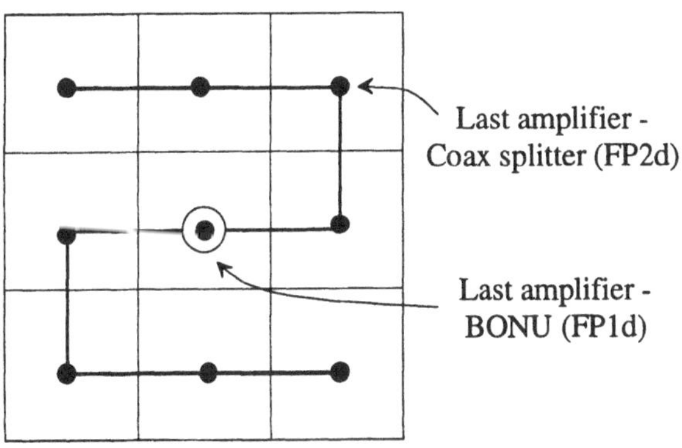

Fig 6.4 The higher bus level, i.e. link level 2d in the bus/star architecture.

The length l_{LL2d} in link level 2d is calculated from:

$$l_{LL2d} = \frac{1+\sqrt{N}}{2} \cdot \sqrt{\frac{N}{d}}\ km \qquad (6.10)$$

where N here is the total number of subscribers fed from FP1d and d is the subscriber density.

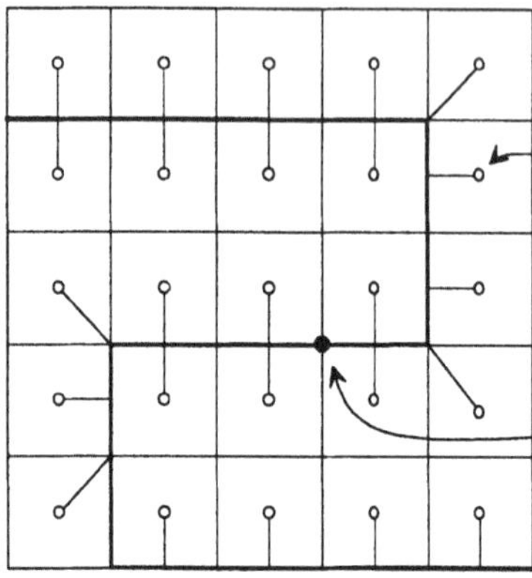

Fig 6.5 The lower bus level, i.e. link level 3d in bus/star architecture.

The length l_{LL3d} in link level 3d is calculated from:

$$l_{LL3d} = \frac{1}{\sqrt{d}} \cdot \left[\frac{1}{2} \cdot \left(N - \left(\sqrt{N} - 2 \right) \right) + \frac{1}{\sqrt{2}} \left(\sqrt{N} - 2 \right) \right] km \tag{6.11}$$

Finally, the total length can be calculated by combining equations 6.3 and 6.4 from the SYNTHESYS model and equations 6.10 and 6.11 from the bus model.

6.5 THE TITAN MODEL

The basic structure of the SYNTHESYS model and its variants is based on a star topology and polygon geometry. In many cases this is a valid assumption, but there are cases where the polygon structure is not ideal. Because of this TITAN has developed a completely new geometric model, namely the TITAN model, which is more flexible in many senses. The TITAN allows modelling of clustered areas where subscribers are not homogeneously distributed. The topology can be either a star, ring or bus or a combination of these. In addition, the shape of the model area and location of flexibility points within the model area are taken into consideration.

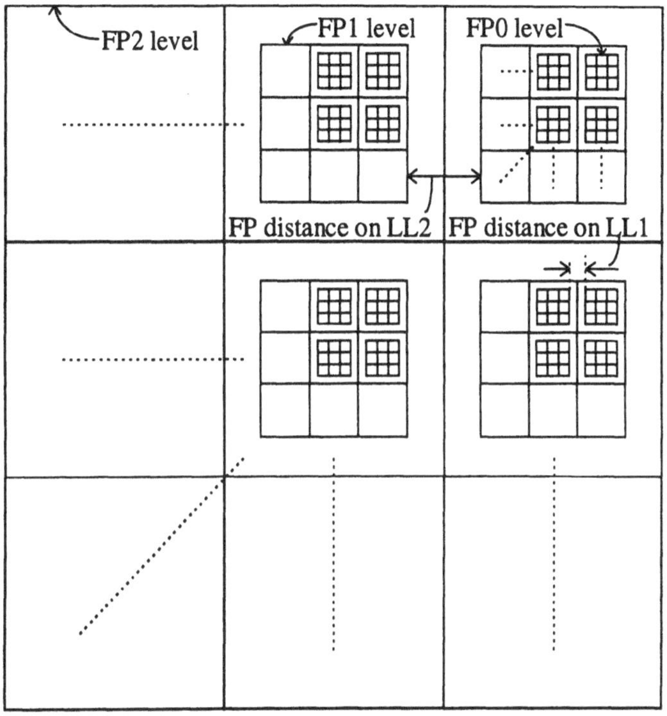

Fig 6.6 The first three levels of the TITAN geometric model.

In Fig 6.6 the basic structure of the TITAN model is shown. The model is based on a layered structure in which each layer uses the same basic TITAN geometric model, but with different parameters. Total cable and duct lengths in the model area are achieved by simply adding cable and duct lengths from different layers. Between the layers there can be a certain amount of empty space. One layer can, for example, represent a village and the next higher layer will be a larger region where there are scattered villages and uninhabited areas in between. The basic model area in each layer for a star topology is rectangular, as depicted in Fig 6.7.

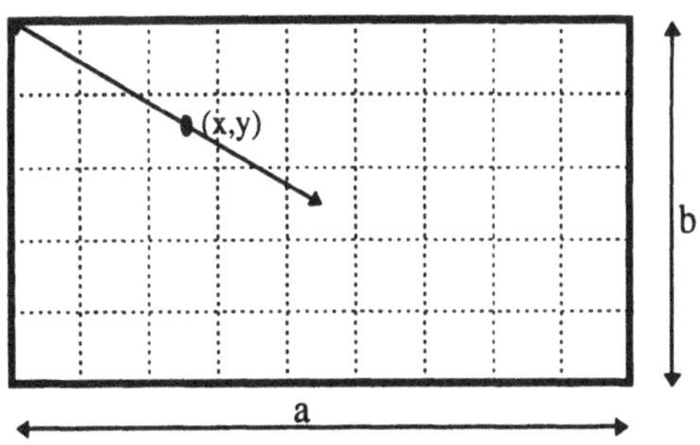

Fig 6.7 The basic TITAN model area for one model layer.

Fig 6.8 shows how the ducts are located within the basic model area. Also one example cable from flexibility point (x,y) to the user or the lower level flexibility point (i,j) is shown.

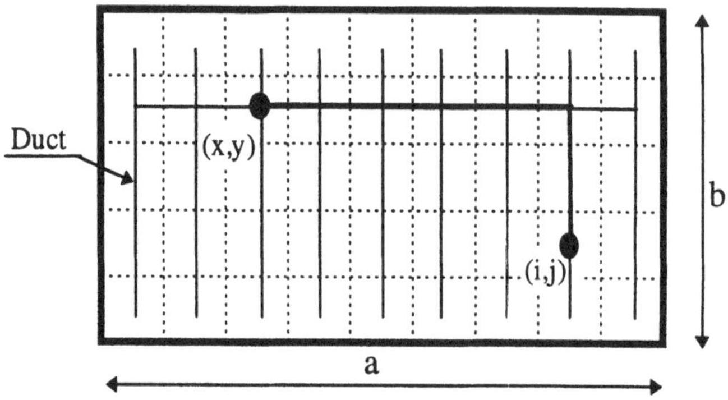

Fig 6.8 Ducts and an example cable route within the basic TITAN model area.

The duct length, l_d, is given by

$$l_d = a \cdot b - 1 = n - 1 \tag{6.12}$$

where n is the distribution ratio at a given network level. Each square in the model area represents a lower model layer or, at the lowest level, a subscriber. The sublayer or subscriber located at point (i,j) will be connected to distribution point (x,y) with a cable that makes one rectangular turn on its way, as shown in Fig 6.8. The total cable length in case of a star topology at any given network layer will be:

$$l_c(a,b,x,y) = \sum_{i=0}^{b-1} \sum_{j=0}^{a-1} \left[|i-y| + |j-x| \right] \tag{6.13}$$

where a is the length of the model area
b is the width of the model area
x is the length co-ordinate of distribution point, $x \in [0,a\text{-}1]$
y is the width co-ordinate of distribution point, $y \in [0,b\text{-}1]$.

After some calculus this yields

$$l_c(a,b,x,y) = \frac{a}{2} \cdot \left[(b-y-1)^2 + y^2 + b - 1 \right] + \frac{b}{2} \cdot \left[(a-x-1)^2 + x^2 + a - 1 \right] \tag{6.14}$$

The variables used in (6.14) are not very suitable for use in the final application. It is more intuitive to express the cable length as a function of distribution ratio *n*, area shape *s* and position *p*. The shape factor *s* indicates the shape of the model area, i.e. if the area is a square or a rectangle ($a = s \cdot b$). The position variable $p \in [0,1]$ equals zero when the distribution point is in the middle of the area. As *p* grows, the distribution point moves along the arrow in Fig 6.7 towards the corner of the area. When *p* equals one, the distribution point is in the corner of the area.

Now *a, b, x* and *y* can be expressed as function of *n, s* and *p* as follows:

$$a = \sqrt{ns}$$

$$b = \sqrt{\frac{n}{s}}$$

$$x = \frac{\sqrt{ns} - 1}{2} \cdot (1 - p) \tag{6.15}$$

$$y = \frac{\sqrt{n/s} - 1}{2} \cdot (1 - p)$$

Substituting the above expression into (6.14) gives the cable length as a function of *n, s* and *p*. The resulting equation is rather complicated and it is not included here. However, the equation can be easily implemented in almost any programming language by implementing (6.14) and equation group (6.15) separately and nesting the results of (6.15) into (6.14)

Previous equations enable the estimation of cable and duct lengths in the case of a star topology. For bus and ring topologies the situation is much simpler. Fig 6.9 shows the case for a ring topology.

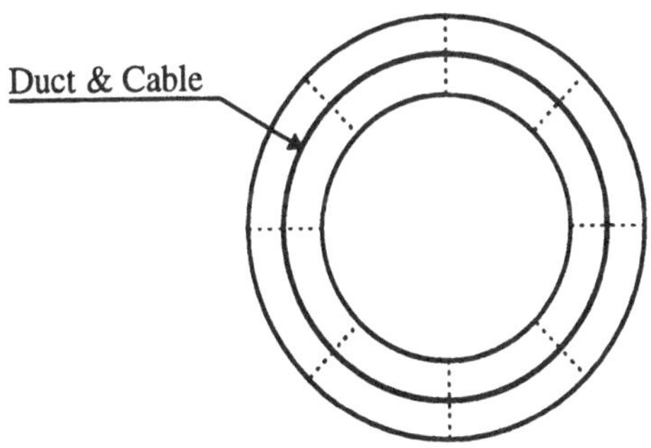

Fig 6.9 Ring topology (n = 8).

$$\begin{aligned} l_{duct} &= n \\ l_{cable} &= n \end{aligned} \tag{6.16}$$

Thus, the cable and duct lengths are the same and equal to the splitting ratio. Fig 6.10 shows the situation for a bus topology.

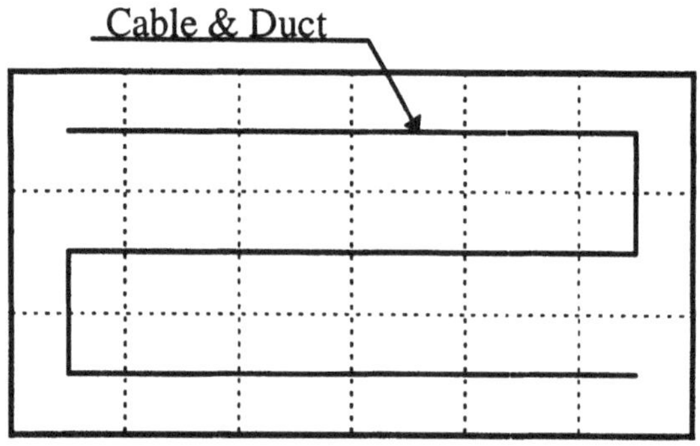

Fig 6.10 Bus topology (n = 18).

$$\begin{aligned} l_{duct} &= n-1 \\ l_{cable} &= n-1 \end{aligned} \tag{6.17}$$

The equations derived here for the TITAN model give the estimates for cable and duct lengths in undetermined *units*. Before these estimates can be used in economics modelling, they must be mapped into real length units, e.g. metres. This is done by calculating the length of one *unit* at the lowest network level from the subscriber density per square kilometre. After this, the length of a *unit* in the second layer can be calculated from the *unit* length at the first layer and from the known geometry between the layers.

During the TITAN project the TITAN model has been developed and used for evaluation purposes, but it has not been implemented in any released version of the TITAN tool. However after proper testing and validation, which may yield further improvements, the model is likely to replace the SYNTHESYS model and its variants.

REFERENCES

[1] Gardan, D, Zaganiaris, A, Madani, A, Madigou, R, Machon, D. Techno-economics of advanced optical subscriber networks. *In: Proc. IEEE Global Telecommunications Conference and Exhibition, Dallas, USA, 1989, 1335-1339.*

[2] Ims, L A, Olsen, B T, Myhre, D, Mononen, J, Lähteenoja, M, Ferrero, U, Zaganiaris, A. Multiservice Access Network Upgrading in Europe: a Techno-economic Analysis. *IEEE Communications Magazine,* 34, (12), 1996, 124-134.

[3] CATV Evolution Case Study. Jewell, S, Aknai, P, Drieskens, M, Fisher, D, Kraushaar, J, Bruyn. S. Brussels, Belgium, 1994. (RACE 2016/STRATOSPHERIC deliverable).

[4] Ramassh, T. Studies of Different Geometrical Models for Future Broadband Access Network. Thesis for Masters of Science, University of Oslo, 1995.

Model for mixed demographic structures

7

Eduardo de Castro, Manuel de Oliveira Duarte

7.1 INTRODUCTION

The economic evaluation of a telecommunication network requires the estimation of the demand for the telecommunication services, as well as its location in space. If the demand for services is a basic input for the estimation of future revenues, its geographical distribution is the key factor for the definition of network topology and, thus, for the evaluation of the length of ducts and cables, a fundamental component of the costs. In this chapter the TITAN (tool for introduction strategies and techno-economic evaluation of access network) geometric model designed to calculate the length of cables and ducts for telecommunication networks is applied to different urban and rural structures, reflecting varied patterns of settlement typical of different European regions [1,2].

The geographical distribution of population and of residential demand for telecommunication services follows a highly complex pattern. Population densities not only reflect the contrast between urban and rural settlements, but also show strong inter-urban and intra-urban variations. Urban densities depend, among other factors, on the cultural and economic conditions of each country and region, on the size and function of urban agglomerations and, inside each city or metropolitan area, on the position relative to the centre of each urban district. Demand for different telecommunication services and applications depends also on cultural and economic conditions, varying according to the characteristics of countries, regions, cities and city areas we are considering.

The capacity to cope with that diversity is a fundamental requirement for any economic evaluation of telecommunication networks and for the TITAN tool, in particular. Concerning the installation costs, it is necessary to define a generic geometric model designed to calculate the length of cables and ducts and to link this model to the particular conditions of each geographic area. Expected revenues depend on the estimated demand for services, a variable which is also spatially contingent.

According to the geometric models presented in the previous chapter, a telecommunication network can be modelled as a sequence of flexibility points linked together. Flexibility points (FP) are hierarchically ordered between the centre of the network (the service access point or FP_n) and the customer premises network interface (FP_0). Each flexibility point of a given order is linked with a flexible point of the order immediately superior and to several flexibility points of the order immediately inferior. In general, the link level *p-1* connects the flexibility points of order *p* and *p-1* (see Fig 7.1).

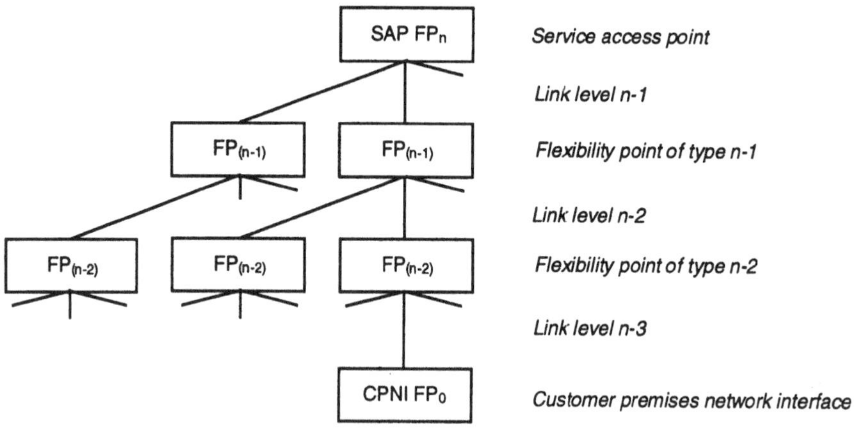

Fig 7.1 Generic access network model.

Thus, any territory served by a telecommunication network is divided into a hierarchical succession of areas and sub-areas, fed by flexibility points of different orders and corresponding to different link levels. Fig 7.2 shows a territorial model with successive levels of areas (L_1, L_2, L_3), each one corresponding to equivalent link levels. The model differs from reality only because it assumes that the areas form a regular pattern of squares with constant sizes and separated by constant distances (d_1, d_2, d_3). However, the losses in accuracy are clearly compensated by the gains in simplicity. The length of ducts and cables of a network can be immediately calculated if one knows, for each kind of link level areas, its topology and the parameter *n*, the number of flexibility points of order *p* contained in each area of link level *p*. As an alternative to *n* the distance *d* between the link level areas can be calculated. These two parameters can be easily related to each other.

This generic square model can be used to represent large and heterogeneous geographical areas. This is true in spite of the fact that for each separate level of the network the distribution of flexibility points (or the area spacing) must be uniform. Calculations can be made for each area type, assuming that it covers all the territory, and then linear combinations can be made by mixing the results obtained for each type of area proportionally to the relative size of the areas. The next section will show how the model can be used to describe different kinds of urban and rural settlements.

Fig 7.2 Territorial model of regularly distributed squares.

7.2 MIXED GEOGRAPHICAL STRUCTURES

In order to apply the geometric model presented above it is necessary to know how complex geographical structures of residential land use can be reduced to combinations of regularly distributed squares. This is a difficult task that requires the previous construction of a land use typology defining the spatial distribution of residential buildings, as well as the gathering of empirical data necessary to the description of real places as combinations of the defined land use types.

It must be stressed that the empirical data used in the following examples, although based on real situations represent a limited set of rural settlements and urban agglomerations. Thus, rather than an accurate and representative sample of European geographical structures, it must be considered an instrument to exemplify the methodology used.

7.3 SPATIAL DISTRIBUTION OF RESIDENTIAL BUILDINGS

The following typology of land distribution for residential use is based on seven types of buildings, described on Table 7.1. Each category of building has a particular number of floors, a particular number of homes per floor (and, thus, a particular number of homes) and their average size.

Relating this typology to the geometric model, it can be assumed that each building is associated to a flexibility point of order 1 (FP_1), linked with all its homes, the flexibility points of order 0 (FP_0). Thus each building is an area of link level 0 (L_0). It should be noticed that, in order to maintain the general framework of the geometric model, the number of floors and the number of homes per floor adopted are all perfect squares.

The next link level area (L_1) corresponds to blocks of buildings. In order to simplify the typology , each type of building corresponds to only one type of block (see Table 7.2). For example, building type A is a tall, large and isolated one, whilst buildings B, C and F are associated in blocks of 16, and building G is a typical single residence.

L_1 areas constitute a good example to show the limitations of the geometric model and its assumption of sets of squares, contiguous or separated by constant distances. For instance, blocks of 16 buildings are in general associated with two different shapes. The first one is shown in Fig 7.3. It does not fit the assumptions of the geometric model because the space in the middle of the square, corresponding to nine buildings, is empty. Comparing the solution shown in Fig 7.3 to a square of 4·4 contiguous buildings, corresponding to the TITAN geometric model, it is easy to see that, for star and ring topologies, the length of cables and ducts calculated by the model is underestimated relatively to the case shown in the figure (for the bus architecture the length is the same in both cases). Another common shape for blocks is the row shown in Fig 7.4. Once more, except for bus topology, the square geometric model, compared with the row shape, underestimates lengths. This problem can be partially solved assuming that the blocks represented in Fig 7.3 are equivalent to perfect squares made up of buildings separated by a given distance. In the case of the upper part of Fig 7.3 that distance is more or less equivalent to a redistribution of the empty space inside the square; in the case of the lower part of Fig 7.3 the distance is absolutely fictitious and is equivalent to a small shrinking of the buildings. The distance between buildings defined in Table 7.2 reflects this approach.

The following link level area (L2) corresponds to residential units with a size of 10 ha, and made up of blocks of buildings in numbers that, for the sake of simplicity, are again perfect squares. It was assumed once more that each type of block corresponds to only one type of residential unit (see Table 7.2). Because the different blocks are separated by streets, small gardens, etc., typical distances between units were defined. Residential units of type A, C and E are sets of blocks of more or less tall buildings, separated by wide intervals. Residential units of types B and D correspond to dense inner city areas, with relatively narrow streets and almost no empty spaces. Finally, units G and F correspond to single homes, isolated or grouped in rows.

The three link level areas presented above constitute our typology of land use for residential buildings, defining seven different densities of homes. Making linear combinations of these seven categories, an almost infinite variety of land use characteristics can be defined.

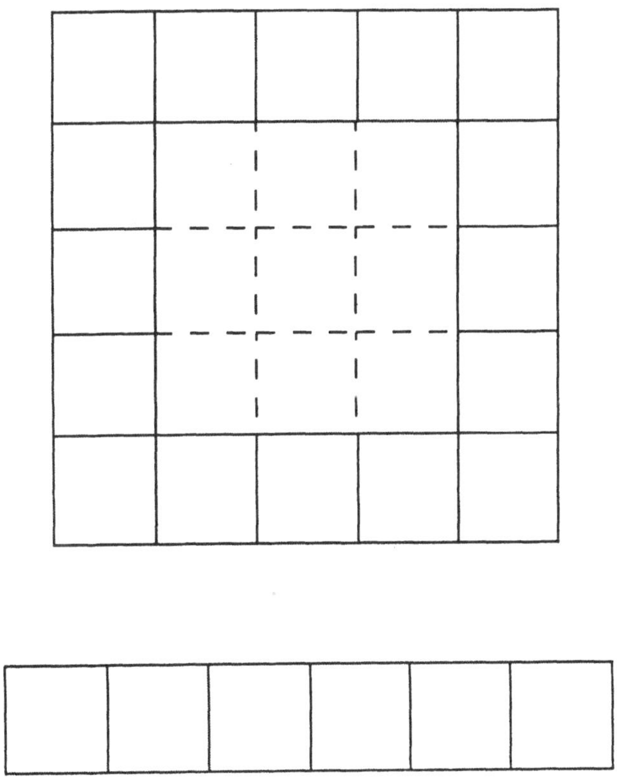

Fig 7.3 Blocks of buildings.

Table 7.1 Level 0, buildings.

Type of building	*Number of floors*	*Number of homes per building*	*Average size of homes (m^2)*
A	16	144 (16·9)	120
B	9	18 (9·2)	80
C	9	36 (9·4)	150
D	4	8 (4·2)	80
E	4	16 (4·4)	150
F	1	1	150
G	1	1	400

Table 7.2 Level 1, blocks of buildings.

Type of block	*Number of buildings per block*	*Number of homes per block*	*Average space between buildings (m)*	*Area of blocks (m^2)*
A	1	144	-	1,080
B	16	288	1.54	3,520
C	4	144	0.00	2,400
D	16	128	0.80	3,040
E	4	64	0.00	2,400
F	16	16	1.89	3,200
G	1	1	-	400

Table 7.3 Level 2, residential units with a size of 10 ha.

Type of residential unit (and respective buildings and blocks)	*Number of blocks per residential unit*	*Separation between blocks (m)*	*Density (homes per ha)*
A	16	46.2	230.4
B	16	19.7	460.8
C	9	56.4	129.6
D	25	8.1	320.0
E	9	56.4	57.6
F	16	22.5	25.6
G	49	25.2	4.9

7.4 MODELLING MIXED GEOGRAPHICAL STRUCTURES

The seven types of areas presented in the previous section are the basis for the description of any kind of residential and urban structure in accordance with the TITAN geometric model. The following examples are based on real data, covering different types of demographic agglomerations located in different European countries.

The first case is a big metropolis in central Europe, with a population of about three million inhabitants. Table 7.4 shows a sample of typical areas of the referred metropolis, which can be described as linear combinations of groups of 16 residential units separated by a given distance. These areas, with a size of 300 ha, constitute the third link level of the geometric model and the distance separating the residential units reflects the existence of a fraction of the area which is devoted to non-residential buildings specialised in service activities, factories, empty spaces, roads, etc.

The distances between residential units were calculated using data defining the fraction of built areas occupied with residential units (see Table 7.4). Representing this fraction by symbol f, d being the distance between residential units and a the area of each residential unit, it is easy to see that:

$$d = \sqrt{a}\left(\frac{1}{\sqrt{f}} - 1\right) \tag{7.1}$$

It should be noticed that each type of area is made up of residential units with different sizes rather than with the constant size of 10 ha assumed in Table 7.3. This is necessary in order to make compatible all the conditions implicit in the table. Consequently, the application of the geometric model to this link level, as an aggregation of residential units, requires the re-scaling of the size of units from 10 ha to the values corresponding to each type of area.

The fourth link level adopted describes the overall metropolitan area and is a linear combination of groups of four built-up areas (see Table 7.5). The distance between urban areas, corresponds to urban parks and forests, big avenues and motorways, railroads, airports, rivers and channels, etc. Similarly to the previous level, it was necessary to adopt different sizes for built areas, rather than the 300 ha assumed in Table 7.4.

Finally, the fifth link level, corresponding to the overall metropolitan area, was defined as a linear combination of groups of 36 urban areas. The empty space separating these areas was considered to be null.

The other cases, presented below were modelled according to the same methodology. They represent the following situations:

- a rural area of central Europe with around 3,000 inhabitants (Table 7.7)
- a medium size city of southern Europe, with more or less one hundred thousand inhabitants (Table 7.8 and Table 7.9).

The only differences relative to the previous case concern the number of link levels adopted. For the rural region a level three area was enough to describe it. A level four area was used to represent the medium size city.

Table 7.4 Level 3, metropolitan area of central Europe (built areas with 16 residential units and a size of 300 ha).

Type of area	*Description*	*Combination of residential units*	*Percentage of the area occupied with residential units*	*Size of each residential unit (ha)*	*Separation between residential units (m)*	*Average number of POTS subscr. per home (%)*	*Percentage of residential POTS relative to all POTS subsc.*	*Density (homes per ha)*
a	Central business area	0.4B+ 0.3C+0.3D	35	6.56	176.8	93	41.2	111.7
b	Central business area	0.15A+0.25D+0.6E	50	9.38	126.9	80	46.1	74.6
c	Central area, mainly residential	0.4B+0.3C+0.3D	50	9.38	126.9	93	79.6	159.6
d	Central area, mainly residential	0.15A+0.25D+0.6E	70	13.12	70.7	80	82.3	104.4
e	Residential area of inner periphery	0.3A+0.5C+0.2E	55	10.31	111.9	93	89.3	80.0
f	Peripheric residential area	0.3E+0.2F+0.5G	85	159.4	33.8	89	90.7	21.1
g	High income peripheric residential area with some service development	0.3C+0.5E+0.2G	60	11.25	97.6	105	77.1	41.2
h	Peripheric residential area with some service development	0.15C+.35E+.5F	65	12.19	83.9	93	79.4	34.1
i	Outer peripheric residential area	0.35E+0.35F+0.3G	65	12.19	83.9	84	92.9	19.9
j	Outer peripheric residential area	0.4E+0.3F+0.3G	65	12.19	83.9	103	93.7	20.9

Table 7.5 Level 4, metropolitan area of central Europe (urban areas with 4 built-up areas and a size of 2,500 ha).

Type of urban area	*Description*	*Combination of built areas*	*Percentage of the urban area occupied with built areas*	*Size of each built area (ha)*	*Separation between built areas (m)*	*Average number of POTS subscribers per home (%)*	*Percentage of residential POTS relative to all POTS subscribers*	*Density (homes per ha)*
A_1	*Central business and residential areas*	0.2 a+ 0.4 c+0.4 e	45	281.25	823	93	74.9	53.3
A_2	*Central business and residential areas*	0.15b+0.3d+0.55f	55	343.75	646	82	76.9	29.7
A_3	*Peripheric area with some service and industrial development*	0.35 g + 0.65 h	45	281.00	823	98	78.4	16.4
A_4	*Residential peripheric areas*	i	20	125.00	1382	84	92.9	4.0
A_5	*Residential peripheric areas*	j	25	156.25	1250	103	93.7	5.2

Table 7.6 Level 5, metropolitan area of central Europe (overall metropolitan area, with a size of 90,000 ha and 36 urban areas).

Combination of urban areas	*Separation between urban areas (m)*	*Average number of POTS subscribers per home (%)*	*Percentage of residential POTS relative to all POTS subscribers*	*Density (homes per ha)*
$0.15\ A_1+0.2A_2$ $+0.25A_3+0.25A_4$ $+0.15A_5$	0	91	77.6	19.8

Table 7.7 Level 3, rural area of central Europe with a size of 8,600 ha and 1,200 homes spread by 16 villages (residential units).

Combination of residential units	*Percentage of the area occupied with residential units*	*Size of each residential unit (ha)*	*Separation between residential units (m)*	*Average number of POTS subscribers per home (%)*	*Percentage of residential POTS relatively to all POTS subscribers*	*Density (homes per ha)*
0.05 E +0.95 G	18.6	10	2000	126	94.4	0.14

Table 7.9 Level 4, medium size city of southern Europe (overall area, with a size of 12,820 ha and 25 built-up areas of 200 ha).

Combination of urban areas	*Separation between urban areas (m)*	*Percentage of the metropolitan area occupied with built areas*	*Average number of POTS subscribers per home (%)*	*Percentage of residential POTS relatively to all POTS subscribers*	*Density (homes per ha)*
0.03a+0.01b +0.30c+0.13d +0.09e+0.44f	850	39	87.9	76.28	3.56

7.5 GEOGRAPHICAL DIFFERENCES IN DEMAND FOR SERVICES

The availability of information concerning the expected residential demand for the different advanced telecommunication services and applications is a fundamental requirement for the assessment of the economic viability of networks. Knowing the demand, expressed as penetration rates, the number of homes defined in the geographical models should be multiplied by the rates, in order to calculate the actual number of residences that will be connected to the network.

Penetration rates below 100 % mean that not all residences are expected to be linked to the network. However, if the penetration is sufficiently high, it may be assumed that all the buildings will be connected and thus, the penetration rate will only define the fraction of homes, inside each building, expected to be linked. If the penetration rate is much smaller than 100 %, it is reasonable to expect that a number of buildings or even blocks of buildings will not be connected. In this last case, the calculations of cable length corresponding to the areas of level zero link and even level one link will be affected by penetration rates.

Forecasts of penetration rates concerning different applications and services in different groups of European countries, provided by the detailed Delphi survey performed by TITAN, have been presented in various publications [3] and in chapter 3 of this book. However, these forecasts do not provide information about geographical variation of demand, at the level of geographical detail adopted here.

The problem can be solved with the use of information concerning disposable household income, at the level of urban areas, since TITAN forecasts provide information about the expected income elasticity of demand for different services. However, for most of European countries, the required data is not directly available and its estimation is a difficult task.

A simplified methodology was adopted to solve the problem. It consists of the assumption that demand for advanced services will be proportional to the present rate of residential penetration of plain old telephony service (POTS). Then, it is possible to adopt Delphi forecasts for future demand for advanced services and assume that the variation of their demand inside each metropolitan area will have the same pattern as the present variation of POTS penetration. It is also necessary to assume that each studied urban and metropolitan area will have an aggregate behaviour identical to the corresponding European division adopted in the Delphi survey.

A final comment refers to the mixed effects of business and residential demand for advanced telecommunications. If, for a given area, corresponding to a given link level, business demand is geographically mixed with residential demand, the costs of the network would be shared by both kind of users. Thus, it can be assumed that, for each area, the costs input to the residential network would be a fraction of the total network costs, corresponding to the weight of residential demand in relation to the overall demand. In the absence of more detailed data it can be accepted that the above mentioned weight is identical to the present proportion between residential and business POTS demand.

REFERENCES

[1] Gardan, D, Zaganiaris, A, Madani, A, Madigou, R, Machon, D. Techno-economics of advanced optical subscriber networks. *In: Proc. IEEE Global Telecommunications Conference and Exhibition, Dallas, USA, 1989, 1335-1339.*

[2] Olsen, B T, Zaganiaris, A, Stordahl, K, Ims, L A, Myhre, D, Øverli, T, Tahkokorpi, M, Welling, I, Drieskens, M, Kraushaar, J, Mononen, J, Lähteenoja, M, Markatos, S, De Bortoli, M, Ferrero, U, Ravera, M, Balzaretti, S, Fleuren, F, Gieschen, N, De Oliveira Duarte, M, de Castro, E. Techno-economic evaluation of narrowband and broadband access network alternatives and evolution scenario assessment. *IEEE Journal on Selected Areas in Communications,* 14 (6), 1996, 1184-1203.

[3] Stordahl, K, Murphy, E. Methods for forecasting long term demand for wide and broadband services in the residential market. *IEEE Communications magazine*, 13, (2), 1995, 44-49.

Component price versus time

8

Borgar Tørre Olsen

8.1 INTRODUCTION

Some of the equipment and components in the TITAN (tool for introduction strategies and techno-economic evaluation of access network) cost database have been on the market for some years. In this case traditional forecast methods can be applied to directly estimate the evolution of cost over time [1]. In studying future strategic network projects such as an upgrade of the present narrowband access network towards a broadband access network, new components and equipment with different degrees of maturity are considered. In some cases only target costs can be obtained. The RACE (research for advanced communications in Europe) 2087/TITAN project has developed a new methodology [2,3] for estimating the cost evolution of those components which combine information from the production process expressed in the learning curves [4-6], and an estimating procedure for the volume forecast of the equipment and components.

The cost evolution of every component in the TITAN cost database is described by the following parameters:

- *Price* (cost) at a given reference year
- *Reference year*
- *Learning curve class*, which gives the cost of the component as a function of produced volume with one parameter: *K* (optimistic or pessimistic)
- *Volume class*, which gives volume as a function of time (forecasted volume) with three parameters: Δt, $n(0)$ and γ
- Δt, which is the time it takes for the accumulated production volume to increase from 10 % to 90 % of the total accumulated production volume
- $n_r(0)$, which is the accumulated production volume by t = 0 (relative to total accumulated volume)
- γ, which describes the asymmetry of the applied logistic curve (set to 1 in the TITAN Tool).

In the TITAN methodology the description of the growth over time of the accumulated volume of every component in the cost database is a standard demand logistic curve:

$$n(t) = M\left[1 + e^{(a+b\cdot t)}\right]^{-\gamma} \tag{8.1}$$

By combining this standard demand logistic curve with a learning curve, the cost evolution in time of every component can be easily described [2-6].

8.2 LEARNING CURVE COEFFICIENT

T. P. Wright first proposed the concept of learning curves [4]:

$$T_n = n^{-\alpha} \cdot T_0 \tag{8.2}$$

where T_n is the average production time for n units, given by

$$T_n = \frac{t_1 + t_2 + \ldots + t_n}{n}$$

where t_n is the time to complete the nth unit, T_0 the time to complete the first unit and n is the number of completed units.

J. R. Crawford applied the same formula, but interpreted T_n as the completion time for the nth unit [5]. Wright's law describes the cumulative effect of learning, while Crawford's formula only refers to scale effects. A disadvantage of Wright's law is the appearance of strong autocorrelation, affecting the statistical estimation of its parameters, a problem which always arise when trying to correlate accumulated values.

In the TITAN methodology we have assumed that component cost (price) P_n is proportional with production time T_n for the nth component, which gives:

$$P_n = n^{-\alpha} \cdot P_0 \tag{8.3}$$

Now we can write

$$P_{2n} = (2n)^{-\alpha} \cdot P_0 \tag{8.4}$$

which may be expressed as

$$P_{2n} = (2)^{-\alpha} \cdot n^{-\alpha} \cdot P_0 \tag{8.5}$$

and

$$P_{2n} = (2)^{-\alpha} \cdot P_n \tag{8.6}$$

and

$$P_{2n} = K \cdot P_n \tag{8.7}$$

K is the factor by which the price is reduced when the production volume is doubled. K is called the learning curve coefficient and is related to α by

$$K = (2)^{-\alpha} \tag{8.8}$$

or

$$\alpha = -\log_2 \cdot K \tag{8.9}$$

$$\alpha = \frac{-\log_x \cdot K}{\log_x \cdot 2} \tag{8.10}$$

The K factor is a parameter which may be obtained from the production industry.

8.3 COMPONENT COST AS A FUNCTION OF TIME

We know that n and P_n is a function of time. Thus

$$P(t) = n(t)^{-\alpha} \cdot P_0 \tag{8.11}$$

In principle $n(t)$ is the global volume (for the world production of a component) and P_0 is the cost of the very first component. Both of those inputs are sometimes difficult to obtain. What possible solutions are there? One solution is to use relative values, in which case P_0 may be omitted:

$$\frac{P(t)}{P(0)} = \left[\frac{n(t)}{n(0)}\right]^{-\alpha} \tag{8.12}$$

Furthermore the global volume $n(t)$ may be removed from the expression, by observing that

$$\frac{P(t)}{P(0)} = \left[\frac{n_r(t)}{n_r(0)}\right]^{-\alpha} \tag{8.13}$$

(8.13) holds true since

$$\left[\frac{n(t)}{n(0)}\right]^{-\alpha} = \left[\frac{n_r(t)}{n_r(0)}\right]^{-\alpha} \tag{8.14}$$

where $n_r(t)$ and $n_r(0)$ are relative values (i.e. normalised to 1). Rearranging the expression gives:

$$P(t) = P(0) \cdot \left[\frac{n_r(t)}{n_r(0)} \right]^{-\alpha} \tag{8.15}$$

or

$$P(t) = P(0) \cdot \left[n_r(0)^{-1} \cdot n_r(t) \right]^{\log_2 \cdot K} \tag{8.16}$$

$n_r(0)$ is the relative component volume at time t = 0. The relative accumulated growth $n_r(t)$ of the component volume in TITAN is described by the logistic curve in formula (8.1) normalised to 1.

$$n_r(t) = \left[1 + e^{(a + b \cdot t)} \right]^{-\gamma} \tag{8.17}$$

8.4 THE COST EVOLUTION IN THE SYMMETRIC CASE ($\gamma = 1$)

In the TITAN cost database the parameter γ is fixed to 1. Let us express a and b with parameters which are more meaningful:

$$n_r(t) = \left[1 + e^{(a + b \cdot t)} \right]^{-1} \tag{8.18}$$

By setting $t = 0$

$$n_r(0) = \left(1 + e^{a} \right)^{-1} \tag{8.19}$$

and rearrange

$$a = \ln\left[n_r(0)^{-1} - 1 \right] \tag{8.20}$$

a may be expressed by the relative component volume in year 0.
What about b? Let

$$n_r\left(t_1\right) = 0{,}1 \tag{8.21}$$

$$n_r\left(t_2\right) = 0{,}9 \tag{8.22}$$

Thus

$$\left[1+e^{\left(a+b\cdot t_1\right)}\right]^{-1} = 0{,}1 \quad \& \quad \left[1+e^{\left(a+b\cdot t_2\right)}\right]^{-1} = 0{,}9 \tag{8.23}$$

$$1+e^{\left(a+b\cdot t_1\right)} = 10 \quad \& \quad 1+e^{\left(a+b\cdot t_2\right)} = \frac{10}{9} \tag{8.24}$$

$$e^{\left(a+b\cdot t_1\right)} = 9 \quad \& \quad e^{\left(a+b\cdot t_2\right)} = \frac{1}{9} \tag{8.25}$$

$$e^{a}\cdot e^{b\cdot t_1} = 9 \quad \& \quad e^{a}\cdot e^{b\cdot t_2} = \frac{1}{9} \tag{8.26}$$

By dividing the right side by the left side we get

$$e^{b\cdot(t_2 - t_1)} = \frac{1}{9^2} \tag{8.27}$$

$$b\cdot\left(t_2 - t_1\right) = -2\cdot \ln 9 \tag{8.28}$$

By definition

$$\Delta T = t_2 - t_1 \tag{8.29}$$

$$\Delta T = \frac{-2\cdot \ln 9}{b} \tag{8.30}$$

$$b = \frac{-2\cdot \ln 9}{\Delta T} \tag{8.31}$$

If we substitute the expression for a and b into the logistic curve, we get

$$n_r(t) = \left(1+e^{\left\{\ln\left[n_r(0)^{-1} - 1\right] - \left[\frac{2\cdot \ln 9}{\Delta T}\right]\cdot t\right\}}\right)^{-1} \tag{8.32}$$

where the only inputs are $n_r(0)$ and ΔT. Fig 8.1 illustrates this relation.

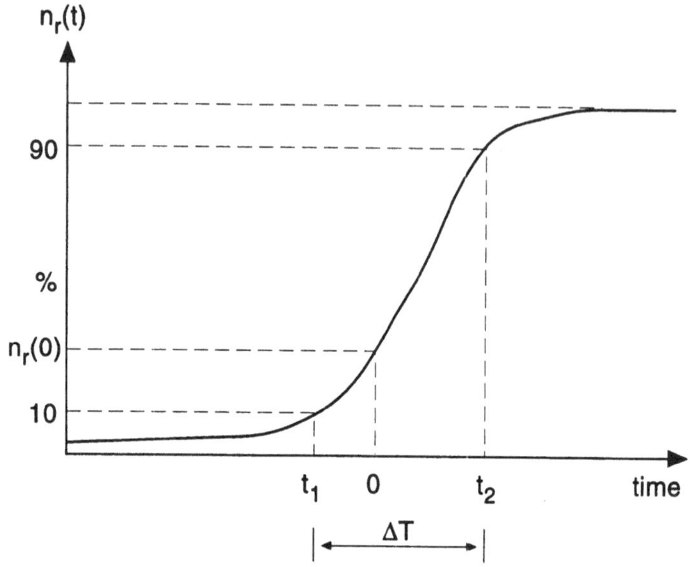

Fig 8.1 The forecast function for the evolution of the relative accumulated volume $n_r(t)$.

The expression for $n_r(t)$ may now be substituted into the learning curve formula (8.16):

$$P(t) = P(0) \cdot \left[n_r(0)^{-1} \cdot n_r(t) \right]^{\log_2 \cdot K} \tag{8.33}$$

yielding the final expression for price versus time in the TITAN cost database

$$P(t) = P(0) \cdot \left[n_r(0)^{-1} \cdot \left(1 + e^{\left\{ \ln\left[n_r(0)^{-1} - 1 \right] - \left[\frac{2 \cdot \ln 9}{\Delta T} \right] \cdot t \right\}} \right)^{-1} \right]^{\log_2 \cdot K} \tag{8.34}$$

Only four parameters are input to the formula:
$P(0)$, the price in the reference year 0,
$n_r(0)$, the relative accumulated volume in year 0,
ΔT, the time for the accumulated volume to grow from 10 % to 90 %,
K, the learning curve coefficient.

In order to illustrate this relation, and to get a normalised component price, we put *P(0)* = 1 and *K* = 0.95 (Fig 8.2). By keeping $n_r(0)$ = 0 .001 constant and letting the parameter ΔT range from two to 20 years, we can illustrate the evolution of the normalised price versus time for different ΔT, as shown in Fig 8.2. Fig 8.3 shows the impact of *n(0)* on the normalised price, keeping ΔT = 5 years as a constant.

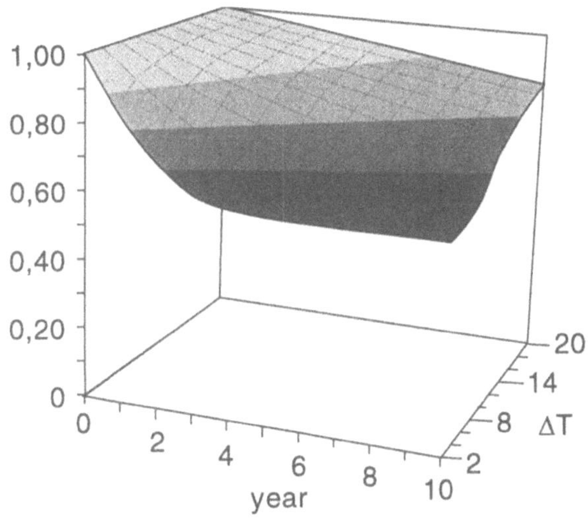

Fig 8.2 The impact of ΔT on the normalised price, keeping $n_r(0)$ = 0,001.

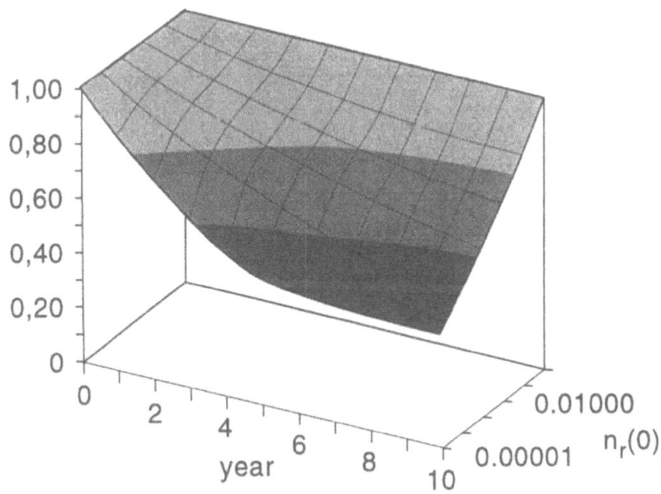

Fig 8.3 The impact of $n_r(0)$ on the normalised price, keeping ΔT = 5 years as a constant.

From the previous expression it is clear that the asymptotic price level when t approaches ∞ does not depend on ΔT, and is given by:

$$P(\infty) = P(0) \cdot \left[\frac{1}{n(0)}\right]^{\log_2(K)} \tag{8.35}$$

In addition, for small t the slope of the price curve is proportional to ΔT^{-1}.

8.5 THE COST EVOLUTION IN THE ASYMMETRIC CASE ($\gamma \neq 1$)

The logistic curve for γ different from 0 is illustrated in Fig 8.4.

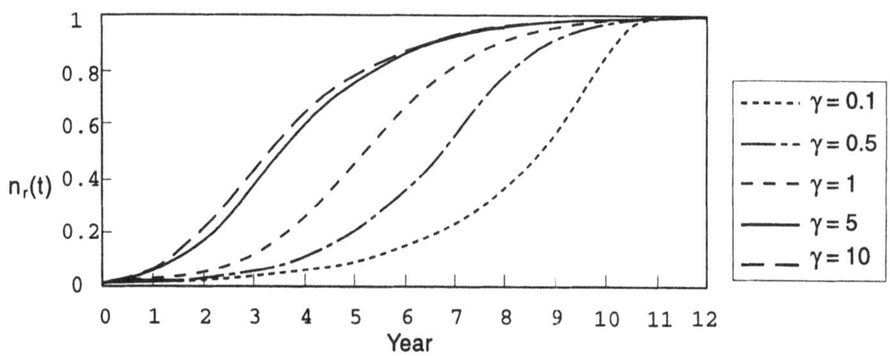

Fig 8.4 The impact of ΔT on the normalised price, keeping ΔT = 5 years as a constant.

Let us express a and b and γ with parameters which are more meaningful:

$$n_r(t) = \left[1 + e^{(a + b \cdot t)}\right]^{-\gamma} \tag{8.36}$$

$$n_r(0) = \left(1 + e^{a}\right)^{-\gamma} \tag{8.37}$$

$$a = \ln\left[n_r(0)^{\frac{-1}{\gamma}} - 1\right] \tag{8.38}$$

Let

$$n_r(t_1) = 0{,}1 \tag{8.39}$$

$$n_r(t_2) = 0{,}9 \tag{8.40}$$

Thus

$$\left[1+e^{(a+b\cdot t_1)}\right]^{-\gamma} = 0{,}1 \quad \& \quad \left[1+e^{(a+b\cdot t_2)}\right]^{-\gamma} = 0{,}9 \tag{8.41}$$

$$1+e^{(a+b\cdot t_1)} = 10^{\frac{1}{\gamma}} \quad \& \quad 1+e^{(a+b\cdot t_2)} = \left(\frac{10}{9}\right)^{\frac{1}{\gamma}} \tag{8.42}$$

$$e^{(a+b\cdot t_1)} = 10^{\frac{1}{\gamma}} - 1 \quad \& \quad e^{(a+b\cdot t_2)} = \left(\frac{10}{9}\right)^{\frac{1}{\gamma}} - 1 \tag{8.43}$$

or

$$e^{a} \cdot e^{b\cdot t_1} = 10^{\frac{1}{\gamma}} - 1 \quad \& \quad e^{a} \cdot e^{b\cdot t_2} = \left(\frac{10}{9}\right)^{\frac{1}{\gamma}} - 1 \tag{8.44}$$

By dividing the right side by the left side we get

$$e^{b\cdot(t_2 - t_1)} = \frac{\left[\left(\frac{10}{9}\right)^{\frac{1}{\gamma}} - 1\right]}{\left[10^{\frac{1}{\gamma}} - 1\right]} \tag{8.45}$$

Defining δ by

$$\partial = \frac{\left[\left(\frac{10}{9}\right)^{\frac{1}{\gamma}} - 1\right]}{\left[10^{\frac{1}{\gamma}} - 1\right]} \tag{8.46}$$

yields

$$b = \frac{\ln \partial}{\Delta T} \tag{8.47}$$

$$\Delta T = \frac{\left(\ln\left\{\frac{\left[\left(\frac{10}{9}\right)^{\frac{1}{\gamma}} - 1\right]}{\left[10^{\frac{1}{\gamma}} - 1\right]}\right\}\right)}{b} \tag{8.48}$$

Hence, the final expression for price versus time in general is:

$$P(t) = P(0) \cdot \left[n_r(0)^{-1} \cdot \left(1 + e^{\left\{ \ln\left[n_r(0)^{\frac{-1}{\gamma}} - 1 \right] + \left[\frac{\ln \partial}{\Delta T} \right] \cdot t \right\}} \right)^{-\gamma} \right]^{\log_2 \cdot K} \tag{8.49}$$

Where δ is a function of γ only. Five parameters are input to the formula:
P(0), the price in the reference year 0,
$n_r(0)$, the relative accumulated volume in year 0,
ΔT, the time for the accumulated volume to grow from 10 % to 90 %,
K, the learning curve coefficient,
γ, the asymmetry of the logistic curve.

This relation is demonstrated in Fig 8.5, in which the price versus time is shown for $P(0) = 1$, $n(0) = 0{,}01$, $\Delta T = 5$, $K = 0{,}8$ and different γ

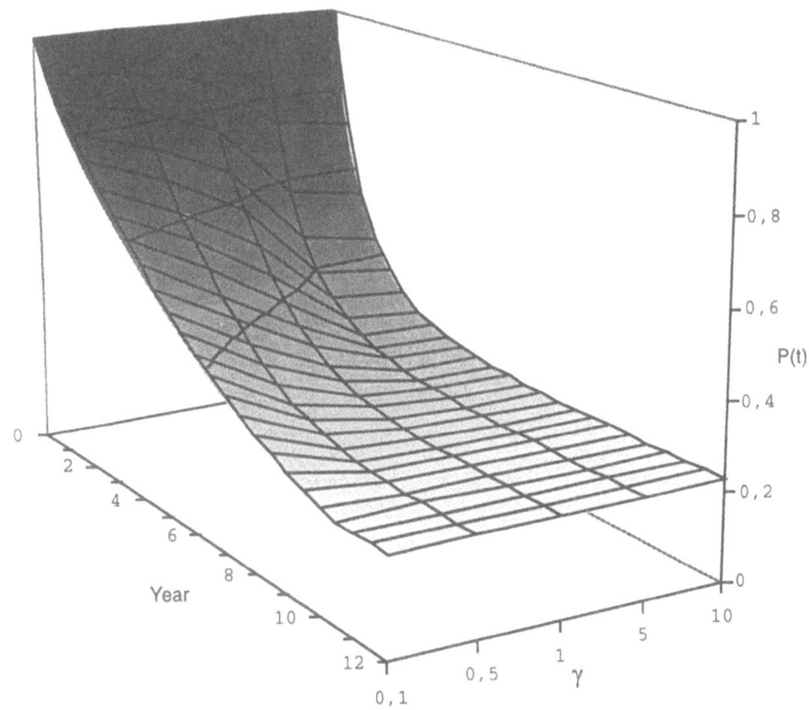

Fig 8.5 The price versus time for different γ.

REFERENCES

[1] CCITT. GAS 10 Handbook. Data acquisition and forecasting methods for telecommunication planning. 1987.
[2] Olsen, B T, Zaganiaris, A, Gieschen, N, Tahkokorpi, M, Drieskens, M, Markatos, S. RACE 2087/TITAN: Tool for Introduction scenarios and Techno-economic studies for the Access Network. *In: Proc. RACE Open Workshop on Broadband Access, Nijmegen, the Netherlands, June 7-8, 1993.*
[3] Olsen, B T, Stordahl, K. Forecast of price development of network components, based on an extension of the learning curve model. *Telektronikk,* 90, (1), 1994, 166-172.

[4] Wright, T P. Factors affecting the cost of airplanes. *Journal of Aeronautic Science*, 3, (4), 1936, 122-128.
[5] Crawford, J R. Learning curve, ship curve, ratios, related data. Locheed Aircraft Corporation, 1944.
[6] Yelle, L E. The learning curve: histrorical revew and comprehensive survey. *Decision Science*, 10, (2), 1979, 302-328.

Sensitivity analysis and risk assessment 9

Kjell Stordahl

9.1 INTRODUCTION

The development of the access network presents telcos with difficult and important strategic problems, both technically and financially. Technically because of the need to accommodate network evolution with changing and still uncertain technologies; financially because of the large cost and economic implications of replacing the present extensive access network. Decisions related to the introduction of new switching and transmission technologies for the core network are generally met within a ten years time frame, compared to 20 - 30 years for the access network. Although residential fiber in the loop will be a key element in the telco network strategy, its introduction seems to be slower and more complex than foreseen some years ago.

The new entrants to the local loop market introduce unprecedented parameters which lead to uncertainty in component costs and demand for services, civil works and operation, administration and maintenance costs of the technology to be employed, and uncertainty in the economic, political and fiscal environments in which development and operation will take place. The risks are a tough combination of a bewildering choice of possible technologies. Copper, fibre, radio and various combinations are all possible. Investment costs are high and making the wrong choice at the wrong moment can be an expensive mistake. The key to success is matching the capabilities of the technology and the customer requirements. Despite the risks, there are potentially large rewards in this market. In the presence of uncertainty factors, a thorough risk and sensitivity analysis is likely to substantially speed up the implementation of the project, since it clarifies the profitability of the investment to the operator, under various conditions.

9.2 AREA TYPES AND STRATEGIES

The access network areas are characterised by loop length, subscriber density, size, proportion between business and residential customers, existing infrastructure and type and quality of existing network components. The same strategies will not necessarily be used in each area, hence they will be evaluated in different types of areas.

The planning activity of a network operator usually starts from the definition of the target of the company and from the definition of the strategies to achieve such a target in different areas where it operates the network. The assumed target of the network operator is the maximisation of the present value of cash flow. To achieve this target, the network operator will undertake a number of actions that will eventually modify the network structure, the set of services provided to the customers and the penetration among the potential customers. Such set of actions can be generally termed as *"strategy"*.

The strategy should allow the network operator to match the forecast demand of services in a certain geographical area. In the following we will assume that in the selection of the strategy a number of initial conditions already have been defined. In addition, regulatory aspects and degree of competition have to be predicted. Among these initial conditions the subscriber density and type can vary in time and be a cause of uncertainties in the evaluation of operator strategy. Similarly, the service penetration and the number of competitors are not fully predictable: these two parameters will be further considered in the analysis of the potential risks for the operator. On the basis of this information, the network operator will define a number of strategies for achieving its target. In particular it has to define the following items:

- a network architecture capable of supporting the selected services
- the set of subscribers to be served
- the service tariff
- a plan to schedule investments in time.

The sources of uncertainty in the selected strategy mainly arise from the subscription of potential customers in a competitive situation, by real service penetration and by the costs of installing and operating a network compliant to the selected architecture. The evaluation of the impact of such uncertainties on the possibility of reaching the predefined target will constitute the main part of the work.

Comprehensive work has been done in the project RACE (research for advanced communications in Europe) 2087/TITAN (tool for introduction strategies and techno-economic evaluation of access network) in modelling and calculating the cost for various network structures [1-4]. The network architectures considered during the TITAN project were:

- twisted pair with asymmetric digital subscriber line (ADSL) and/or high-speed digital subscriber line (HDSL) between the local exchange and the customer
- fibre to the curb (FTTC) with twisted pair and coaxial cable between the optical network unit (ONU) and the customer
- FTTC with coaxial cable between the ONU and the customer
- fibre to the building (FTTB) with twisted pair and coaxial cable between the ONU and the customer
- radio in the local loop (RLL) with wireless link between the ONU and the customer.

The different architecture scope with different requirements for service transport and their selection will form a major part of the overall strategy definition. Additionally the operator should decide whether to implement the new architectures as:

- an integrated network to provide all types of services to the selected customers
- an overlaid network to provide a specific set of advanced services to selected customers.

A number of alternatives are available for subscription selection:The tariffs for the offered services will be influenced by the cost/revenue balance of the operator, by the willingness of the subscribers to pay and by competition with other operators. Due to the complex relationship with other variables, the tariff values will not be considered as a source of uncertainty.

The investment plan is determined by the financial budget of the company, by the evolution of the service demand in time and by the ageing of the installed network. Simplifying the problem, the investment can be related to the increase of subscriber numbers and to the replacement of aged components as it is presently done in the TITAN tool. The only causes of uncertainty associated with the investments plan will arise from the variations in component costs and forecast service demand.

In every area, a set of strategies denoted by:

$$S_1, S_2, \text{---------}, S_p \tag{9.1}$$

are examined to find the impact on the roll-out the access network.

9.3 GAIN CRITERIA IN RISK ASSESSMENT

Two main criteria commonly used in investment planning, are considered: The net present value (NPV) and the internal rate of return (IRR). For a given strategy, these two figures will tell us whether it is worth investing or not.

Net present value (NPV): for a strategy S and a period of study [0,n] where n defines the years, R_t the return , C_t the costs and I_t the investments at year t. RV_n is the residual investment at the end of the study period and α is the discount factor.

$$NPV(S,n,\alpha) = \sum_{t=0}^{n} \frac{R_t + C_t + I_t}{(1+\alpha)^t} + \frac{RV_n}{(1+\alpha)^n} \tag{9.2}$$

Internal rate of return (IRR): IRR(S,n) = α such that NPV(S,n,α) = 0 An investment is tolerable when NPV > 0 at the end of the period and is profitable when IRR(s,n) > α. Then the invested funds are covered and the explotation generates a positive return. Since the above two indicators have different meanings, two strategies A and B may lead to opposite results NPV(A,n,α) > NPV(B,n,α) while IRR(A,n) < IRR(B,n). NPV is linked to the amount of the expected returns whereas IRR emphasises the investment efficiency. An IRR based methodology favours a project with rather low investments whereas a NPV based methodology favours a project with rather high investments. None of these two indicators can be left aside The final decision will consider both of them and will be linked to other strategic or political criteria.

9.4 CRITICAL VARIABLES FOR ACCESS NETWORK EVALUATIONS

9.4.1 Important variables and their dependencies

The most important variables identified and their interdependencies are as follows:

- types of services
- type of network architecture
- service demand
- competition
- cost of components
- tariffs
- type of area
- network quality
- discount rate.

9.4.2 Distillation of critical variables

In order to perform a reasonable cost assessment, it is necessary to concentrate the analyses on a limited number of variables, which are chosen according to their influence on the results:

1. Critical network components. The network component is specified after the network architecture is defined.
2. Demand for services and network competition.
3. Running costs and new technology.

9.5 APPLICATION OF RISK METHODOLOGY

9.5.1 Probability distributions and time dependency

The critical variables listed in the previous sections are likely to have significant impact on the gain *G*, depending on the strategy. The size of the variable will change with time. In the risk model the variables are defined as *parameters* and denoted by:

$$\theta_1(t), \theta_2(t), ------, \theta_m(t) \tag{9.3}$$

Each parameter $\theta_i(t)$, $i = 1, 2, ----$ m has a probability distribution denoted by:

$$p_{it}(\theta_i(t)) \quad i = 1, 2, ------ m \tag{9.4}$$

Suppose that parameter number two is the price of a specific network component. Because of mass production the price of the component will decrease with time. Unfortunately we do not know the exact price. Therefore price is described by a probability distribution. Since the price changes with time, also the probability distribution will change with time.

In 1998 we know the exact price for a given network component. In 1999 the price will decrease, but still our estimate of the expected price will be reasonably good. Hence the probability distribution describing the price variation has a small standard deviation. The uncertainty in the price estimate will increase as a function of time. At the same time the expected value of the prices decreases as a function of time.

Usually the Normal probability distribution describes these situations very well. The Normal distribution is uniquely defined by two related parameters: the expectation μ and the standard deviation σ. Fig 9.1 gives an illustration of how the probability distribution changes with time.

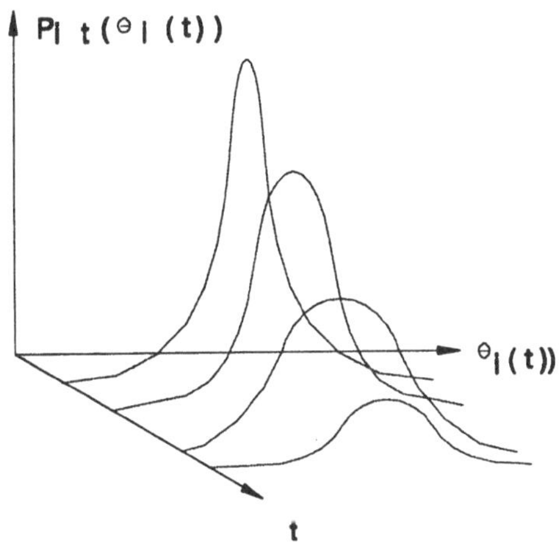

Fig 9.1 The probability distribution as a function of time.

In a complete risk analysis every important parameter has to be described by a probability distribution which changes with time. Because of the complexity of the modelling and volume of calculations only parameters (variables) which have a significant impact on the gain should be included in the analysis. It is also important to start with a few variables and examine their influence to understand the model behaviour.

9.5.2 Specification of the Normal distribution

Estimates or predictions of the expected value of the parameter must be performed with the benefit of earlier experience. For network components the predictions will be based on learning curve models like Wright's and Crawford's formula [5-7]. The expected value of other important parameters must be predicted using other methodologies based on the specific situation. The expected values must be predicted for each year over a ten year period.

As mentioned earlier the Normal distribution is uniquely defined by the expectation and the standard deviation. The standard deviation must be estimated in a subjective way. Guidelines for prediction of the standard deviation σ is as follows:

Let the parameter be θ. Then the value of σ according to tables of the Normal distribution, must be chosen such that:

$$P(\mu - \sigma < \theta < \mu + \sigma) \approx 0.68 \tag{9.5}$$

$$P(\mu - 2\sigma < \theta < \mu + 2\sigma) \approx 0.95 \tag{9.6}$$

$$P(\mu - 3\sigma < \theta < \mu + 3\sigma) \approx 0.997 \tag{9.7}$$

Equation (9-7) shows that μ and σ should be predicted such that the probability for θ to exceed the interval $\mu \pm \sigma$ is 0.68 or 68 %. In the same way the probability for θ to exceed two times the interval is 95 % etc.

9.5.3 Segmentation of the Normal distribution

The question is how calculations should be performed for each parameter and for a set of parameters. If the effect of the variation in one parameter only is analysed at a time, then it is possible to use more discrete values than when the effect of variation of a set of parameters is analysed simultaneously.

Three alternatives for segmentation of the Normal distributions in intervals and middle fractiles are presented. The different alternatives are based on the normal distribution with expectation 0 and standard deviation 1, which is called the standard normal distribution $N(0,1)$. This is in fact the general case since it is always possible to transform a Normal distribution $N(\mu,\sigma)$ with expectation μ and standard deviation σ to the standard normal distribution.

The relation between the general Normal distribution, $N(\mu,\sigma)$ and the standard Normal distribution $N(0,1)$ is as follows:

Let t_α be the α fractile in the standard Normal distribution. Then

$$P\left(\theta < t_\alpha\right) = \alpha \tag{9.8}$$

Now if θ is normal distributed, $N(\mu,\sigma)$, then

$$P\left(\theta < \left(\mu + t_\alpha \cdot \vartheta\right)\right) = \alpha \tag{9.9}$$

Fig 9.2 shows a segmentation of the Normal distribution and illustrates how the risk assessment is performed

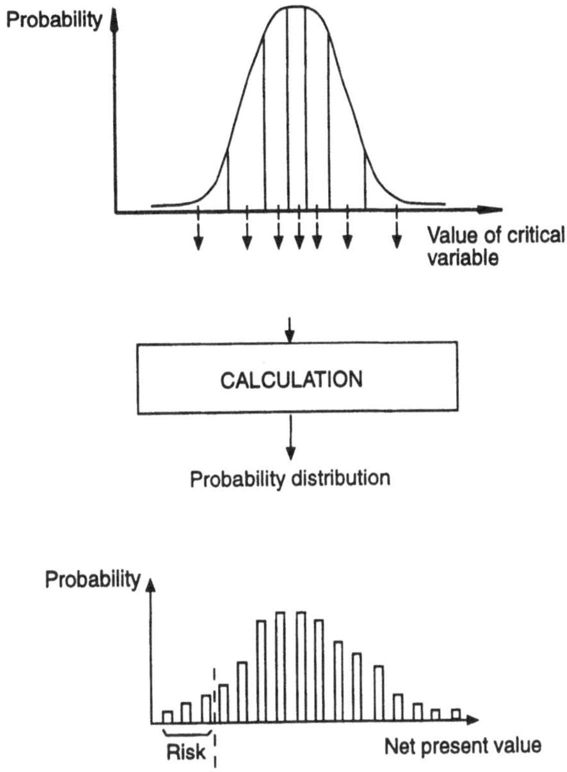

Fig 9.2 Segmentation of the Normal distribution and risk assessment.

9.5.4 Specification of asymmetric distributions

In some cases the size of the standard deviation is not equal on both side of the expectation. This is usually the case when the expectation is close to zero. To describe the situation we need an asymmetric probability distribution. The Gamma distributions are used for their purpose.

9.5.5 Simplification by introduction of relative uncertainty

A simplification of a combined function of the logistic model and Wright's formula is used to predict the expected costs for the years following a given reference year. The simplified model predicts the future costs as a function of time. Moreover, in risk analysis it is important to make simplifications to avoid unnecessary work and to save computing time. Usually there is a relation between the size of the expectation and the size of the standard deviation. In statistical theory the concept relative standard deviation is used. The reason is that when the expectation increases the standard deviation also increases. Often there is a linear relation between the increase in the expectation and the standard deviation.

When it is possible to use the relative standard deviation of every important or critical parameter, the prediction of the standard deviation is performed by the application of equations (9.5) - (9.7) together with the estimation of the standard deviation as a fraction of the expected value.

$$\sigma = f(\mu) = \kappa\mu \qquad (9.10)$$

Now we have to predict the fraction κ. By using this approach it is possible to decrease the volume of input data.

As mentioned previously, the prediction of the expected value $\mu(t)$ as a function of time is performed by a combined simplified function of the logistic model and Wright's formula. To decrease the volume of input data the standard deviation is also predicted by the time function $\kappa(t)$. Equation (9.10) is then extended to:

$$s(t) = \kappa(t) \cdot \mu(t) \qquad (9.11)$$

Additional work has to be carried out to determine the $\kappa(t)$ function for the various important or critical parameters. In the TITAN cost database the network components are classified in five confidence classes according to their relative standard deviation.

9.6 EVALUATION CRITERIA

Suppose now that we want to examine strategy S_j. The gain by using strategy number j is G_j, $j = 1, 2, \text{------} p$. Various expressions may be used for the gain such as NPV and IRR. As mentioned earlier, the gain is dependant on a set of parameters and also the distinct strategy. Hence the gain can be expressed as a function of the strategy and the parameters:

$$G = G_j(\theta_1, \theta_2, \text{------}, \theta_m) \qquad (9.12)$$

In Fig 9.3 the gain G_j is illustrated in an m+1 dimensional space.

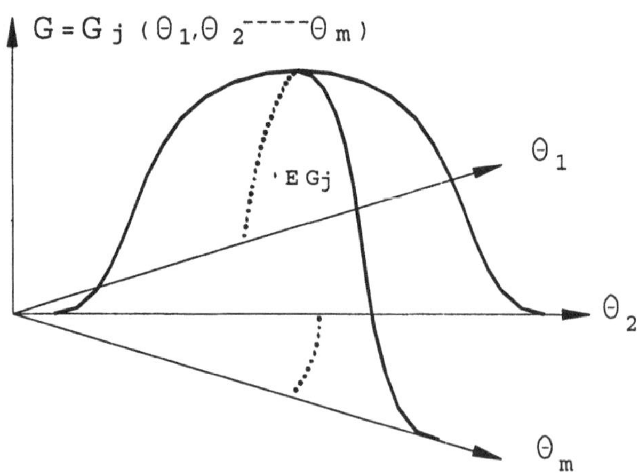

Fig 9.3 The gain G_j using strategy j as a function of the parameters.

We need to have a set of criteria in order to examine the different strategies. The most important criterion is the expected gain. Secondly some criteria concerning the risk have to be included. It is possible to look at the estimated standard deviation of the gain. Another possibility is to calculate the probability that the gain is greater than 0, or the probability that the gain is less than a critical value Δ. The criteria for evaluating strategies S_j, $(j=1, 2, - - - - - , p)$ are then statistically defined by:

$$EG_j \tag{9.13}$$

$$\sigma_j (G_j) \tag{9.14}$$

$$P(G_j > 0) \tag{9.15}$$

$$P(G_j < \Delta) \tag{9.16}$$

P( ) denotes the probability.

9.7 TWO APPROACHES FOR RISK ASSESSMENT

The risk assessment may be performed either by simulation or by ordinary calculation. However, the volume of calculations requires programs for calculation of joint probabilities when risk assessments are performed and when the critical variables are assigned values according to their probability distributions. Alternatively, simulations may be used. The Crystal Ball© simulation program is used in connection with the TITAN tool.

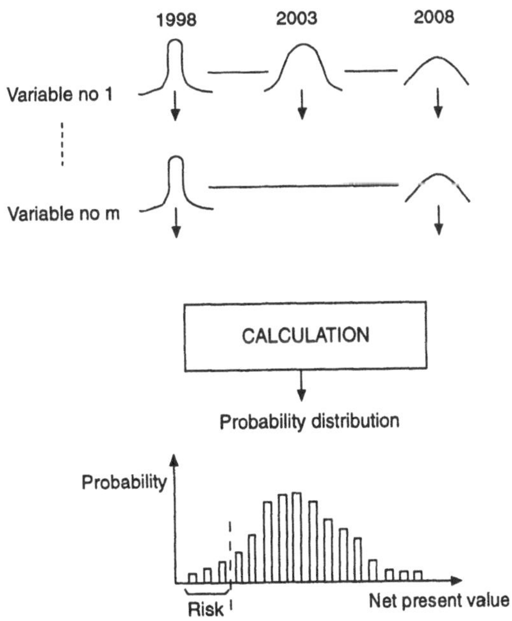

Fig 9.4 Simulation of a set of critical variables and risk assessment.

Furthermore, two risk assessment methodologies may be applied for simulation and calculation. Both require well defined parameter distributions. The first methodology consists of analysing the impact of each parameter on the forecast values. This approach is very similar to a sensitivity analysis.

In the second approach all the assumptions or input parameters are changed simultaneously in order to get the distribution of the forecast values, which typically may be the present value of the cash flows and the internal rate of return. The distribution of input parameters or assumptions is split into equal probability sets. The number of sets has to be defined initially. In Crystal Ball© random numbers are generated for each set of the distributions before the calculations. This sampling method is precise, but the implementation takes time. Fig 9.4 illustrates how risk assessment is performed based on simulation of a set of probability distributions.

9.8 SENSITIVITY ANALYSIS

Sensitivity analysis is a simplification of risk analysis, in which only some values for each variable are chosen, instead of introducing the probability distribution for the various variables. The intervals constituted by the chosen values of each variable are determined in the same way as the probability distributions were constructed. The effect of each variable can be evaluated separately or simultaneously.

The common approach is to start with one variable and calculate the effect of different values of the variable within an a priori probable interval. If the values do not have significant influence on the results, one option is to test how large the variation must be in order to influence the results, even if the values are not probable. The same procedure is applied to all the variables separately. Thus, it is possible to study the effect of the variables and rank the variables according to their influence. Usually the sensitivity analysis stops with an evaluation after the calculation, based on variation of each variable separately, but it is also possible to study the effect of some variables simultaneously. However, the calculation can not be covered in an representative way with a simultaneous sensitivity study. Some effects can be calculated, but it is difficult to draw unique conclusions.

It is recommended to use sensitivity analysis in the first part of a study. The sensitivity analysis will tell which variables are of most importance. If the study does not focus on the risk, then a sensitivity analysis will be acceptable. It is time consuming to carry out a complete risk analysis. First of all the preparations in order to determine the probability distributions and to plan the calculations and/or simulations take time. Secondly, it is time consuming to perform the calculations and/or simulations. Hence, the time available for evaluating the various case studies will determine what type of methodology is to be chosen to predict the risk.

REFERENCES

[1] Olsen, B T, Zaganiaris, A, Gieschen, N, Tahkokorpi, M, Drieskens, M, Markatos, S. RACE 2087/TITAN: Tool for Introduction scenarios and Techno-economic studies for the Access Network. *In: Proc. RACE Open Workshop on Broadband Access, Nijmegen, the Netherlands, June 7-8, 1993.*

[2] Zaganiaris, A, Ferrero, U, Balzaretti, S, Tahkokorpi, M, Kalervo, M, Olsen, B T, Stordahl, K. Methodology for risk assessment and techno-economic evaluation of optical access networks. *In: Proc The 20th European Conference on Optical Communications (ECOC '94), Florence, Italy, September 25-29, 1994, 3, 83-90.*

[3] Zaganiaris, A, Olsen B T, Stordahl, K, De Bortoli, M, Tahkokorpi, M. Life-cycle cost and risk assessment of the access network. From RACE 2087/TITAN. *In: Proc. 6th International IEEE workshop on Optical Access Networks (OAN '94), Kyoto, Japan, Oct. 16-19, 1994, 1.4-1 - 1.4-10.*

[4] Stordahl, K, Murphy, E. Methods for forecasting long term demand for wide and broadband services in the residential market. *IEEE Communications magazine*, 13, (2), 1995, 44-49.

[5] Wright, T P. Factors affecting the cost of airplanes. *Journal of Aeronautic Science*, 3, (4), 1936, 122-128.

[6] Crawford, J R. Learning curve, ship curve, ratios, related data. Locheed Aircraft Corporation, 1944.

[7] Yelle, L E. The learning curve: histrorical revew and comprehensive survey. *Decision Science*, 10, (2), 1979, 302-328.

Tool implementation 10

Markku Tahkokorpi

10.1 INTRODUCTION

This chapter gives a brief description of the general structure of TITAN (tool for introduction strategies and techno-economic evaluation of access network) and how the methodology described in the previous chapters has been implemented in Excel® spreadsheet software. The development of the TITAN tool started from the SYNTHESYS tool, which had been developed by the evolution prospects and frameworks (EPF) group of the RACE (research for advanced communications in Europe) project 1044, entitled Integrated broadband communications development and implementation strategies. The SYNTHESYS tool contained the core functionality of the TITAN methodology, including the SYNTHESYS geometric model. However, at that stage the whole tool was implemented as one very large Excel© 3 sheet, which made its use fairly cumbersome.

The tool development within RACE 2087/TITAN was aimed at implementing the methodology in a user friendly and open way as possible. The openness has been dealt with by using Excel© as the programming platform, which enables the use of the tool both in PC and Macintosh environments with only minor adaptations required [1].

Furthermore, the initial goal was a methodology and a tool for evaluating introduction scenarios for fibre in the loop systems. However, during the project, the technology scope was widened to include any access network technology [2-5].

10.2 THE TOOL STRUCTURE

The user interface has been implemented using the Excel© macro language, with the macros included in one kernel file. The database part of the tool is implemented in a separate Excel spreadsheet file. Hence, the tool has two working modes: The *main mode* is used for the actual modelling and calculations. The *database mode* is used for the editing of the database contents. In the *main mode*

After the calculations, output data as well as the input data can be extracted to a separate log file. The data in this file can then be further formatted and processed using standard Excel© functions. Several *main sheets* (architecture models) may be open simultaneously during the running of the tool, but the number of simultaneous database files is limited to one. Fig 10.1 shows the file structure of the TITAN tool.

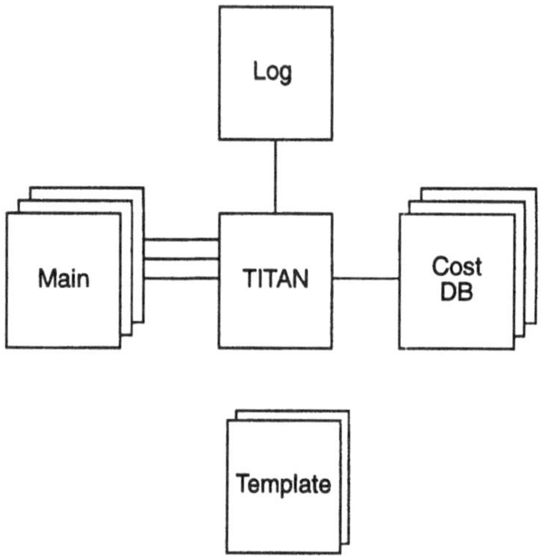

Fig 10.1 The file structure of the TITAN tool.

In the figure *TITAN* represents the program core, which consists of TITAN macros, approximately 2,000 lines of Excel© 3 and Excel© 4 macro language code. The *TITAN main file* implements the user interface and parts of the calculation processes. Most of the TITAN calculations are common Excel© functions included in the main sheet cells. In addition, template files for creating new main files and cost database files are included in the tool. Chapter 5 contains additional information on the TITAN methodology.

10.2.1 The structure of the main sheet

The *main sheet* consists of several sections. Information notes about the model are found in the first section. The second section contains general inputs such as discount rate, tax rate and other financial inputs. The third section of the *main sheet* defines the geometric model and its input parameters.

In the fourth section the services are defined as a time series specifying the penetration percentages and tariffs (annual revenues) per user over the study period. The study period can be in the range from two to 15 years and the start year can be any year.

The architecture is defined in section five, the so called *shopping list*, in which the investments made each year of the model run are specified. The investment may be linked to the outputs of the geometric model and/or the service penetrations. References to the values within the geometric model are made by the use of pre-defined model names, whilst the references to the services are manually established using standard Excel© references. In the *shopping list* section cumulative investments, operation, administration and maintenance (OAM) costs, depreciation and rest value for each cost component are calculated.

All the investment related expenses are calculated within the *shopping list* section. Then, in the final section the previously described inputs and calculation results are combined in order to provide various economic outputs. The majority of the values in this section are time series, as for instance the annual investments or running costs. However, some values are scalar values which are calculated from these vectors. Total investments, running costs per user, NPV and IRR are examples of these scalar outputs. The *main sheet* is described in more detail in appendix C.

10.2.2 The structure of the database sheet

The database sheet consists of several tables. The main table is the components database in which each cost component is described by its *reference price, reference year, price evolution class, OAM class* and *write-off class*. The classes themselves are also defined in the database file. An additional table describing services in terms of penetration and tariff evolution is included in the database file. The services may either be defined by the user manually within the main sheet, or by the use of the included database, which holds pre-defined services, comprising default service evolutions and characteristics. Further information on the database is found in chapter 5 and in appendix C.

10.2.3 Models implemented in the tool

The basis of the TITAN modelling and the tool is a layered network model where the network consists of a hierarchy of flexibility points and link levels connecting these flexibility points. Geometric models are included in the TITAN tool for calculation of cable and duct lengths. The SYNTHESYS model, the modified SYNTHESYS model, the bus/star model and the TITAN model are described in chapter 6.

The component prices for each year are calculated from the cost database data using the price evolution formula described in chapter 8. The component prices are updated every time the database link is updated.

10.2.4 The calculation flow

Every time a network architecture is calculated, the input data may be updated initially from the database, if needed. Following the optional update of these inputs, the complete main sheet is calculated in order to update all inputs. Then the depreciation is calculated by a separate macro, since the depreciation calculation requires iterative summing of depreciation figures from preceeding years. Finally, the main sheet is re-calculated in order to include the calculated depreciation values in the results.

10.2.5 Risk assessment

The TITAN tool has been slightly modified to enable risk assessment by the use of the commercial risk assessment tool package Crystal Ball®, which is an add-in to Excel® and Lotus 123®. Crystal Ball® enables a group of inputs in an Excel® spreadsheet to be defined as assumptions. A probability distribution is defined for these assumptions, and a set of outputs are selected as forecasts. Crystal Ball® then runs a Monte Carlo or Latin Hypercube simulation with the assumption values, and calculates the model hundreds or thousands of times with different input values drawn from defined distributions. The forecasts then show the probabilities of selected outputs in the case under study. The risk assessment methodology developed in TITAN is described in more detail in chapter 9.

REFERENCES

[1] Olsen, B T, Zaganiaris, A, Stordahl, K, Ims, L A, Myhre, D, Øverli, T, Tahkokorpi, M, Welling, I, Drieskens, M, Kraushaar, J, Mononen, J, Lähteenoja, M, Markatos, S, De Bortoli, M, Ferrero, U, Ravera, M, Balzaretti, S, Fleuren, F, Gieschen, N, De Oliveira Duarte, M, de Castro, E. Techno-economic evaluation of narrowband and broadband access network alternatives and evolution scenario assessment. *IEEE Journal on Selected Areas in Communications,* 14 (6), 1996, 1184-1203.

[2] Olsen, B T, Zaganiaris, A, Gieschen, N, Tahkokorpi, M, Drieskens, M, Markatos, S. RACE 2087/TITAN: Tool for Introduction scenarios and Techno-economic studies for the Access Network. *In: Proc. RACE Open Workshop on Broadband Access, Nijmegen, the Netherlands, June 7-8, 1993.*

[3] Zaganiaris, A, Gieschen, N, Olsen, B T, Stordahl, K, Tahkokorpi, M, Drieskens, M, Ajibulu, A. A methodology for achieving life-cycle costs of optical access networks-from RACE 2087/TITAN. *In: Proc. 11th Annual Conference European Fibre Optic Communications and Networks (EFOC '93), the Hague, the Netherlands, June 30-July 2, 1993.*

[4] Zaganiaris, A, Tahkokorpi, M, Drieskens, M, Van Hoecke, J, Olsen, B T, Markatos, S, De Bortoli, M. Life-cycle costs and economical budget of optical and hybrid access networks. *In: Proc. 5th IEEE Conference on Optical/Hybrid Access Networks, Montreal, Canada September 7-9, 1993.*

[5] Tahkokorpi, M. TITAN methodology. *In: Proc. Eurescom Workshop on Optical Multiservice Access Networks (WOMAN), Turin, Italy, September 23-24, 1996.*

Section 3

Case studies and application results

Analysis of narrowband technology options

11

Borgar Tørre Olsen, Kjell Stordahl, Alcibiade Zaganiaris, Nikolaus Gieschen, Markku Tahkokorpi, Marc Drieskens, Sotirios Markatos, Leif Aarthun Ims

Access network architectures for the short and medium term narrowband needs have been analysed with TITAN (tool for introduction strategies and techno-economic evaluation of access network). All technical solutions are assumed deployed in a greenfield environment, in which there is no existing infrastructure, apart from ducts. Single home areas and apartment block areas have been studied. For these particular studies the theoretical outputs of the tool have been validated by comparisons to real operating networks in Europe.

11.1 INTRODUCTION

Accelerated competition and dramatic technological breakthroughs focus the attention of the operators on satisfying new customer needs, on cutting costs and on expanding their revenue base. With increasing pressure to minimise costs and to maximise revenues, a large variety of access network architectures must be rigorously examined in order to determine the most appropriate ones for the different area types and service demand profiles. A first step in assessing the economic viability of different access networks consists of a comprehensive evaluation of installed first costs (IFC), operation, administration and maintenance (OAM) costs and potential revenues, given the service provision capability of the system. For provision of the current residential narrowband services, several technology options exists. The chapter presents a techno-economic analysis of the short and medium term major technology alternatives for a greenfield situation, carried out with the tool developed in RACE (research for advanced communications in Europe) 2087/TITAN [1-3].

11.2 NARROWBAND TECHNOLOGY ARCHITECTURES

The four different narrowband (narrowband capacity = n·64 kbit/s, i.e. capacity < 2 Mbit/s) architectures considered in this study are:

1. Conventional copper solution, taken as a yardstick for comparison
2. Fibre to the curb (FTTC) with 32 subscribers per optical network unit (ONU)
3. Fibre to the building (FTTB) with both 32 and eight subscribers per ONU
4. Radio in the local loop (RLL).

These architectures were mapped into the SYNTHESYS double-star geometric model [4]. All architectures are assumed to serve 1,024 subscribers per service access point (SAP) [5].

An ad hoc copper architecture was derived, called copper to the building (CTTB), in which the drop cable is independent of user density and is replaced by 20 meters of indoor cable. This enabled a comparison of a copper network to FTTB architectures. In the conventional copper architecture two twisted pairs serve every home. 30 % of the installed pairs are assumed to be unused. The copper cables are dispatched and spliced at the flexibility points. The only electronics included in this architecture are line cards for plain old telephone service (POTS) and line cards for narrowband integrated services digital network (N-ISDN). The copper architecture studied is shown in Fig 11.1.

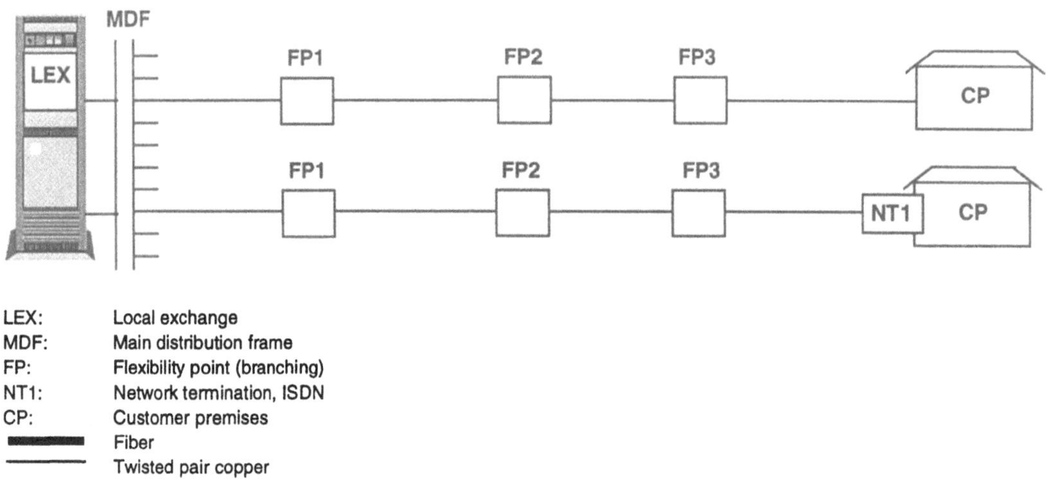

Fig 11.1 The reference copper architecture studied, CTTB.

The copper architecture was compared to two fibre solutions and one radio solution. The fibre alternatives are telephony over passive optical network (TPON) architectures, in both FTTC and FTTB configurations. The fibre architectures studied are shown in Fig 11.2. The FTTC solution has 1:2 splitters at the exchange site and a 1:4 splitter in a central flexibility point. The system has separate fibres for upstream and downstream signals, adding up to a total of four fibres to each ONU. The maximum capacity of the ONU used in this architecture is 64·64 kbit/s, whilst the maximum capacity of the optical line terminal (OLT) is 64·2 Mbit/s. The number of line cards used depends on the penetration of services.

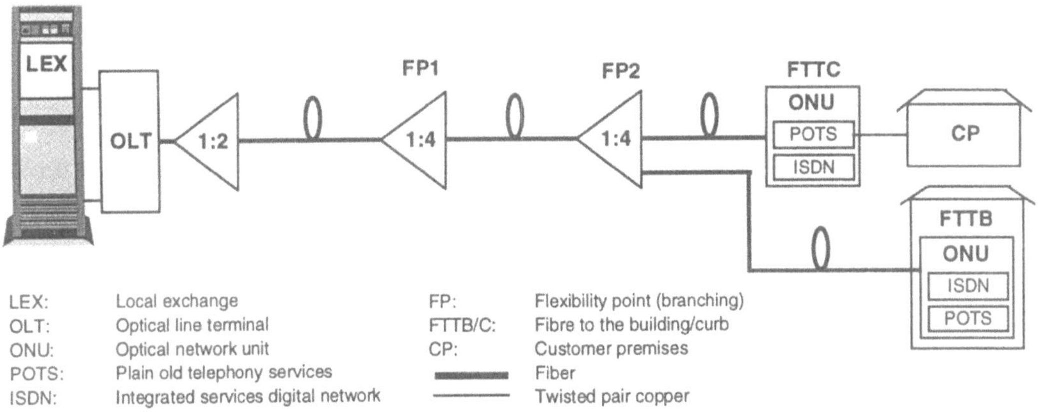

Fig 11.2 The reference fibre architectures studied. The architectures are based on TPON technology, in both FTTC and FTTB configurations.

The FTTB alternative is a PON solution similar to the one described above for FTTC. The main difference is that the copper cable from the ONU to the subscriber is replaced by a constant length of 20 meters indoor cable. Two architecture versions are studied:

1. ONUs in buildings with up to 32 subscribers
2. ONUs in buildings with up to eight subscribers.

The RLL architecture is a digital European cordless telecommunication (DECT) architecture on a copper infrastructure. Fig 11.3 depicts the RLL architecture studied.

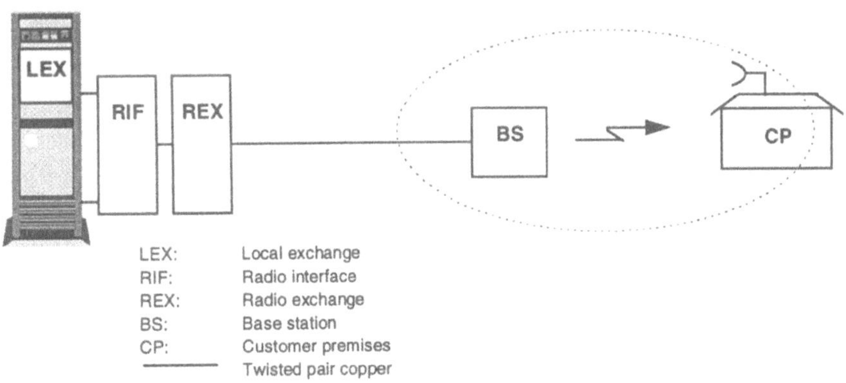

Fig 11.3 The reference RLL architecture studied.

Base stations covering a maximum of 200 meter radius are placed in different flexibility points depending on subscriber density. For low density areas with less than 560 subscribers/km^2, directional antennas are used to extend the range of the base stations up to 800 meters. The number of base stations is determined by the range and traffic capacity of the base stations.

11.3 RESULTS AND DISCUSSION

The above architectures were analysed and compared to the conventional copper network. In addition, the impact of the feeder part is calculated. The results presented must be related to the specific input parameters and depend on the 1994 cost knowledge. The key assumptions are listed in Table 11.1:

Table 11.1 The key assumptions used in the analysis.

Civil works	*Duct availability (when included), 90 % (fibre), 50 % (copper)*
Services	POTS and N-ISDN; demand according to forecast.
Depreciation	Straight Line
Discount factor	7.5 %
Cable overlength	40 %
Copper pair occupation	70 %
Central management OAM savings	30 % (for fibre based systems only)
Revenue per user	POTS: 160 ECU/year ISDN: 530 ECU/year

11.3.1 Installed first costs

The impact of the subscriber density on the installed first costs (IFC) is seen in Fig 11.4. The handset is not included in the RLL costs. With the given duct availability both the RLL and the FTTC solutions are significantly cheaper than conventional copper solutions for low subscriber densities. RLL turns out to be a promising solution, from the point of view of the operators, especially in low density areas where directional antennas can be effective.

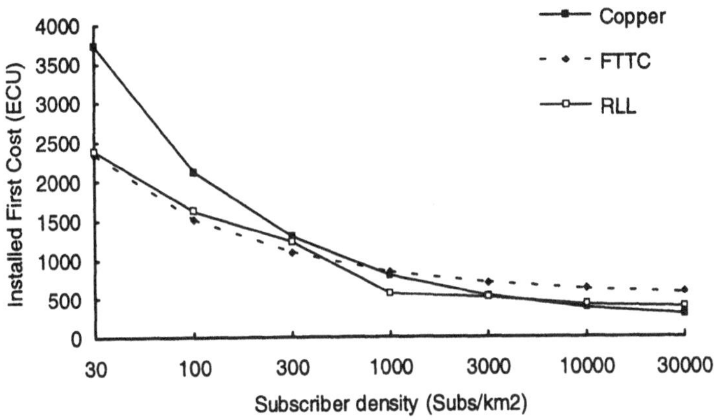

Fig 11.4 The impact of subscriber density on the IFC.

11.3.2 Impact of feeder cost

The impact of the feeder on the IFC of some architectures was studied by comparing the FTTC32 architecture with a feeder section both to a simple copper network and to a network consisting of a remote subscriber system (RSS) with fibre feeder and copper drops. Fig 11.5 depicts the impact of the feeder length on three different technical solutions. As shown in the figure, for short feeder lengths simple copper solutions are the cheapest ones, while the FTTC alternative is comparable or even cheaper than conventional fibre driven RSS architectures for feeder lengths beyond four km.

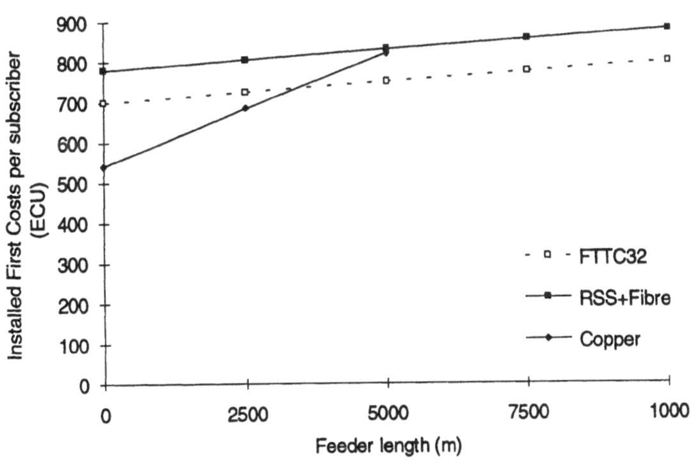

Fig 11.5 The impact of the feeder length on the IFC.

Fig 11.6 and Fig 11.7 illustrate the effect of civil works in the distribution part on the IFC for the fibre and copper architectures. In Fig 11.6 an FTTC architecture is compared to a copper network. Above the solid line in Fig 11.6 the FTTC solution is cheaper and below the solid line the copper solution is cheaper. The two dotted lines represent a 10 % cost change (tolerance). In Fig 11.7 the FTTB32 and CTTB32 architectures are compared. Above the solid line in Fig 11.7 the FTTB32 architecture is cheaper and below the solid line the CTTB32 architecture is cheaper. The figures illustrates that high civil works costs make fibre alternatives preferable. The ONU cost is one of the critical cost elements in the fibre solutions. The significant difference between FTTB with eight or 32 users per ONU is mainly attributed to the larger number of ONUs in the former alternative.

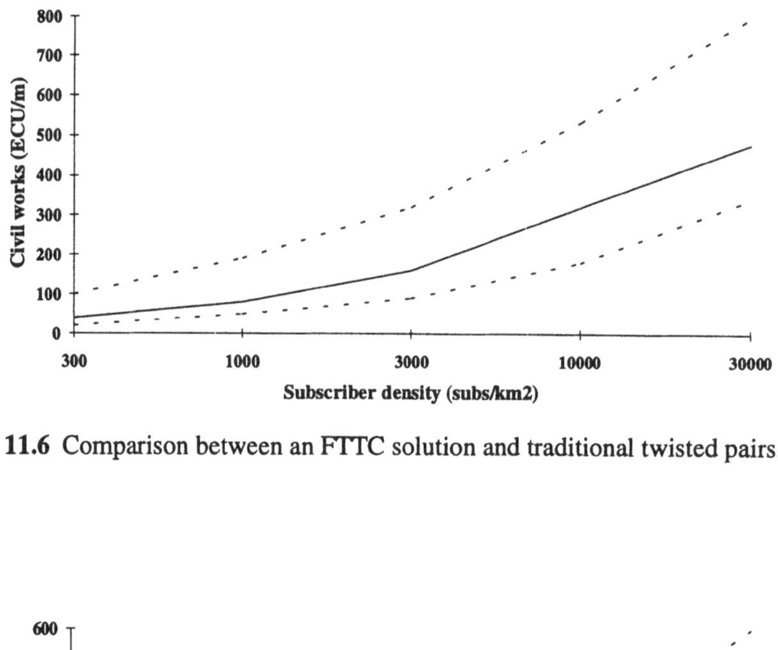

Fig 11.6 Comparison between an FTTC solution and traditional twisted pairs.

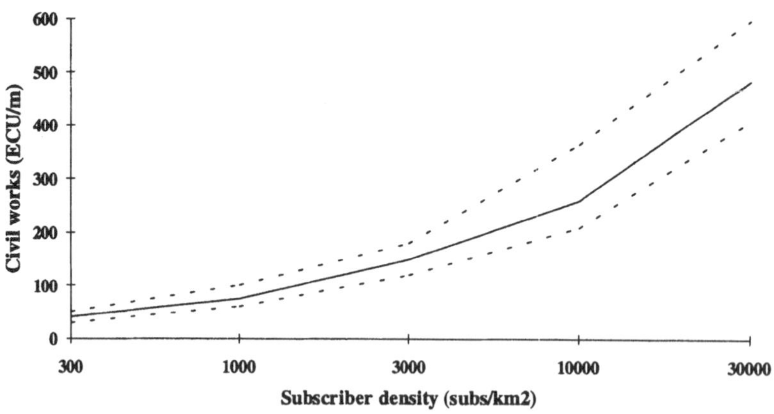

Fig 11.7 Comparison between FTTC32 and CTTB32 solutions.

11.3.3 Life cycle cost

The life cycle cost (LCC), which is the sum of the investments and the running costs, have been calculated for systems using the three relevant transmission media. The results in Fig 11.8 and Fig 11.9 show for the two areas the LCC of the different systems. LCC are shown in black. The white columns in the figures represent the IFC excluding civil works, and the hatched columns represent the IFC including civil works, minimum case.

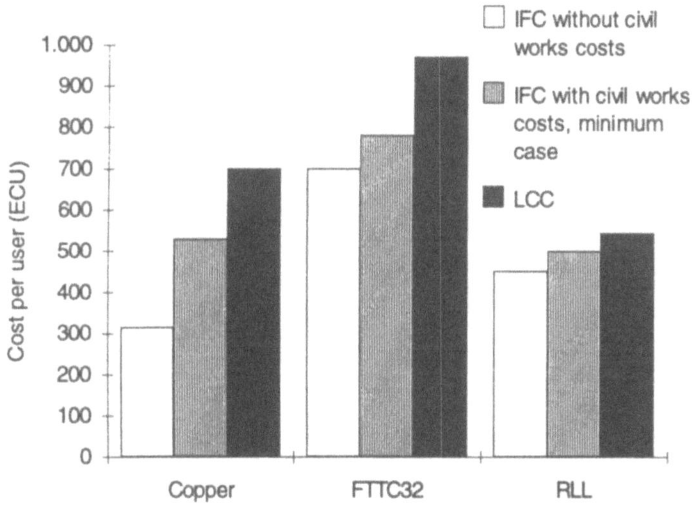

Fig 11.8 The IFC of the different systems for single homes and apartment building areas with a subscriber density of 3,000 subscribers/km^2.

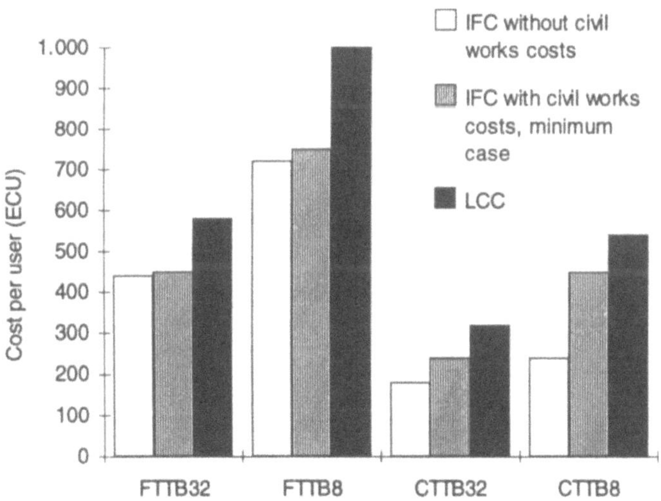

Fig 11.9 The IFC of the different systems for single homes and apartment building areas with a subscriber density of 3,000 subscribers /km^2.

Fig 11.10 and Fig 11.11 show the LCC without civil works for the architectures studied. Fig 11.12 and Fig 11.13 show the LCC including civil works.

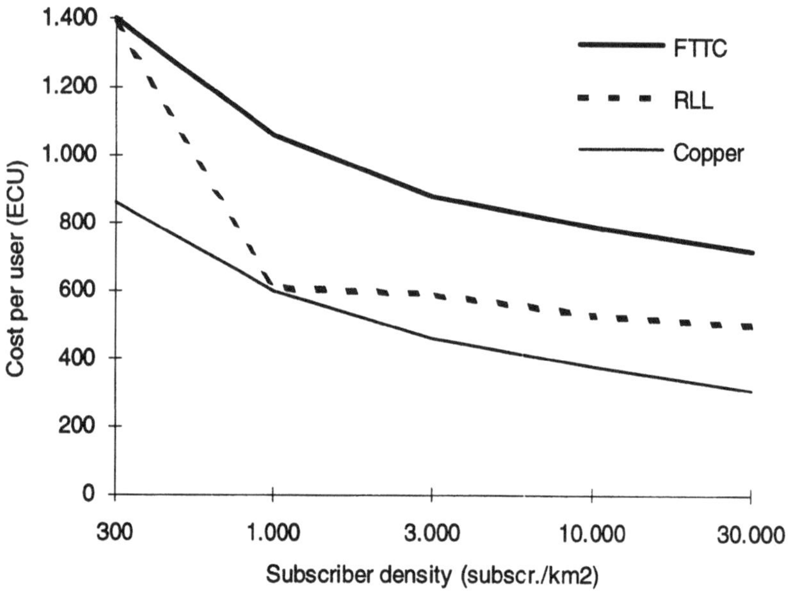

Fig 11.10 LCC excluding civil works for copper, FTTC and RLL architectures.

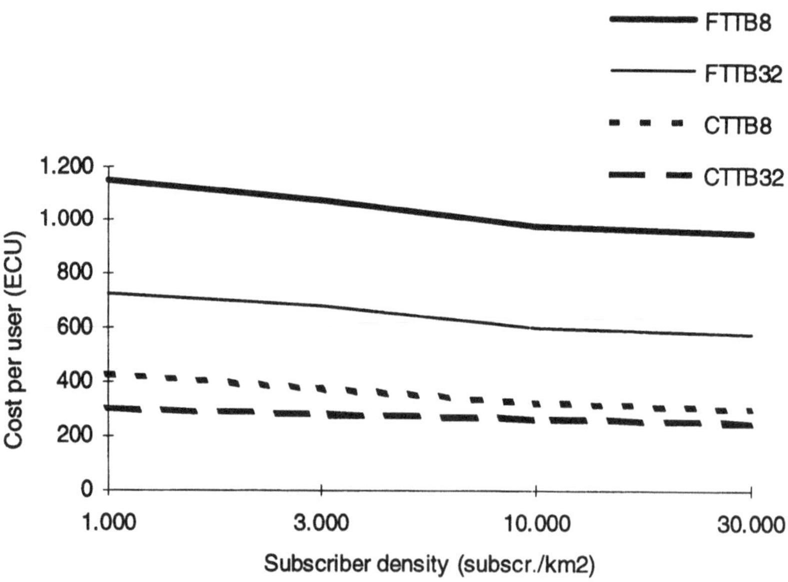

Fig 11.11 LCC excluding civil works for CTTB32, CTTB8, FTTB32, FTTB32 architectures.

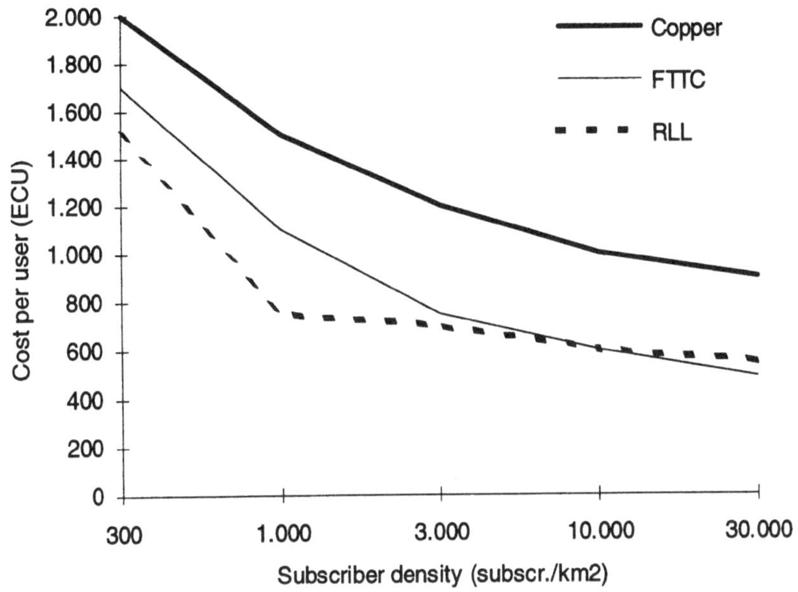

Fig 11.12 LCC including civil works for copper, FTTC and RLL architectures.

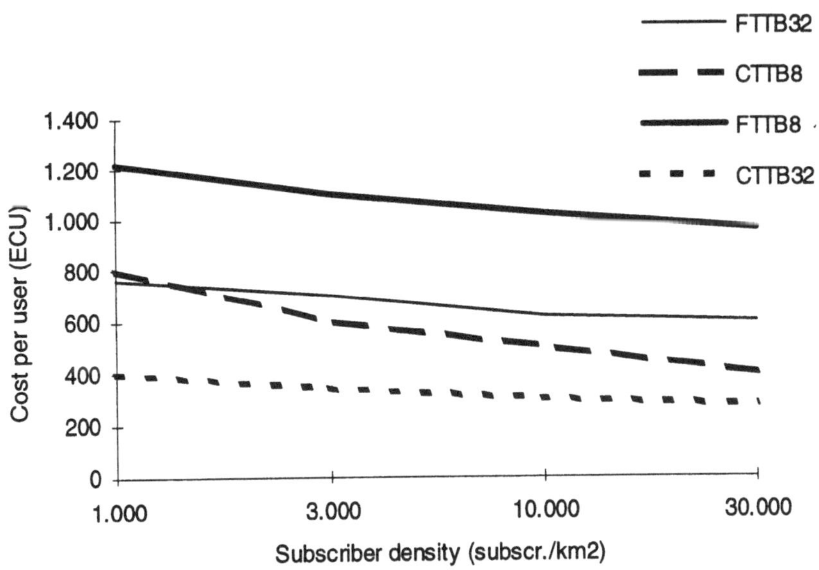

Fig 11.13 LCC including civil works for CTTB32, CTTB8, FTTB32, FTTB32 architectures.

The results show the sensitivity of LCC with respect to subscriber density. Fig 11.8 - Fig 11.13 illustrate that the LCC of the fibre systems is closer to the copper solution compared to IFC calculation results. This is due to reduced OAM costs.

The fibre and radio based networks benefit from automated management, here yielding 30 % savings on the total OAM costs. The maintenance part of the OAM cost is related to the environmental conditions at the ONU location [6]. Therefore, significant cost savings may be achieved when the ONU is located inside a building.

11.3.4 Project calculations

The cash balance (cumulative sum of the annual cash flows) and the payback periods of the projects have been calculated for the given penetration and revenue assumptions in order to obtain an indication of the overall economics of such access network projects. Fig 11.14 shows the cash balance of the FTTB32 and CTTB32 architectures including civil works and with an average urban subscriber density of 3,000 subs/km^2. Fig 11.15 shows the payback periods of the FTTB32 and CTTB32 as a function of the average distribution loop length. The results show payback periods of less than five years for FTTB32, and between 1.5 years and 3.5 years for CTTB32.

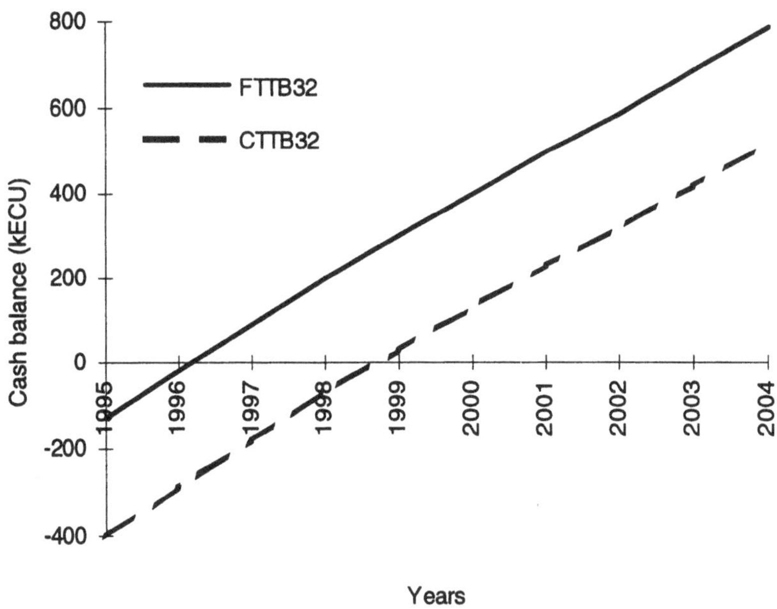

Fig 11.14 Cash balance for FTTB32 and CTTB32 architectures. The results are shown for a subscriber density of 3,000 subscribers/km^2.

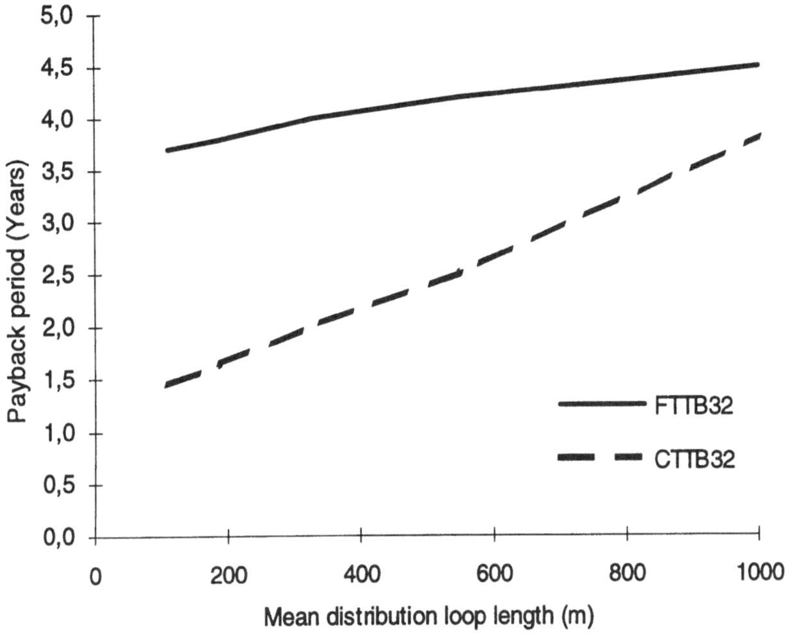

Fig 11.15 The payback periods of the FTTB32 and CTTB32 architectures as a function of the average distribution loop length.

11.4 SUMMARY

The results shown here imply that, if only POTS and N-ISDN services are taken into account, the copper network is still the cheapest solution under conventional conditions. If the civil works cost is very high, FTTB32 becomes the most cost effective alternative. In low density areas (densities less than 3,000 subscribers/km²) the RLL seems to represent an interesting alternative [6,7]. However, several major benefits from the optical networks are difficult to quantify and are not fully accounted for in this study:

- civil works de-congestion
- upgradeability to broadband access
- shorter delay for enabling/disabling a customer with a service, which constitutes a competitive advantage
- switching cost savings due to longer spans.

REFERENCES

[1] Olsen, B T, Zaganiaris, A, Stordahl, K, Ims, L A, Myhre, D, Øverli, T, Tahkokorpi, M, Welling, I, Drieskens, M, Kraushaar, J, Mononen, J, Lähteenoja, M, Markatos, S, De Bortoli, M, Ferrero, U, Ravera, M, Balzaretti, S, Fleuren, F, Gieschen, N, De Oliveira Duarte, M, de Castro, E. Techno-economic evaluation of narrowband and broadband access network alternatives and evolution scenario assessment. *IEEE Journal on Selected Areas in Communications,* 14 (6), 1996, 1184-1203.

[2] Olsen, B T, Zaganiaris, A, Gieschen, N, Tahkokorpi, M, Drieskens, M, Markatos, S. RACE 2087/TITAN: Tool for Introduction scenarios and Techno-economic studies for the Access Network. *In: Proc. RACE Open Workshop on Broadband Access, Nijmegen, the Netherlands, June 7-8, 1993.*

[3] Zaganiaris, A, Gieschen, N, Olsen, B T, Stordahl, K, Tahkokorpi, M, Drieskens, M, Ajibulu, A. A methodology for achieving life-cycle costs of optical access networks-from RACE 2087/TITAN. *In: Proc. 11th Annual Conference European Fibre Optic Communications and Networks (EFOC '93), the Hague, the Netherlands, June 30-July 2, 1993, 2, 136-141.*

[4] Gardan, D, Zaganiaris, A, Madani, A, Madigou, R, Machon, D. Techno-economics of advanced optical subscriber networks. *In: Proc. IEEE Global Telecommunications Conference and Exhibition, Dallas, USA, 1989, 1335-1339.*

[5] Zaganiaris, A, Tahkokorpi, M, Drieskens, M, Van Hoecke, J, Olsen, B T, Markatos, S, De Bortoli, M. Life-cycle costs and economical budget of optical and hybrid access networks. *In: Proc. 5th IEEE Conference on Optical/Hybrid Access Networks, Montreal, Canada September 7-9, 1993.*

[6] Zaganiaris, A, Olsen B T, Stordahl, K, De Bortoli, M, Tahkokorpi, M. Life-cycle cost and risk assessment of the access network. From RACE 2087/TITAN. *In: Proc. 6th International IEEE workshop on Optical Access Networks (OAN '94), Kyoto, Japan, Oct. 16-19, 1994, 1.4-1 - 1.4-10.*

[7] Olsen, B T, Zaganiaris, A, Markatos, S, Ferrero, U, Tahkokorpi, M, Schjetne, I, Øverli, T. Access network alternatives from RACE 2087/TITAN. *In: Proc. 12th Annual Conference European Fibre Optic Communications and Networks (EFOC '94), Heidelberg, Germany, June 21-24, 1994.*

Radio in the local loop for rural and suburban areas

12

Rolf B. Haugen, Borgar Tørre Olsen, Bjørn Erik Eskedal

The chapter analyses various technologies for radio in the local loop (RLL)[1]*. Special attention is paid to rural and suburban areas. Techno-economic calculations are carried out for the digital European cordless telecommunication (DECT) system, demonstrating the competitiveness of wireless solutions as compared to traditional copper wire architectures. The calculations have been carried out with the tool developed within the project RACE* (research for advanced communications in Europe) *2087/TITAN (tool for introduction strategies and techno-economic evaluation of access network) [1].*

12.1 INTRODUCTION

Wireless communications have demonstrated a remarkable growth over the last years and continue to grow by 25 % year, as compared to a growth of 7-8 % for long distance calls and 3-4 % for local telephone services. Predictions indicate that by the end of this decade the penetration of cellular communications in Europe will reach 20-25 % and that 50 % of all telephones bought will be mobile phones.

The success of cellular communications has led to a renewed interest in radio communications as such, including *radio access technologies,* or radio in the local loop (RLL). This interest is further furnished by social changes, like the ones in Eastern Europe, and the fact that the new mobile operators are now competing in the huge revenue public network market. Wireless solutions give a rapid provision of communications links in countries, or areas, that are lacking telecommunications infrastructure. Furthermore they provide low-cost

[1] Based on "Radio in local loop for rural and suburban environments," by R. B. Haugen, B. T. Olsen, B. E. Eskedal. © 1994, IEEE. Reprinted with permission, from Proceedings of IEEE International Conference on Personal Wireless Communications, Bangalore, India, 1994.

investments, flexibility, and provision of basic services, all important factors for a rapid build up of new infrastructure.

In the previous East Germany more than 20 billion DM have been invested the last four years for a massive build up of the network infrastructure. More than two million new telephones have been installed, doubling the previous number of subscriptions. Since the capacity of radio networks is limited due to restricted spectrum availability, business customers, who are the main contributors to the economic upturn in East Germany, were connected first. The number of wireless accesses is thus also limited, somewhat less than 100,000 subscribers in Germany are by 1994 connected to the network via radio

The limited availability of frequency spectrum implies capacity limitations for the various radio solutions. One way of increasing capacity, is to achieve enhanced frequency reuse, i.e. to introduce smaller cells. *Microcell is* understood to be cells with local coverage of a few hundred meters, up to about or slightly above one kilometre (km). This is in contrast to the larger cells of nordic mobile telephone (NMT) and, often, global system for mobile communication (GSM) that are denoted m*acrocells.* Microcellular networks emerged in mobile systems with the third generation systems, first the personal communications network (PCN), and later on the universal mobile telephony system (UMTS). The GSM based solution in the 1.8 GHz band, the so called digital cellular system (DCS) 1800, is a microcell system tailored for high traffic and small handsets. PCN does not imply coverage over large areas as is the case for ordinary mobile communications. However, handover between adjacent cells is possible within certain technical limitations.

Microcell structures may also be combined with optical fibres to reduce the dimension and number of radio equipment. Fibre links to the radio sites may be used to increase the capacity (by increased exploitation of frequency reuse) or permit radio access to specific areas like subways and airways. Hence, optimal deployment of radio and fibre in the access network, is strategically very important for the public network operators (PNOs).

Cordless systems like DECT are also based on microcells. DECT was proposed by Swedish Telecom and Ericsson in 1985, was endorsed by conférence Europeenne des administrations postes et de télécommunications (CEPT) in 1988 and later finalised by the European telecommunications standards institute (ETSI) in 1992. It is a two-way time division multiple access (TDMA) system originally designed for cordless private branch exchange (PBX) applications. The present realisation operates with ten carriers in the 1,880-1,900 MHz frequency band and has dynamic channel allocation as an important feature. It has a 50 metres reach indoors and possibly 300 metres outdoors with omni-directional antennas. With directional antennas ranges exceeding a kilometre are within reach

DECT is a means of obtaining *mobility* in the fixed network, both locally in a community or around a PBX in the initial phase, and at a later stage with roaming possibilities. Roaming can be achieved via the concept of universal personal telecommunications (UPT) and use of the intelligent network or by interworking with GSM. But DECT may also be used as a means of accessing the local network.

12.2 MOTIVATION FOR USING RADIO ACCESS

There are several reasons for telecommunication operators to access public networks via radio. Among the various motivations we find:

- rapid and cheap build up of areas with not fully developed infrastructure
- a means of introducing mobility into the network
- possibility for the mobile operators to capture customer from the PNOs.

New developments was discussed in the introduction. Wireless access is taken into use to rapidly connect new customers to the network in many developing countries, as is also the case in East Germany.

Introduction of *mobility* is a means for PNOs to give their fixed network subscribers a service that otherwise is provided by the mobile operators. The subscribers are offered an advanced cordless phone service within the local area, which later may be extended to countrywide coverage. Hence, DECT is an extension of the fixed network that gives the subscribers mobility as an added service. The mobility is a bit limited since DECT cannot be used from (fast) moving vehicles. Nevertheless, this may be sufficient for most subscribers.

The *mobile operators* are of course interested in wireless access since that gives them a way of capturing customers from the PNOs. The former operators will argue that wireless access systems are simplified cellular systems and might be included into their general licence. What the regulatory authorities will decide remains to be seen.

In Norway we have a DECT pilot network in the small community of Førde on the west coast of the country [2]. The DECT pilot network in Norway is motivated by the possibility of introducing mobility into remote communities. Førde is situated in a mountainous area with a population of about 9,000 inhabitants. 230 subscribers are connected to the DECT network in the trial. was The relative short distance between residential areas and the business centre was one of the reasons for choosing Førde. This enables the handsets to be used within the entire local community.

The system in Førde consists of small portable handsets (PH) that communicate with a maximum number of 160 radio base stations, so called radio fixed points (RFPs), connected to a central control unit (CCU), as shown in Fig 12.1.

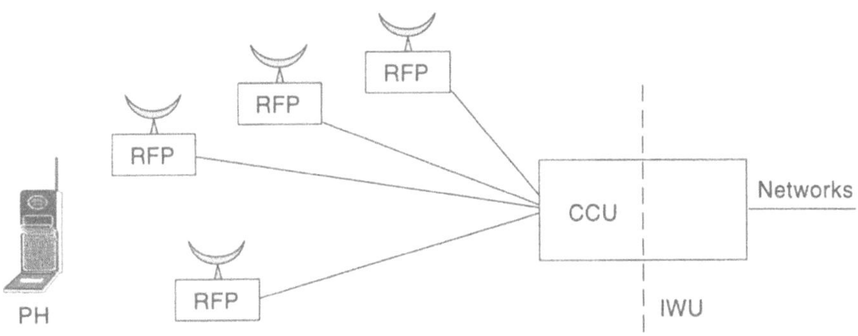

Fig 12.1 Generic DECT system.

The TDMA system allows each base station to use 12 duplex time slots for the calls i.e. each handset can choose among 12 time slots. In the Førde project only eight slots can be simultaneously taken. Totally the system can carry 40 simultaneous calls. In the trial all of the 160 RFPs have been put into service. The large number of base stations is partly due the coverage of the residential area as well as the central business area, and partly due to indoor coverage of some community houses.

12.3 TECHNOLOGY CHOICES

Various technologies can be employed to provide radio access. However, one should bear in mind that radio systems originally created for different applications are not necessarily suited for replacement of copper unless modifications or redesigns are done. The main differences between radio and copper lie in transmission capability, robustness of the radio link and radio range interference.

The alternative radio access systems are *cordless systems*, modified *cellular* systems, *point-to-multipoint microwave* systems as well as additional proprietary systems. None of these systems has however yet emerged as the clear choice. The optimised system choice will depend on which RLL approach the network operators have in mind for the required service and the type of area to be covered. In this paper, the following subdivision is done:

- RLL as replacement for conventional copper loop
- RLL for providing mobility to the users.

In the first situation a small box, the subscriber radio terminal, is typically mounted on the outside wall of the home of the subscriber. A wire connects the subscriber equipment, i.e. telephone and telefax, to the radio terminal. It communicates across the air interface via a directional antenna to the remotely located radio base stations. No mobility is provided, and the RLL decision is likely to be based on potential line cost saving.

In the second approach, the operators provide the users with mobility. Several mobility levels are possible, ranging from limited mobility at the customer premises to wide area coverage, like the cellular mobile systems of today. The latter solution is the most flexible one, as it starts with limited mobility, but with upgrade possibility. Different tariff structures will emerge, giving the users various coverage choices at differentiated prices.

In East Germany three types of RLL are in use, all of which are replacement for the conventional copper loop:

- trunked mobile radio
- modified analogue cellular system
- point-to-multipoint microwave systems.

The *trunked system* is installed in more than 40 locations, each with a maximum number of 180 users. The frequency range is 400-450 MHz, with a maximum reach of between eight and ten km. The features include the basic services of telephone, telefax and data transmission. The other two systems are described in more detail below.

12.3.1 Digital cordless systems

Current cordless systems like DECT and cordless telephony, second generation (CT2) have been designed to provide low power cordless access between small handsets and fixed points for distances up to, typically, a few hundred meters with a standardised radio interface. They serve only as an access network specified for connection to target networks like the public switched telephone network (PSTN), integrated services digital network (ISDN) and local area networks (LANs). Since the systems define only the access part without switching functionality, the system complexity is limited. Both DECT and CT2 have been designed for use in a variety of environments, like domestic, public access, cordless PBX and RLL. The growing interest for using cordless standards for RLL applications is reflected by the number of trials already going on with DECT and CT2.

The speech channels in DECT are based on 32 kbit/s. The communication between the RFP and the CCU is a proprietary half-duplex ISDN 2B+D transmission using two twisted pairs. Each of the pairs carries four adaptive differential pulse code modulation (ADPCM) speech channels. In the CCU these channels are converted to ordinary 64 kbit/s channels, but due to proprietary standards, a line interface module (LIM) is needed as a connection between the public switch and the CCU. With a V5.1 interface incorporated into the digital exchanges the LIM interface will be superfluous.

In DECT, control and data information are multiplexed together into one packet, which is transmitted over the air interface. A packet starts with a 32 bit synchronisation field, the S-field, followed by a 64 bit A-field containing control and signalling information. The B-field carries user data of 320 bits during a call, providing 32 kbit/s of user data. With this TDMA frame structure DECT offers simplex and duplex communications as well as provides symmetric and asymmetric links. The simplest duplex service uses a single pair of evenly spaced time slots. Higher data rates are achieved by the use of more time slots, lower rate services are achieved by the use of half a time slot. The advantages of DECT as an RLL are:

- highly flexible standard, high speech quality, good security and authentication
- dynamic channel allocation, which simplifies network radio planning
- support for conventional PSTN switch interfaces
- supports for basic telecommunication services and enhanced ISDN services.

A drawback of both DECT and CT2 is the limited range due to low RF power. In addition, DECT is sensitive to multipath reflections. It still remains to be seen how suitable cordless technologies will be in an RLL application. However, the use of directional antennas and repeaters will increase the radio range significantly (towards five km for DECT).

12.3.2 Mobile based systems

The *cellular* RLL systems installed in Germany are based on modifications of NMT 900 supplied by Nokia and Ericsson. The frequency range(s) in use are four windows between 812 and 873 MHz, with a maximum range of several kilometres and a maximum number of users per system is >>10,000. By now there are about

50,000 access points for this system. The RAS 1000 system works without the mobile switching centre (MSC), provides more complex speech scrambling, larger audio bandwidth, faster telefax (4.8 kbit/s) and has a 2 Mbit/s interface

Experience in operating *analogue* cellular systems has given insight into their strengths and weaknesses. The strengths of RLL applications are large area coverage, low delay, high degree of transparency for dual tone multi-frequency (DTMF) tones, and telefax and voice band data up to 4.8 kbit/s. Their main limitations are the poor security, the lack of high speed data or ISDN support, and variable speech quality at the cell boundaries.

Fixed RLL systems both based on NMT and total access communication system (TACS) already exist and are used in several locations. Especially for remote rural areas with low user densities and a poor existing infrastructure, analogue systems may for a long time still be a very attractive solution.

The *digital* mobile system GSM has already proven to be a success for cellular mobile subscribers. This makes it, and the GSM based DCS 1800 system, strong RLL candidates. As with analogue systems, GSM gives a wide area coverage, typically 30 km using 16 W base stations and an outdoor 8 W mobile terminal. Secure authentication and encryption algorithms are defined. Consistent speech and service quality within the whole cell and a comprehensive set of standardised services, including ISDN supplementary and signalling services, are provided. Below are listed some drawbacks of GSM used as an RLL [3]:

- speech quality and delay does not meet fixed network standards
- limited data throughput (e g. GSM does not support ISDN 2B+D capacity)
- no transparency to voice band data signalling tones (e.g. DTMF in downlink) due to voice codes, which prevents the use of standard telephony products at the subscriber end
- no network interface to the PSTN local exchange.

The lack of a network interface to a PSTN local exchange is a significant disadvantage of GSM/DCS 1800 used for RLL applications, since this requires that local calls are routed through the complete mobile network via its MSC. This increases the cost and congestion in the GSM network. Several possible solutions to this problem are considered. One possibility is to use the GSM air interface for RLL and to have a combined local switch integrating the switching functionality of GSM with that of the fixed network, and thus providing a direct access to a local switch.

12.3.3 Microwave point-to-multipoint

Microwave point-to-multipoint (MPMP) systems are *designed* for local loop replacement and therefore meet the RLL service requirements. The systems support conventional PSTN switch interfaces and give good radio range. On the other hand, poor spectrum efficiency and cellular inflexibility indicate that most MPMP systems only are economical for long range applications, in which telecommunications are provided to isolated subscribers or small clusters in rural areas. However, new ideas to combine MPMP and DECT replacing the last drop to the subscriber with a DECT cordless connection, may lead to the beneficial introduction of local mobility.

The MPMPs in Germany are digital systems working in the 2.5 GHz range. The service radius is up to two km and each system can support 300 customers. A typical configuration involves a central station connected to the main distribution frame and a subscriber station to which the telephone set is connected. The system supports 64 kbit/s leased lines, ISDN basic access in addition to ordinary telephone services. Economic considerations show that the system is cost effective if every subscriber station supports at least ten telephone lines or other accesses.

12.3.4 Proprietary standards

A number of proprietary standards are being developed both in Europe and the USA for RLL. Most of these systems are fixed, in the sense that they do not provide the users with mobility. Especially interesting are fixed radio systems using spread spectrum techniques. It is claimed that CDMA has several advantages compared to TDMA and frequency division multiple access (FDMA), including

- high immunity to interference sources
- robust operation in fading channel due to inherent frequency diversity
- high spectrum efficiency.

In a fixed RLL environment spread spectrum may become a strong candidate enabling the use of directional antennas at the subscriber end and thereby providing sufficient inter-cell isolation, which normally is a weakness of such systems.

12.3.5 Technology comparisons

An overview of the strengths and weaknesses of the various RLL technologies are given in Table 12.1 [3].

12.4 RLL CONFIGURATIONS FOR RURAL AND SUBURBAN AREAS

12.4.1 Rural areas

A typical rural area in Scandinavia is sparsely populated with subscriber densities of 1,100 subscribers/km^2 and lower. The terrain is often rough, hilly and forested, and people tend to live in 1-3 km narrow lines along rivers, fjords, roads or valleys. Normally 80 % of the subscribers live in small villages of 50-500 subscribers, with a radius of about 1-9 km, and a typical density of 100 subscribers/km^2. The more dense centres are spread over large areas with significantly lower density and located 5-20 km apart. The remaining 20 % are evenly distributed in small clusters varying from 1-50 houses. The upper part of Fig 12.2 shows a typical rural area, illustrating some possible network architectures and indicating different technology implementations.

Table 12.1 Strengths and weaknesses for the various RLL alternatives.

Feature	*GSM*	*Analog cellular*	*DECT*	*CT2*	*MPMP*
Strengths	• range 10-35 km, robust radio interface • standardised • security • multitude of GSM services	• range 35 km • Voice band (VB) transparency up to 4.8 kbit/s • low cost and complexity	• capable up to 2B+D • VB transparency up to 4.8 kbit/s • PSTN interface • security • simplicity • 20 MHz spectrum	• low delay • VB transparency up to 4.8 kbit/s • PSTN interface • low complexity	• range 50 km • 64 kbit/s capability • relay operation • PSTN interface
Weaknesses	• speech quality and delay • VB transparency • no PSTN interface • max 9.6 kbit/s data	• low security level • no digital services • poor performance at cell borders	• short range • sensitivity to multipath intersymbol interference	• short range • limited data capacity	• spectrum inefficient • requires line of sight • non-standard air interface
Improvement required in	• speech quality • VB transparency • PSTN interface	• authentication and security • PSTN interface	• increased range	• increased range • additional privacy	• good where economical

Due to the low population density, the number of base stations is determined by range limitations, and not by capacity. Using repeaters to communicate across rough terrain increases the radio range considerably. In more concentrated areas, short range radio technologies, such as cordless systems with repeaters, can economically justify local mobility. This is aspect is addressed in section 12.5.

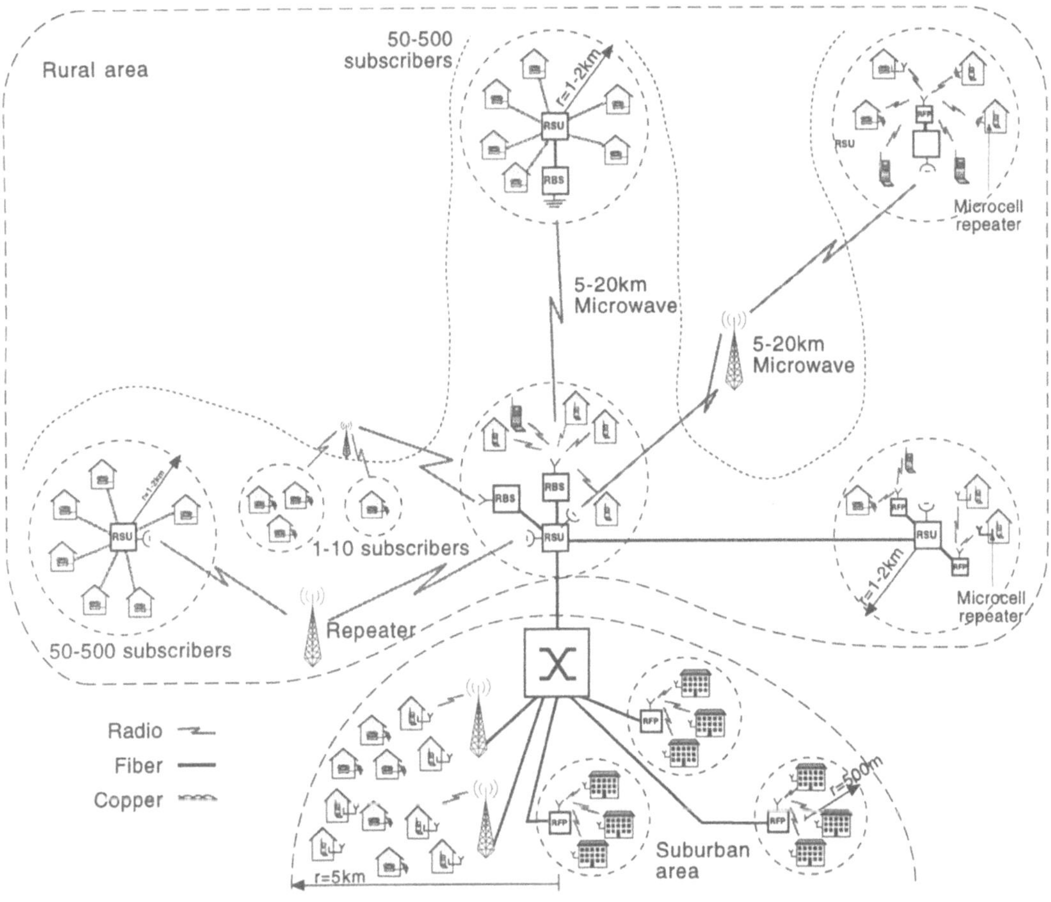

Fig 12.2 Example of RLL in suburban and rural areas.

12.4.2 Suburban areas

Suburban residential areas are moderately populated areas with an average density of 100 - 1,000 subscribers/km^2, extending over a radius of five km. Such areas often surround a dense inner city ring. Local variations in housing density within the area exist, resulting in a typical pattern as shown in the lower part of Fig 12.2. In high density suburban areas (radius 0.5 km) user densities often exceed 1,000 subscribers/km^2. The customers to be served in these dense local areas live in blocks or apartment houses or they may be small businesses or industrial customers. At the outskirts, a lower density area with approximately 500 subscribers/km^2 is found. This lower density area typically consists of detached houses, often with small gardens.

12.5 ECONOMIC ANALYSIS OF RLL

The TITAN tool has been applied in the techno-economic assessment [5-8]. In this study a preliminary forecast on future penetration of POTS and N-ISDN is used, which assumes an exponential growth of N-ISDN up to 20 % in a ten year period. A slight rise followed by a small decline in the penetration of POTS is foreseen. The average annual tariff of POTS and N-ISDN are fixed at 160 ECU and 560 ECU respectively.

An RLL solution (DECT) is analysed and compared to an conventional copper access network. The architectures are mapped into a geometric model, as shown in Fig 12.3.

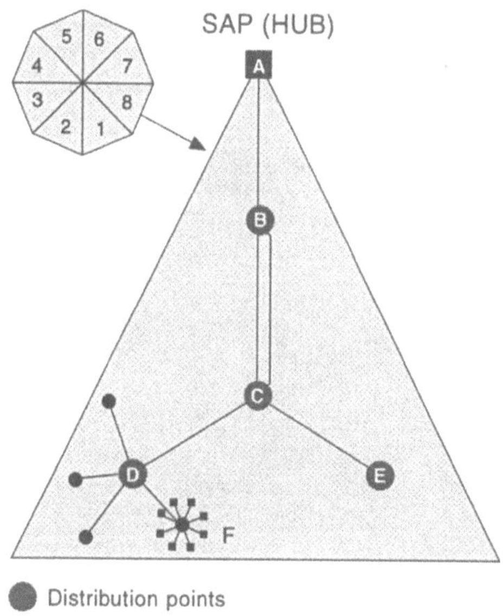

Fig 12.3 Geometric model used in the calculations.

The model represents a double star structure with two flexibility point levels in addition to the hub, flexibility point 0. The first level of flexibility points include the points B, C, D and E. The second level of flexibility points is represented by the point F. Network components can be placed in any of the flexibility points, depending on subscriber density or distribution loop length. The architectures are assumed to serve 1,024 subscribers per hub. The copper architecture consists of two twisted pairs to every subscriber premises. The copper cable is dispatched and spliced at the flexibility points. We assume that 30 % of the installed pairs are unused. The only electronics included in this architecture are the POTS and the N-ISDN line cards. The RLL architecture on copper infrastructure has base stations covering a maximum of 200 metre radius, located in different flexibility points, depending on subscriber density. For low density areas with densities less than 560 subscribers/km^2, directional antennas are used to extend the range of the base stations up to 800 meters. The number of base stations is determined by the range and traffic capacity of the base stations. In the calculation we have not included the feeder part of the network, because the feeder network will be very similar for copper access networks and RLL solutions. The handsets are not included in the calculation.

Civil works encompass digging, ducting and surface reinstatement. The cost per unit length depends on the area type, the distance from the exchange and the technology used. Conventional techniques consist of ducting and subducting for subsequent pulling of the cables through the conduits. The branching part of the network, between the subscriber premises and the nearest flexibility point, is of paramount importance since it is not shared by several subscribers. We assume that the civil works in this part are carried out using direct burial or aerial drop cable along the outer walls of the houses. This yields a price of 2 ECU/m for civil works, plus 2 ECU/m for the installation. For conventional ducting 25 ECU/m is taken as an average figure.

Fig 12.4 shows the comparison between the installed first cost (IFC) as a function of subscriber density for the DECT radio access network and the conventional copper access network. The different topologies shown in Fig 12.2 represent local subscriber densities from 50 subscribers/km^2 in rural areas (mainly single unit houses) to about 3,000 subscribers/km^2 in suburban areas (buildings containing several flats).

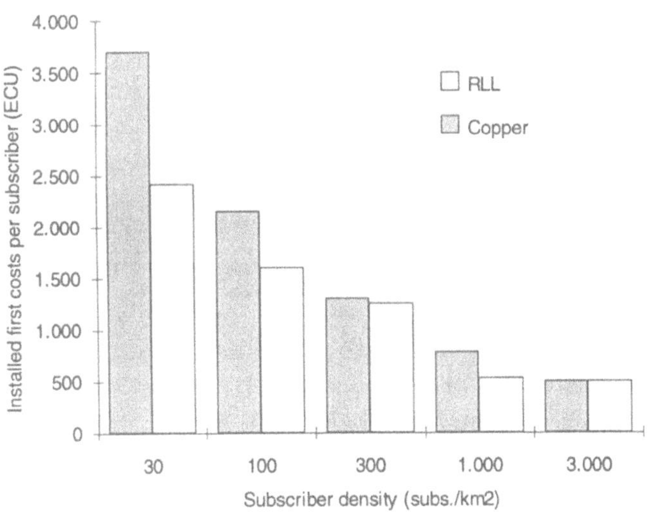

Fig 12.4 IFC as a function of subscriber density for DECT and copper architectures.

Fig 12.5 depicts the same results, illustrating the relation between IFC and average loop length for DECT and copper architectures.

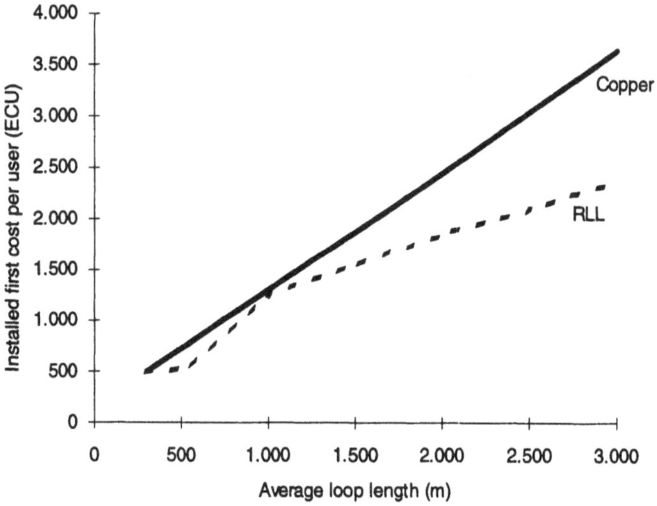

Fig 12.5 IFC as a function of average loop length for DECT and copper architectures.

12.6 SUMMARY

The paper has discussed various technologies for RLL systems and techno-economic evaluations have been carried out. The calculations show that radio based solutions are cost effective compared to copper for low to medium high subscriber densities. This holds even if the handset is included.

For low *densities* DECT gives limited mobility due to the required use of directional antennas. In suburban areas with subscriber densities >300 subscribers/km^2, the cost of RLL access is close to the cost of copper access. However, in these areas, the RLL solution provides the additional feature of mobility.

In the present calculation, we have assumed passive antennas. In some of the rural areas, with small clusters of houses widely dispersed, the use of repeaters, will increase the efficiency of the DECT solution.

Since DECT is an advanced RLL-system originally designed with local mobility, the system is not optimised for low cost radio connections. Simplified radio solutions, like RAS 1000, will in many cases be a cheaper solution. Hence, in our calculations DECT represents an "upper limit" for wireless access to public networks [1].

REFERENCES

[1] Haugen, R B, Olsen, B T, Eskedal, B E. Radio in local loop for rural and suburban environments. *In: Proc. IEEE International Conference on Personal Wireless Communications, Bangalore, India, 1994.*

[2] Eskedal, B E. The DECT system. *Telektronikk,* 91, (4), 25-29, 1995.

[3] Rapeli, J. Application of technologies for Wireless Local Loop.

[4] ETR on Radio Local Loop. ETSI, November 1993.

[5] Olsen, B T, Zaganiaris, A, Gieschen, N, Tahkokorpi, M, Drieskens, M, Markatos, S. RACE 2087/TITAN: Tool for Introduction scenarios and Techno-economic studies for the Access Network. *In: Proc. RACE Open Workshop on Broadband Access, Nijmegen, the Netherlands, June 7-8, 1993.*

[6] Stordahl, K, Murphy, E. Methods for forecasting long term demand for wide and broadband services in the residential market. *IEEE Communications magazine,* 13, (2), 1995, 44-49.

[7] Zaganiaris, A, Gieschen, N, Olsen, B T, Stordahl, K, Tahkokorpi, M, Drieskens, M, Ajibulu, A. A methodology for achieving life-cycle costs of optical access networks-from RACE 2087/TITAN. *In: Proc. 11th Annual Conference European Fibre Optic Communications and Networks (EFOC '93), the Hague, the Netherlands, June 30-July 2, 1993, 2, 136-141.*

[8] Zaganiaris, A, Tahkokorpi, M, Drieskens, M, Van Hoecke, J, Olsen, B T, Markatos, S, De Bortoli, M. Life-cycle costs and economical budget of optical and hybrid access networks. *In: Proc. 5th IEEE Conference on Optical/Hybrid Access Networks, Montreal, Canada September 7-9, 1993.*

European multiservice access network upgrades 13

Leif Aarthun Ims, Borgar Tørre Olsen, Dagfinn Myhre, Markku Lähteenoja, Jari Mononen, Umberto Ferrero, Alcibiade Zaganiaris

This chapter presents a techno-economic evaluation of strategies for developing the traditional access network along a cost effective path towards a future-proof multiservice broadband network. A broad range of appropriate network alternatives have been assessed for each of five representative European upgrade cases. The technology options that have been evaluated range from conventional copper based networks to emerging new technologies such as hybrid fibre coax (HFC) networks and asynchronous transfer mode (ATM) based broadband passive optical networks (PONs)[1].

13.1 INTRODUCTION

The quality of existing infrastructure, average subscriber capacity demand, demography and technology choice are among the key elements likely to affect the economic viability of access network upgrade strategies under varying boundary conditions. Currently, there is no single technology choice or strategy envisaged for access network upgrading, and this may be the case for several years [1,2]. The EURESCOM P306 project "Access network evolution and preparation for implementation" has had as one of its objectives the strategic evaluation of the access network alternatives for the residential and small business market in Europe [3-7]. The scenarios studied include gradual provisioning from 1995 to 2005 of services like plain old telephony service (POTS), narrowband integrated services digital network (N-ISDN), asymmetric switched broadband (ASB), symmetric switched broadband (SSB) and broadcast broadband. The demand forecasts for the selected bearer services are extracted from the TITAN Delphi survey [8,9]. The techno-economic assessment was carried out by the

[1] Based on "Multiservice Access Network Upgrading in Europe: a Techno-economic Analysis," by L. A. Ims, B. T. Olsen, D. Myhre, J. Mononen, M. Lähteenoja, U. Ferrero, A. Zaganiaris. from *IEEE Communications Magazine*, vol. 34, no. 12, 124-134, Dec. 1996.

EURESCOM P306 project using TITAN (tool for introduction strategies and techno-economic evaluation of access network).

13.2 SET OF BEARER SERVICES

The total western European market demand for fixed voice telephony (POTS and N-ISDN) is expected to increase only marginally over the next few years. However, the demand in the developing European countries is expected to increase significantly. The two major factors affecting the new demand for POTS are the economic growth and demographic changes. Commercial ISDN facilities are offered in an increasing number of European countries.

The bearer services considered are: POTS, N-ISDN, ASB (2 Mbit/s downstream, 16 kbit/s upstream), SSB (2 Mbit/s bi-directional) and broadcast broadband, i.e. community antenna television (CATV). The following annual tariffs have been assumed: 160 ECU for POTS, 530 ECU for N-ISDN, 250 ECU for CATV, 600 ECU for ASB and 1000 ECU for SSB. 1 ECU ~ 1,3 $US. Traffic income has not been accounted for, since this is commonly considered to be trunk network specific revenue.

13.3 CASE STUDIES

Five cases representing situations common among several European operators have been studied, in order to map the diversity in existing infrastructure, short and medium term needs, cultural and social acceptance of new services and economic capabilities exhibited in Europe. A case in this context refers to a possible European telecommunications market situation related to the access network for residential and small business customers in areas with an already established infrastructure. The cases reflect European diversity and are linked to clusters of countries rather than to specific areas. Each case is described by

1. The area type
2. The housing structure
3. The quality of the existing infrastructure
4. The expected service evolution from the point of view of the public network operator (PNO)
5. The operational environment: regulation, competition, political and economic aspects.

For all cases a ten year study period from 1995 to 2005 has been analysed. The case specifics are summarised in detail in Table 13.1 and Table 13.2.

13.3.1 Case 1, Eastern Europe, urban

An urban, apartment block dominated area is analysed. The existing infrastructure is based on copper, most of which is obsolete and insufficient for the present and future demand. The ducts are underdimensioned for additional copper cables, hence fibre or enhanced copper technologies are considered highly relevant solutions in this case. Even the demand for new POTS can not be met without sizeable capital investment.

The penetration of POTS is low, but increasing rapidly towards a level which is in compliance with the telecommunications requirements of the European Union dynamic development programs. The access network evolution will be driven mainly by high demand for new POTS lines, whereas low competition on POTS will have only marginal effect. The PNO is expected to extend its set of services to include CATV, which is growing rapidly, and switched broadband services, both with high penetrations.

13.3.2 Case 2, Northern Europe, urban

The urban area analysed has the majority of its potential subscribers living in apartment blocks whilst the minority is living in individual homes. The existing twisted pair based infrastructure is modern and capable of providing the present and foreseeable demand for POTS. Optical fibres are already used in the feeder section of the network in concentrator and multiplexer configurations.

The penetration of POTS is almost at saturation. POTS will be open to competition at the turn of the century. However, the PNO will remain the dominant operator, both for POTS and CATV in the coming decade. The demand for CATV, ASB and SSB services is the main reason for replacing the existing access network. Competition is expected to have a low effect on the residential and small business market in the timescale 1995-2005 and, hence a negligible influence on the access network evolution. The service penetrations are shown in Fig 13.1.

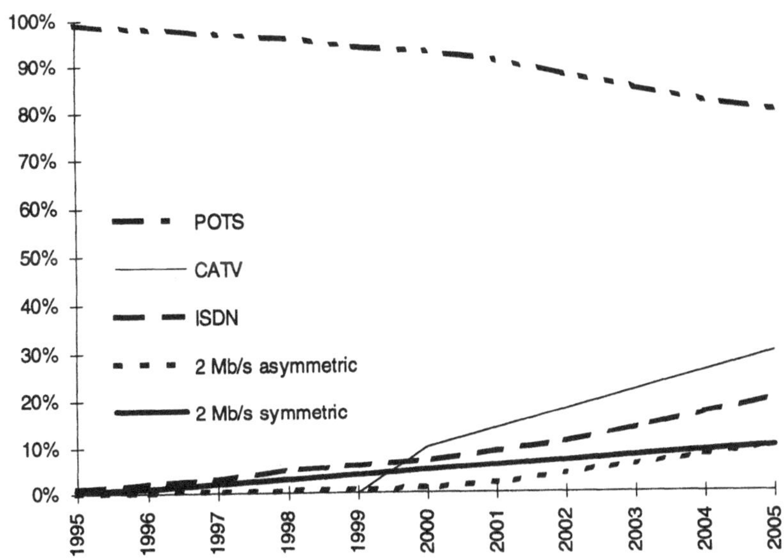

Fig 13.1 Penetration of services, case 2.

13.3.3 Case 3, Northern Europe, suburban

The majority of the subscribers in the analysed suburban area live in individual homes or in buildings with few apartments. The rest of the potential users live in apartment blocks. The existing access network infrastructure is modern, and the present and foreseeable demand for POTS can be accommodated by the network, which for the residential and small business customer segment is based on the use of copper pairs. Optical fibre is already used in the feeder section.

The PNO is the main provider of telecommunication services with the penetration of POTS being almost at the saturation level. It will be allowed to provide CATV from year 2000 on. The need for new service revenues and economic aspects are the main reasons for investing in advanced access network solutions. The PNO is expected to lose a significant number of POTS customers to competitors in the time scale 1995-2005. Therefore, the phased introduction of cost effective, flexible new access network solutions is of importance.

13.3.4 Case 4, Southern Europe, urban

The customers live in an urban area of apartment blocks. The existing copper infrastructure is of good quality and sufficient for the POTS demand. The demand for ASB and SSB services is the main reason for installing fibre in the access network. After the turn of the century POTS will be open to competition, but this will not have significant influence on the POTS penetration. The PNO is not allowed to provide CATV, neither is he intended to do so in the timescale 1995-2005. However, the PNO has a regulatory obligation to rent fibres to CATV operators.

13.3.5 Case 5, Eastern Europe, suburban

The suburban area under study consists of a majority of subscribers living in apartment blocks and a minority of single house dwelling subscribers. The existing infrastructure is based on copper and is modern, less than ten years old. However, the existing narrowband infrastructure is underdimensioned. This is the main driver for a network upgrade. No POTS competition is expected until the turn of the century, and even after year 2000 the competition is not assumed to affect the PNO. The PNO and other operators may provide CATV from 1995.

13.4 ACCESS NETWORK UPGRADE TECHNOLOGIES

The study includes assessment of the most relevant technical alternatives for upgrading the access network to broadband, ranging from copper based solutions to solutions based on emerging new technologies like HFC and ATM based PONs [10-13]. These technologies have been applied to the above five cases. The copper solutions are based on the use of asymmetric digital subscriber line (ADSL) and high bit rate digital subscriber line (HDSL) modems for provision of 2 Mbit/s services. The technology alternatives examined for each of the five cases are summarised in Table 13.1. Illustrated in Fig 13.2 is the enhanced copper alternative considered particularly in case 2.

Table 13.1 The technology alternatives of the five cases.

Case	*Area type*	*Technical alternatives considered*
1	Eastern Europe, urban	1. Enhanced copper network 2. Fibre-based multiplexer 3. PON for new customers 4. PON for all customers 5. ATM PON
2	Northern Europe, urban	1A. Regenerator and enhanced copper 1B. Regenerator, CATV/ASB 2. Enhanced copper network 3. Optical point-to-point, ADSL 4. PON and enhanced copper network 5. ATM PON
3	Northern Europe, suburban	1. Enhanced copper network 2. PON 3. ATM PON (8 or 32 subscribers/ONU) 4. Hybrid fibre coax
4	Southern Europe, urban	1. Enhanced copper network 2. Fibre based multiplexer 3. PON 4. Hybrid fibre coax 5. ATM PON
5	Eastern Europe, suburban	1. Enhanced copper network 2. Enhanced copper network, CATV 3. PON for new customers 4. PON for all customers 5. ATM PON

In addition, narrowband telephony over passive optical networks (TPONs) and broadband PONs have also been studied. The PON solutions are based on the use of fibre to the curb (FTTC) or fibre to the building (FTTB) topologies for provision of POTS, N-ISDN and/or 2 Mbit/s services. The case 3 ATM PON option is depicted in Fig 13.3.

The implementations of the technical alternatives are not the same in the different cases, but in general, for all cases, an initial choice of architecture is made, taking into account the existing infrastructure and the other case characteristics. This initial choice is a key decision, and after that the infrastructure and the architecture selected are assumed to remain fundamentally unchanged during the whole study period. However, throughout the study period new equipment may be installed to accommodate new customers and services.

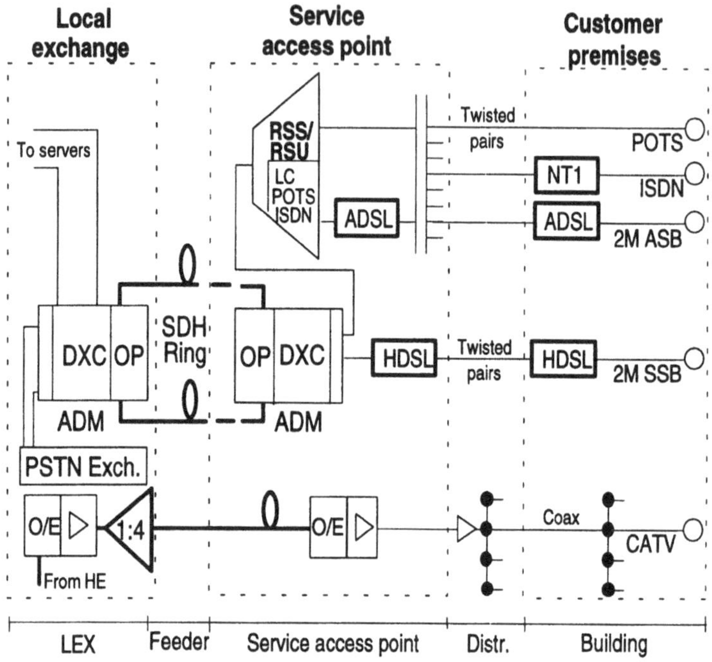

Fig 13.2 Representative copper solution.

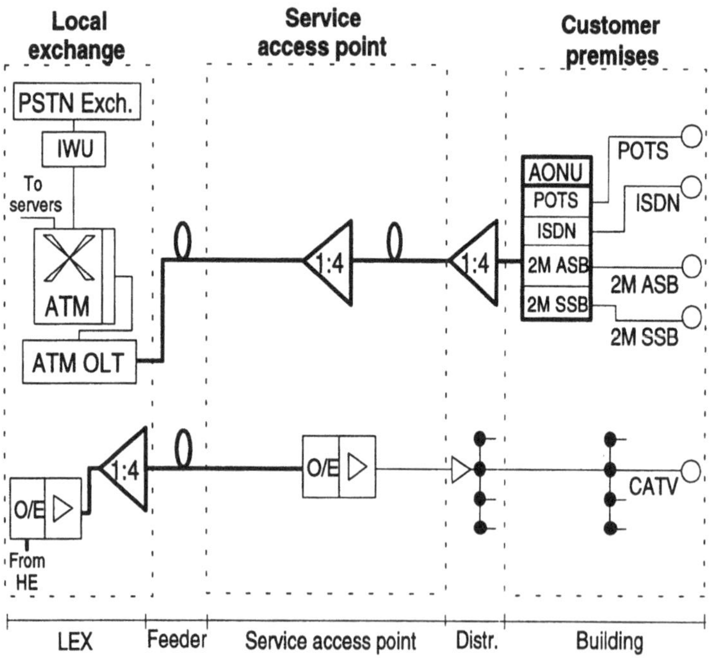

Fig 13.3 Representative ATM PON solution.

All feeder- and distribution cables for all services except for CATV, are installed at the start of the study period, whilst installation of new drop cables - coaxial cable or fibre cable - and service specific equipment for the initially selected architecture is carried out throughout the project period at the time of connection of new service subscribers. The optical cables installed are slightly oversized for future upgrading; the extra cost for this is negligibly small.

13.5 ANALYSIS

13.5.1 The techno-economic methodology

The techno-economic evaluation has been carried out with the TITAN tool. The TITAN network component database was extended by EURESCOM P306. The database contains data gathered from many European sources. It should be noted that the database contains cost elements of different technological maturity and commercial availability. Some of the costs are provided by manufacturers of systems and equipment already available, whilst the costs of components of some emerging technologies such as HFC and ATM based PONs reflect target prices.

13.5.2 General assumptions

The assessment of the upgrade alternatives is based on several key assumptions, summarised in Table 13.2. The potential users are assumed to be uniformly distributed over the access area. Only the part of the network between the local exchange (LEX) and the customer premises is taken into account. The feeder section between the LEX and the service access point (SAP), is included in a ring structure serving four to six SAPs. The SAP, as used in this study, refers to the localisation of the concentrator in the network, as commonly used in some European countries. The SAP in our study has a capacity of 1,024 users.

Civil works costs may contribute significantly to the overall costs and are therefore of paramount importance when considering access network upgrades. Civil works as used here, encompass digging, ducting and surface reinstatement. The cost per unit length depends on the area type, the distance from the exchange and the technology used. These civil works cost consist of ducting and subducting for subsequent pulling of the cables through the conduits. The final drop of the network in case of single homes, between the subscriber premises and the nearest flexibility point, is very cost sensitive since it is not shared by several subscribers. It is assumed that the civil work in this part is carried out using direct burial or aerial drop cable along the outer walls of the houses. This yields a price of 2 ECU/m for civil works plus 2 ECU/m for the installation. For conventional ducting in the distribution part of the network 25 ECU/m is taken as an average figure. This figure is then multiplied by the duct availability.

The amount of civil works required depends on the usability of existing ducts. This duct availability is related to the empty space in existing ducts or the amount of ducts likely to be freed through the replacement of obsolete network. Duct availability is an important input to the TITAN tool. This modelling parameter enables sensitivity analyses with respect to the civil works configuration.

Compared to copper, optical fibre cables need significantly less duct space. This is accounted for by differentiating the duct availability factor accordingly.

Table 13.2 The key assumptions for the assessment of the alternatives of the five cases.

Case	*Area type*	*Duct availability*		*Distribution of users*		*Coverage of existing network*
		Fibre	*Copper*	*Single homes*	*Buildings*	
1	Eastern Europe, urban	60 %	30 %	20 %	80 %	40 %
2	Northern Europe, urban	90 %	50 %	20 %	80 %	100 %
3	Northern Europe, suburban	90 %	50 %	80 %	20 %	100 %
4	Southern Europe, urban	90 %	50 %	20 %	80 %	100 %
5	Eastern Europe, suburban	0 %	0 %	20 %	80 %	75 %

13.5.3 General results

The extensive techno-economic analysis of the above five cases involve dissimilar area types and service patterns, hence all the results are not directly comparable across the cases. Nevertheless, the results highlight some key issues and general aspects described in the following chapters:

1. Upgrade costs compared to narrowband access network costs
2. The effect of dwelling distribution and duct availability on costs
3. The costs of electronics
4. The influence of capacity demand on costs
5. The effect of customer sharing on the costs of optical network termination
6. The effect of competition and penetration on the overall project
7. The timeframe of broadband access network upgrading.

Upgrade costs compared to narrowband access network costs

Fig 13.4 shows the installed first costs (IFC) at the end of the ten year project period of the different architectures realised within each of the cases studied. The IFC is discounted and includes cables, equipment, installation and civil works. Since all cases represent upgrading architectures in a non-greenfield situation, the IFC and life cycle costs (LCC) are only loosely dependent on the subscriber density. In general the fibre based solutions seem to be more expensive than the enhanced copper alternatives. However, the cost difference between optical and enhanced copper technologies turns out to be reasonably small for some of the analysed cases, advocating the introduction of future-proof integrated access technologies.

Fig 13.5 shows the costs per new service connection of the different architectures realised within each of the cases studied. The cost per line for POTS and N-ISDN in the existing access network of northern European countries typically varies between 400-500 ECU in urban areas and 1,000-1,500 ECU in suburban areas. The costs of upgrade in this study vary from just below 260 ECU to 1,100 ECU per new connection, depending on factors like dwelling distribution, the condition of the existing network, choice of technology and the degree of service integration.

These results indicate that an upgrade of the access network to an integrated broadband network will require huge investments, for most areas and technologies analysed. The financial burden of this upgrading is comparable to the one of establishing the narrowband access networks of today.

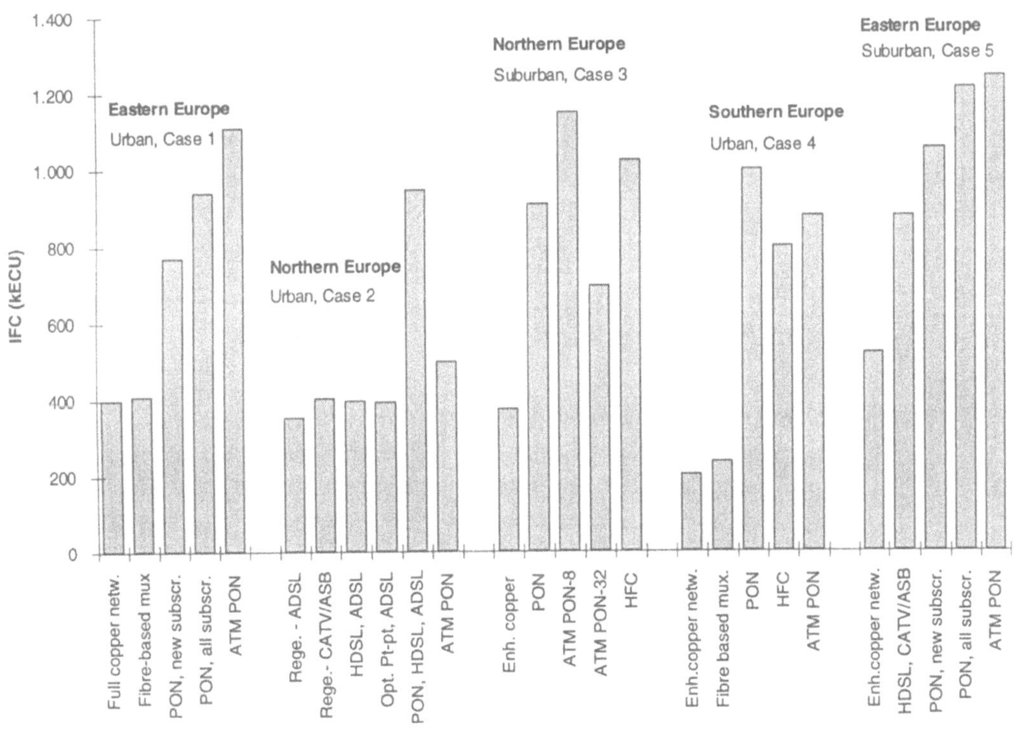

Fig 13.4 The IFC at the end of the ten year project period of the different architectures realised within each of the cases studied.

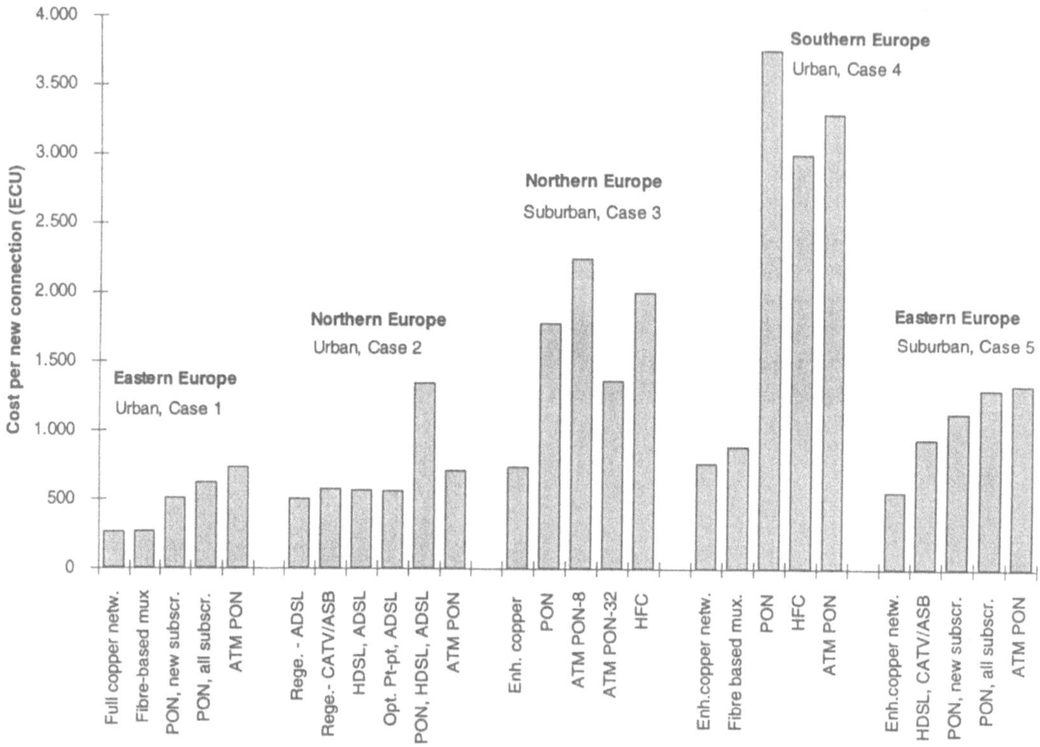

Fig 13.5 The costs per new service connection of the different architectures realised within each of the cases studied.

The effect of dwelling distribution and duct availability on costs

The dwelling distribution greatly influences the range and variation of the costs as shown particularly in case 3 where the majority of the users live in single homes (Fig 13.4). The main differences between cases 2 and 3 are the dwelling distribution and the civil works as described in table 2. Although in case 5 all the ducts have to be created, the fact that the majority of users live in large buildings, compensates for high additional civil works as compared to case 3. Also in case 1 the duct availability is relatively lower, with the housing structure having a similar compensating impact.

Fig 13.6 depicts the cost difference between ATM PON and enhanced copper for each of the cases studied. The cost difference dependence on housing structure and duct availability is clearly shown. The dominant block apartment structure of case 2 leads to low civil works costs and yields a marginal difference between these two technologies. Cases 1 and 5, in which duct availability is low, have significantly larger relative IFC than cases 2 and 3. The cost difference in cases 1 and 5 is also caused by CATV, which is only provided in the ATM PON solutions. Additionally, in cases 1, 3 and 4, the high ATM PON costs reflect the expenses of a complete network upgrade, where also POTS and N-ISDN services are integrated in the broadband PON.

The study emphasises that the dwelling distributions and the condition of the existing network are crucial aspects in choosing the upgrade technology.

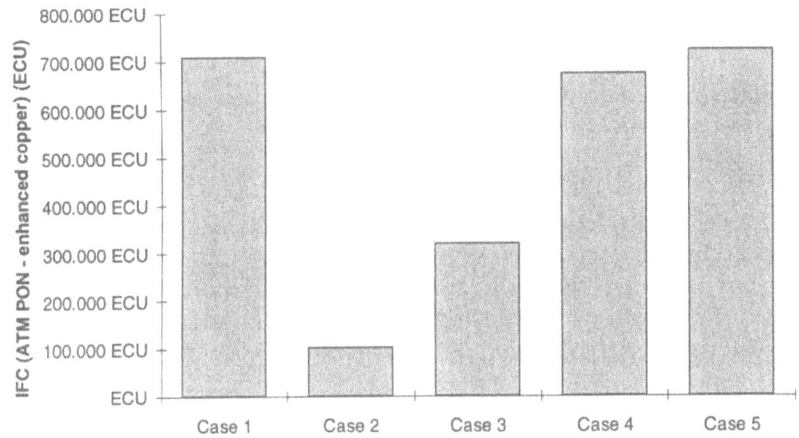

Fig 13.6 Cost difference between the two main technology options, ATM PON and enhanced copper.

The costs of electronics

For all solutions and cases, the costs of electronics are dominant. This is shown for case 3, in particular in the cost breakdown of the different solutions in Fig 13.7.

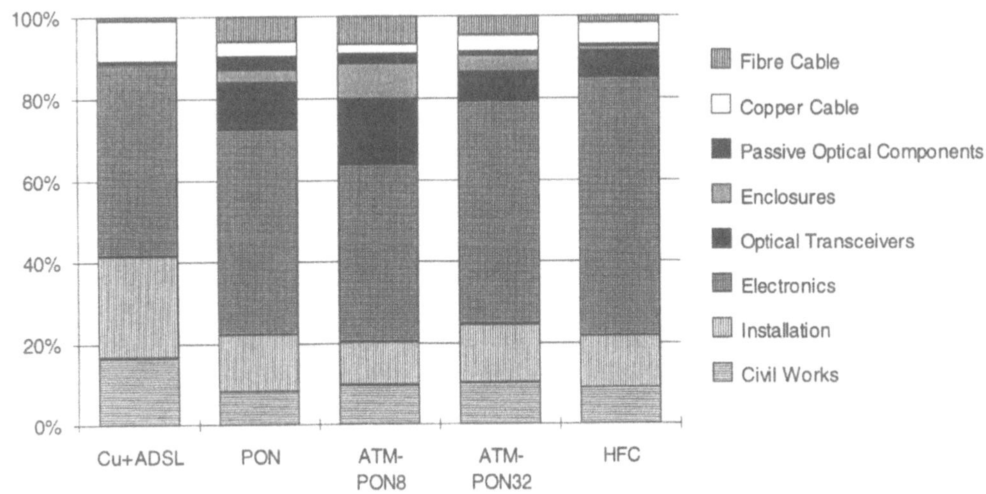

Fig 13.7 Case 3 cost breakdown.

The additional costs of placing these electronic investments in a broadband infrastructure, i.e. deploying fibre, depend on the availability of ducts. For some of these non-greenfield multiservice upgrading cases, the civil works play a less important role compared to greenfield deployment.

This shows that a broadband upgrade requires significant investments in electronics, which are on a similar level regardless the technology. The network operator may then be confronted with the choice of placing these investments in either a narrowband cable infrastructure (twisted pair copper) or broadband cable infrastructure (fibre).

The influence of capacity demand on costs

Fig 13.8 depicts the costs of the two major types of solutions examined (enhanced copper and ATM based PONs) as a function of the average capacity demand at the end of the study period for 1,024 customers. The IFC for ATM PON in fibre to the curb (FTTC) and fibre to the building (FTTB) configurations are included for three degrees of duct availability, 0 %, 50 % and 100 %. The FTTC and FTTB costs are plotted for 8 and 32 customers sharing one optical network unit (ONU) respectively. In addition, the two alternatives for each of the five cases are plotted according to the case specific average subscriber capacity demand: Case 1: 21.5 %, cases 2 and 3: 20 %, case 4: 12 % and case 5: 16 %. The higher slope of the FTTC-8 curves for capacities below 30 % is due to the relatively high number of ONUs compared to broadband subscribers in this region. An FTTC alternative with 32 customers sharing each ONU would have slightly higher IFC than its FTTB counterpart, mainly due to the increased cost of outdoor housing. Similarly configuring FTTB with eight customers per ONU would result in slightly lower cost than the FTTC-8 due to indoor housing savings. The effect of subscriber sharing on costs of optical network termination is discussed in the next section.

Fig 13.8 shows that low demand for broadband services favours the copper based solutions while high capacity demand can favour more aggressive fibre based access network solutions. The operators are currently faced with a choice between optical fibre solutions and enhanced copper. Despite relatively high upfront costs of the former, the lower incremental costs of additional 2 Mbit/s subscribers lead to comparable installed first costs. This is particularly illustrated for case 2 in Fig 13.9 and discussed in more detail later. The break-even points in Fig 13.8 indicate the capacity limit beyond which the demand can justify the deployment of optical solutions in purely economic terms. In general, if the demand is high enough, the fibre-based solutions can be comparable in costs to the copper alternatives. Additionally enhanced copper upgrades may be limited in the number of digital broadband channels per cable of twisted copper pairs. This aspect has however not been addressed in this study.

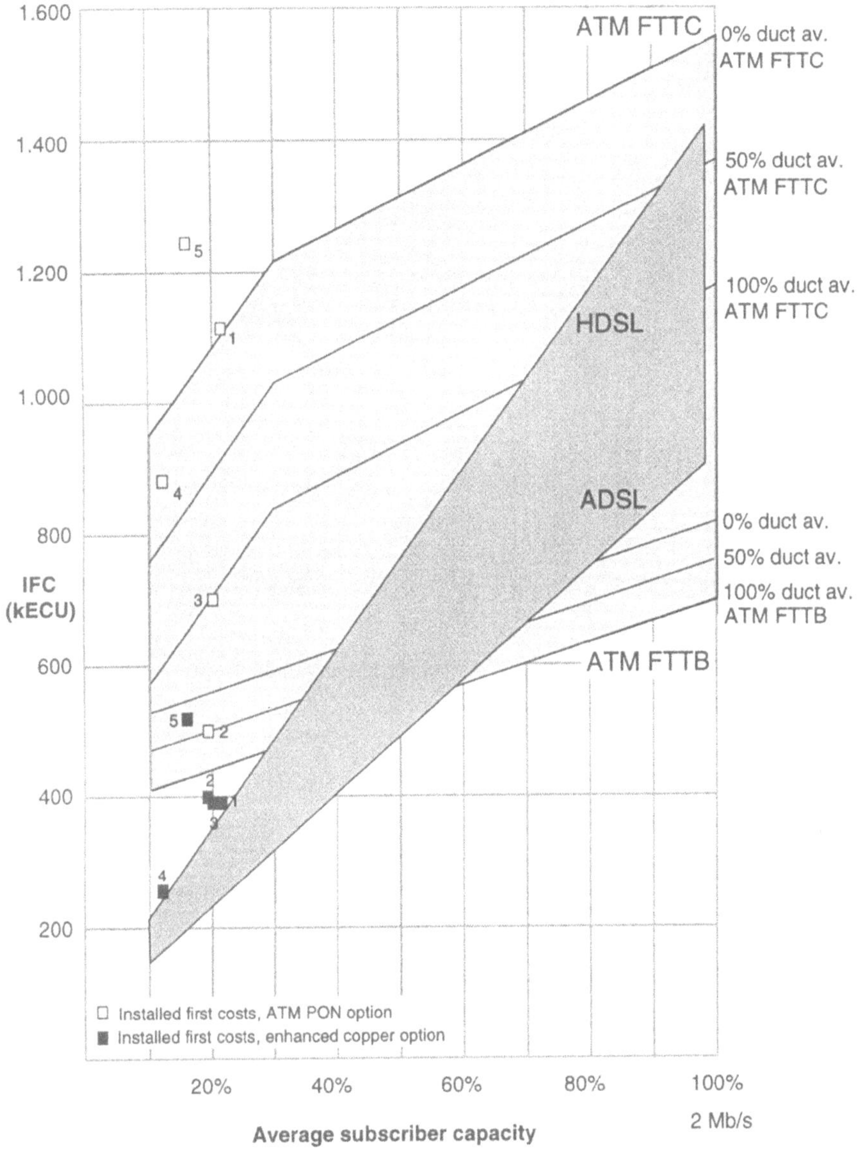

Fig 13.8 The costs of the two major types of solutions examined (enhanced copper and ATM based PONs) as a function of the average subscriber capacity for 1,024 customers.

The overall capacity demand will also be of importance due to the effect of high aggregate capacities on the core network. Even a relatively low penetration of ASB and SSB services with capacities of 2 Mbit/s dramatically influences the dimensioning of the core and feeder network and might lead to saturation. Therefore, network upgradeability will be a key issue in network planning.

In conclusion, the total capacity is a very important parameter to take into account in strategic considerations.

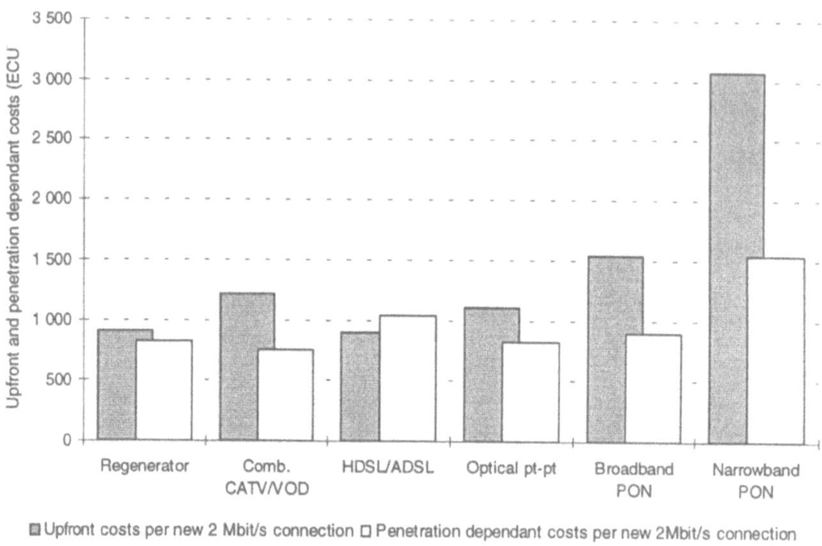

Fig 13.9 Upfront costs and service penetration dependant costs per new 2 Mbit/s connection, case 2.

The effect of customer sharing on the costs of optical network termination

Fig 13.8 depicts the influence on the overall costs of the sharing of network termination equipment as described in the previous section. The IFC for ATM PONs with eight and 32 subscribers sharing one ONU are shown. The cost of the ATM PON is reduced by 50 % for a four fold increase in optical network termination customer sharing, as seen from the respective curves with 100 % duct availability. This illustrates the great benefit with respect to IFC of sharing the ONU costs between a few tens of subscribers.

This confirms that the degree of customer sharing of optical network terminations is an important aspect of a cost effective broadband upgrade with PONs.

The timeframe of broadband access network upgrading

Fig 13.10 shows the net present value (NPV) of the ten year projects. The NPV shown is the cumulative sum of the cash flow from the start of the model year to the final year, i.e. 2005. The residual value of the network is not included.

The NPV divided by the IFC for all cases is shown in Fig 13.11. The ratio of NPV to IFC gives an indication of the level of the yearly internal rate of return (IRR) of the project. A ten year project with a yearly IRR of 10 % will have a NPV/IFC ratio of ~2.6. All of the upgrade projects, except two of the case 4 upgrades, seem to be heavy projects with an IRR comparable to the establishment of the present day narrowband networks. The payback periods of such upgrade projects typically are in the range of five to ten years.

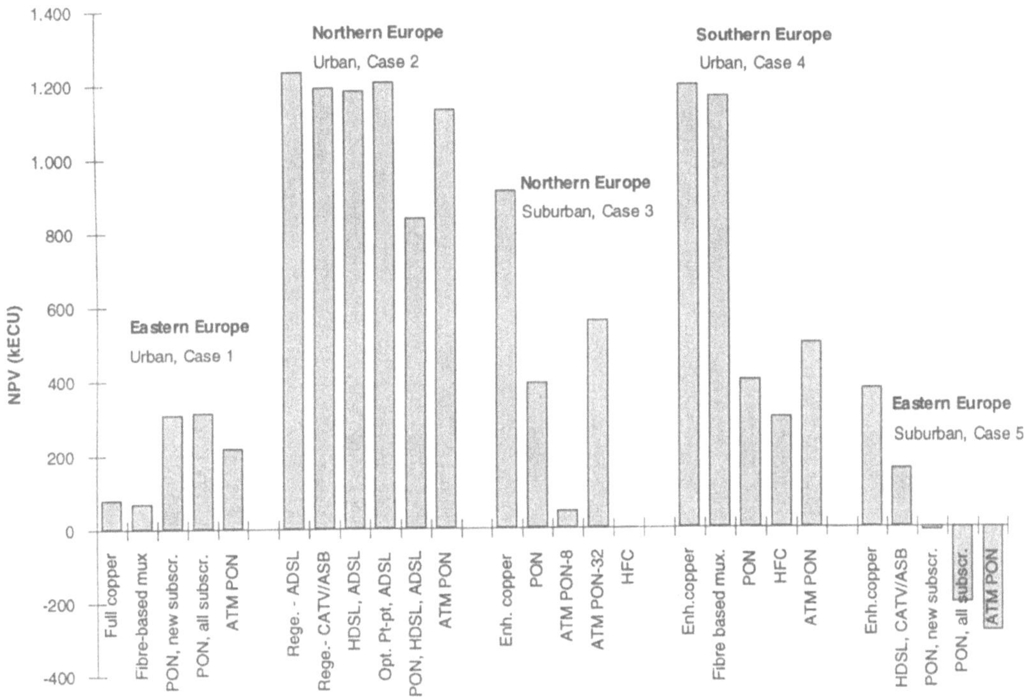

Fig 13.10 The NPV of the different architectures realised within each of the cases studied.

The results show that upgrading relatively modern urban infrastructures (case 2), in a ten year perspective can be quite profitable, provided the operator maintains his POTS and N-ISDN customers. Incomplete existing infrastructures (cases 1 and 5) obviously also require heavy cable and duct layout investments in addition to the capacity upgrade costs, making such upgrades particularly costly.

However, the upgrade effect on the prospects looking beyond the upgrade period is not evident from the NPV. The cash flow in the final year is useful as a long-term predictor. The cash flow in 2005 of the various cases is shown in Fig 13.12. With the given tariffs and penetrations all of the upgrade projects have gained strength through the broadband upgrade compared to the initial annual, cash flow i.e. the final year cash flow is higher than the pre-upgrade cash flow. For a 90% POTS penetration the initial cash flow is ~160 kECU. The upgrade results in a two-fold increase of this annual cash flow for the case 2 operator.

In conclusion, access network broadband upgrades are likely to be long term projects.

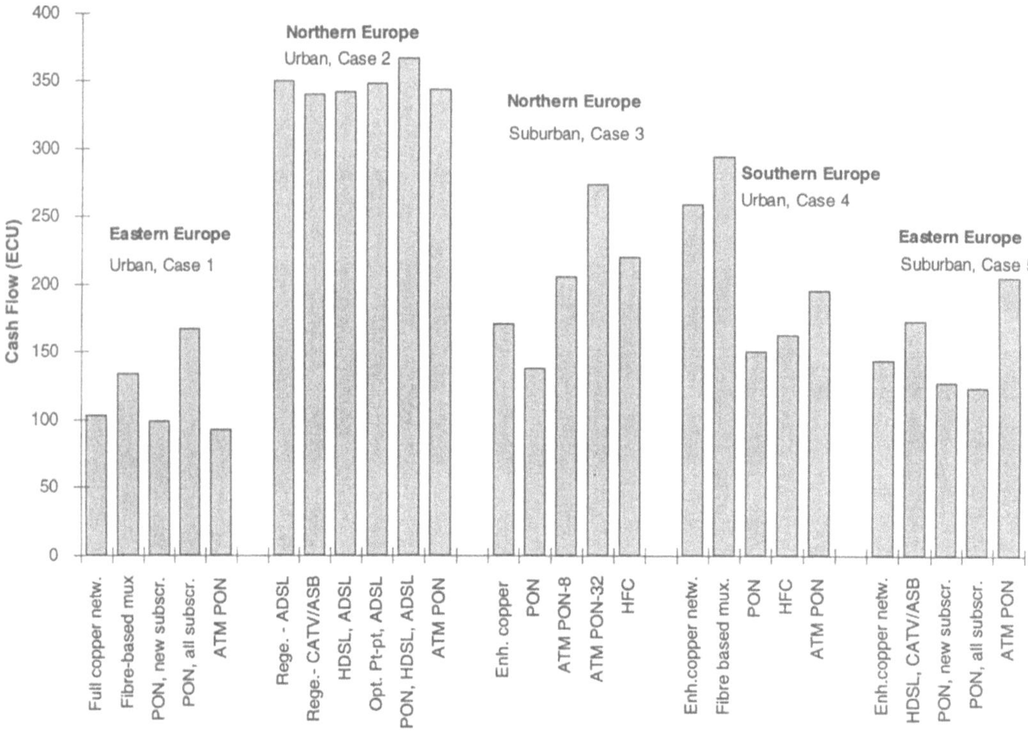

Fig 13.11 The NPV divided by the IFC of the different architectures realised within each of the cases studied.

The effect of competition and service penetration on the overall project

The effect of competition and loss of market shares is clearly demonstrated by the 350 kECU difference in the NPV (Fig 13.10) between the comparable copper based architectures of case 2 (option 2) and case 3 (option 1). This poorer NPV of the PNO in case 3 is caused by the 40 % loss of initial POTS subscribers during the period, as compared to the 20 % customer loss of the case 2 operator. In addition the PNO in case 2 experiences a CATV penetration increase to 30 %, three times as high as the respective penetration of the PNO in case 3. The differences in housing structure have little significance when comparing the enhanced copper alternatives of cases 2 and 3.

Altogether, the PNO in case 4 has an even better net present of his project, mainly due to the fact that CATV is not provided. Even the relatively high CATV penetration (30 %) in case 2 seems to be insufficient for the CATV provision in itself to become profitable with the given tariffs.

The above underlines the importance of the current narrowband customers and revenues, even when upgrading to new broadband services.

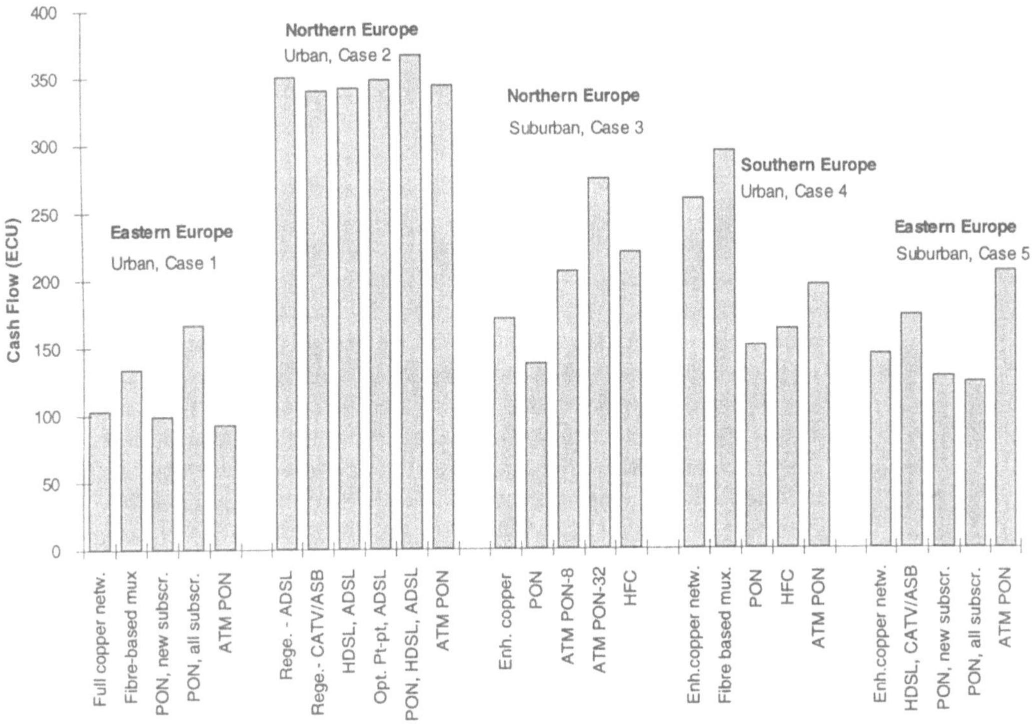

Fig 13.12 The cash flow at the end of the ten year project period of the different architectures realised within each of the cases studied.

13.5.4 Case specific results

Case 1, Eastern Europe, urban

The analysis of case 1 indicates that an upgrade of an existing obsolete access network for new POTS demand probably will represent a significant investment for the PNO (Fig 13.4). The effect of the low duct availability and hence high civil works costs is in this particular case partly compensated by the fact that the majority of users live in large buildings. In implementations 1 and 2 (Table 13.1), HDSL modems and optical point-to-point fibre multiplexers are shared among 30 users respectively. This enables provision of new connections in a rapidly increasing POTS market at a relatively low cost (Fig 13.5).

Low initial POTS and N-ISDN penetrations result in low NPV of the upgrade projects in this case (Fig 13.10). The relatively high investments and long pay-back periods of all solutions assessed (Fig 13.11) necessitate evaluation of alternative access solutions, such as radio based access.

Case 2, Northern Europe, urban

The results show that the cost difference between most of the fibre based solutions and the copper based solutions may be rather low when there is a high percentage of large apartment houses and a good duct layout (Fig 13.4). The installation of TPON for POTS and N-ISDN is examined in particular in case 2 (Fig 13.4). This turns out to be a quite expensive alternative with respect to enhanced copper if the present infrastructure is modern and no additional capacity is provided.

The initial - or upfront - costs of fibre upgrades are higher than for enhanced copper options, mainly because fibre upgrades requires an initial upgrade of the existing infrastructure. There is no significant need for cable infrastructure investments when the upgrade is based on utilisation of the existing copper base. This is illustrated for case 2 in particular in Fig 13.9, which shows the upfront costs and the service penetration dependant, or incremental costs per new 2 Mbit/s connection for the technological options examined. The upfront costs are the costs are the costs of establishing the required common infrastructure. For the fibre optic alternative typically the optical line terminal (OLT), the feeder and distribution fibre cables, the optical couplers and the ONUs which all are shared by all or several customers, are installed in the first year of the upgrade period and hence included in the upfront costs. Typical upfront costs in the enhanced copper architecture are new synchronous digital hierarchy (SDH) add and drop multiplexers (ADM) for increased capacity. The incremental costs are the service penetration dependent costs. Customer specific equipment like ONU interface cards, are installed at the time of connection to the network (according to service penetration) and thus considered incremental costs. The 2 Mb/s ports of these ADMs are together with broadband modems (HDSL, ADSL), line terminals and fibre drop cables considered to be marginal costs. The results in Fig 13.9 show that the HDSL/ADSL alternative has significantly lower upfront costs than the fibre based ATM PON and point-to-point systems. The service penetration dependant costs however, are lower for the fibre solutions.

The cost differences are small and the levels of IRR (Fig 13.11) relatively similar for most of the fibre based solutions and the copper based solutions in this case. This illustrates that the former solutions can be economically viable in apartment block dominated urban areas with good duct layout.

Case 3, Northern Europe, suburban

The analysis underlines that fibre in the loop (FITL) is significantly more expensive than an upgrade of the existing copper network, mainly due to the large portion of single house dwellings.

The issue of customer sharing of network termination equipment as discussed previously, was addressed in particular in case 3. In the case 3 study we found that the costs using ONUs serving eight customers are almost a factor two higher than the costs with ONUs serving 32 customers.

The highly competitive situation might favour the use of FITL systems, which are flexible and may easily be adapted to the varying demand for miscellaneous services. However, the provision of symmetric switched broadband services in addition to POTS and N-ISDN is probably required in order for these solutions to prove feasible.

Case 4, Southern Europe, urban

The results indicate that the introduction of fibre based full service network integration solutions in which all subscribers - new and old - are moved to the new network (implementations 3, 4 and 5, Table 13.1) require very significant investments compared to enhanced copper upgrade solutions for new subscribers only (Fig 13.4). However, the solutions are comparable in costs per new line (Fig 13.5).

The ratios of NPV to IFC of the enhanced copper and fibre based multiplexer solutions of case 4 are high compared to equivalent technological alternatives in other cases. This is mainly due to the beneficial effect of not providing CATV in a situation in which the prospects are only a limited CATV penetration. In contrast, the ratio NPV/IFC of the three other alternatives in case 4 are very low (Fig 13.4), illustrating the heavy burden of moving all customers - both new broadband customers and old narrowband customers - to the new integrated broadband network. Hence, the integrated PON, HFC and ATM PON architectures as studied here may probably only be justified in the long term by prospects such as additional savings due to a high degree of service integration or a rapidly increasing broadband demand.

Case 5, Eastern Europe, suburban

The assessment illustrates that the high overall costs of civil works leads to significant investments in any fixed network upgrade for new POTS lines in an area with an incomplete existing network and no ducts available. However, given that the operator introduces FITL for new POTS customers, the marginal costs of a full rehabilitation, i.e., connecting all customers to the fibre-network, are small compared to the overall costs of the fibre roll-out (Fig 13.4).

The low penetration of POTS and N-ISDN at the start of the study period is one main reason for the low NPV of the upgrade projects in this case (Fig 13.10). The high initial investments (Fig 13.4) and long payback periods (Fig 13.11) for all studied architectures, suggest that alternative access solutions also should be considered.

13.6 SUMMARY

The examination of the above five cases indicates that a further upgrade of the access network to broadband probably for most areas and technologies selected will require equivalent or higher investments than the ones of establishing today's narrowband access networks.

The condition of the existing network infrastructure is of primary importance for an appropriately phased introduction of FITL in areas with already established infrastructures. The introduction strategies greatly depend on dwelling distribution and duct availability.

The costs of electronics dominates the overall costs for all cases, and are on a similar level within the cases, independent of technology.

Low demand for broadband services favours the copper based solutions because they do not require high upfront investments. There is a capacity dependent break-even point beyond which the demand, under some circumstances, can justify the deployment of shared optical access solutions. These represent a paradigm shift with resulting new flexibility for the PNOs, e.g. in terms of bandwidth management and overall capacity.

FITL systems are currently much more advisable for high degree of customer sharing per ONU rather than in configurations with fewer customers per ONU. Since, short time roll-out of FTTH architectures seems very unlikely at present, such solutions have not been considered in this study.

Access network broadband upgrades seem to be quite similar to traditional PNO access network projects in that they are long term projects. The establishment of the existing narrowband networks has relied on long term investments, which most likely will be the case also for a further upgrade of the present network to broadband, independent of technology choice.

The demand for new POTS lines is one of the main driving forces in Eastern European countries and in some other developing areas. Introduction of FITL in developed countries, where a good existing copper network is available, might be advisable starting directly with broadband FITL systems.

This document is based on results achieved in a EURESCOM project; this does not imply that it reflects a common technical position of all the EURESCOM shareholder/parties.

REFERENCES

[1] Cook, A, Stern, J. Optical Fiber Access - Perspectives Towards the 21st Century. *IEEE Communications Magazine*, 32, (2), 1994, 78-86.

[2] Pugh, W, Boyer, G. Broadband Access: Comparing Alternatives. *IEEE Communications Magazine*, 33, (8), 1995, 34-45.

[3] Warzanskyj, W, Baireuther, O. Access Network Evolution and preparation for Implementation: A EURESCOM view. *In: Proc 6th International IEEE workshop on Optical Access Networks (OAN '94), Kyoto, Japan, Oct. 16-19, 1994, 4.1-1 - 4.1-10.*

[4] Warzanskyj, W, Ferrero, U. Access network evolution in Europe: a view from EURESCOM. *In: Proc The 20th European Conference on Optical Communications (ECOC '94), Florence, Italy, September 25-29, 1994.*

[5] Ims, L A, Olsen, B T, D. Myhre, D, Mononen, J, Lähteenoja, M, Ferrero, U, Zaganiaris, A. Multiservice Access Network Upgrading in Europe: a Techno-economic Analysis. *IEEE Communications Magazine,* 34, (12), 1996, 124-134.

[6] Ims, L A, Myhre, D, Olsen, B T, Mononen, J, Lähteenoja, Ferrero, U, Zaganiaris, A, Szomolányi, K, Orfanos, I. Advanced Multiservice Scenarios for Europe. *In: Proc. 13th Annual Conference European Fibre Optic Communications and Networks (EFOC&N '95), Brighton, UK, June 27-30, 1995, 1, 177-183.*

[7] Ferrero, U, de Bortoli, M, Myhre, D, Ims, L A, Olsen, B T, Mononen, J, Lähteenoja, M, Zaganiaris, A, Pousa, M, Szomolányi, K, Orfanos, I. Access

Network Upgrade Strategies: mapping the European Diversity. *In: Proc. 7th International IEEE workshop on Optical Access Networks, Nuremberg, Germany, September 24-28, 1995, 4.2-1 - 4.2-10.*

[8] Stordahl, K, Olsen, B T, Murphy, E, Zaganiaris, A. de Castro, E, Ravera, M. Delphi Survey: Forecasting Demand for Wide- and broadband services. From RACE 2087/TITAN. *In: Proc 6th International IEEE workshop on Optical Access Networks (OAN '94), Kyoto, Japan, Oct. 16-19, 1994, 7.3-1 - 7.3-10.*

[9] Stordahl, K, Murphy, E. Methods for forecasting long term demand for wide and broadband services in the residential market. *IEEE Communications Magazine*, 33, (2), 1995, 44-49.

[10] Mochida, Y. Technologies for Local-Access Fibering. *IEEE Communications Magazine*, 33, (2), 1995, 54-73.

[11] Dyke, P J, Waters, D B. A Review of The Technical Options for Evolving FITL to Support Small Business and Residential Services. *Journal of Lightwave Technology*, 12, (2), 1994, 376-381.

[12] Angenent, J H, Bathe, P. Video-on-demand; implementation in existing and future networks. *In: Proc. 12th Annual Conference European Fibre Optic Communications and Networks, Heidelberg, Germany, June 1994.*

[13] De Prycker, M, Verbiest, W, Mestdagh, D. ATM Passive Optical Networks: Preparing the Access Network for BISDN. *In: Proc. XIV ISS, Yokohama, Japan, October, 1992.*

PNO and cable operator broadband upgrades 14

Leif Aarthun. Ims, Borgar Tørre Olsen,
Kjell Stordahl, Dagfinn Myhre, Alcibiade Zaganiaris

This chapter presents a techno-economic analysis of broadband access network upgrade strategies for public network operators and cable operators in a competitive environment in the residential and small business market. The study focuses on network evolutionary paths from the existing network towards a target broadband architecture. The intermediate steps and final network solutions include the technology options of broadband twisted pair modems, hybrid fibre coax networks and asynchronous transfer mode (ATM) based passive optical networks (PONs) [1].

14.1 INTRODUCTION

The access network is expected to be one of the major battlegrounds of the telecommunication network providers as a full deregulation of the services draws near in Europe. However, the difference in existing access networks of competitors like public network operators (PNOs) and cable operators call for quite different upgrade strategies. This advocates an in-depth study of the existing competing infrastructures, and technology alternatives, which might be the starting point for upgrades to broadband [1-6].

This paper addresses some of the different challenges faced by the PNO and cable operators in adapting their present fixed network to competition. The analysis of broadband upgrade alternatives focuses on one of the most competitive market segments: an urban, apartment block area with short outdoor average loop lengths, in which both the PNO and the cable operator have an established infrastructure for their own subscribers.

The methodology and tool developed within the project RACE (research in advanced communications in Europe) 2087/TITAN (tool for introduction scenario and techno-economic evaluation of access network) has been used to evaluate the selected set of network upgrade alternatives and strategies [7].

14.2 THE CASE STUDY

A network evolution during a 15 year period from 1996 to 2010 has been examined. An urban, residential and small business area with the customers living in apartment blocks with an average of 32 dwellings per block has been considered. The average outdoor loop length from the network concentrator location to the buildings is 250 metres. The PNO and the cable operator have already established network architectures in the area, and a significant part of the area have ducts available for new cables (90% duct availability for fibre and coaxial cable). The cable operator is assumed to have an existing 450 MHz distribution coax network in the initial year, whilst the PNO has an existing twisted pair infrastructure for delivering plain old telephony service (POTS) and narrowband integrated services digital network (N-ISDN). All of the existing cable operator and PNO infrastructures have been fully amortised.

14.3 THE MARKET FORECAST

In [8] new services forecasting and demand projections are reported, derived from current spending patterns of households. Alternatively, market forecasts may be developed from market surveys. In this study we assume that a common set of services is provided by both operators: POTS, N-ISDN, community antenna television (CATV), 2 Mbit/s asymmetric switched broadband (ASB) and 2 Mbit/s and 8 Mbit/s symmetric switched broadband (SSB). The CATV service penetrations used are established based on European averages. The forecasts of the total market of the broadband bearer services are extracted from the Delphi survey on broadband demand, carried out by TITAN in ten European countries [9,10].

Competition is assumed from 1998 on all services. Hence, the market will be shared between the PNO and cable operator. The competition has been modelled by appropriate adjustments of the service penetrations. The PNO starts with the majority of the market share of the high-revenue narrowband switched services, whereas the cable operator initially has the majority of the market share of the low-revenue CATV service. In Fig 14.1 the evolutions of the service penetrations are shown for three selected years. Equal market shares are assumed during the whole period for the broadband switched services.

Only the annual subscription tariffs for bearer services have been considered in the analysis. Traffic income is not accounted for, since this is assumed to be revenue allocated to the trunk network. The tariffs used are based on European averages from the TITAN Delphi survey and other sources. The tariff structure incorporated in the economic analysis includes an evolution of the tariffs during the study period. Competition and penetration may have a significant impact on the evolution of tariffs. The tariff elasticities for the broadband switched services are derived from the Delphi survey, as described in chapter 3. The CATV tariff elasticity is modelled by a similar approach. The POTS and N-ISDN tariffs decrease annually by respectively 1 % and 2 %.

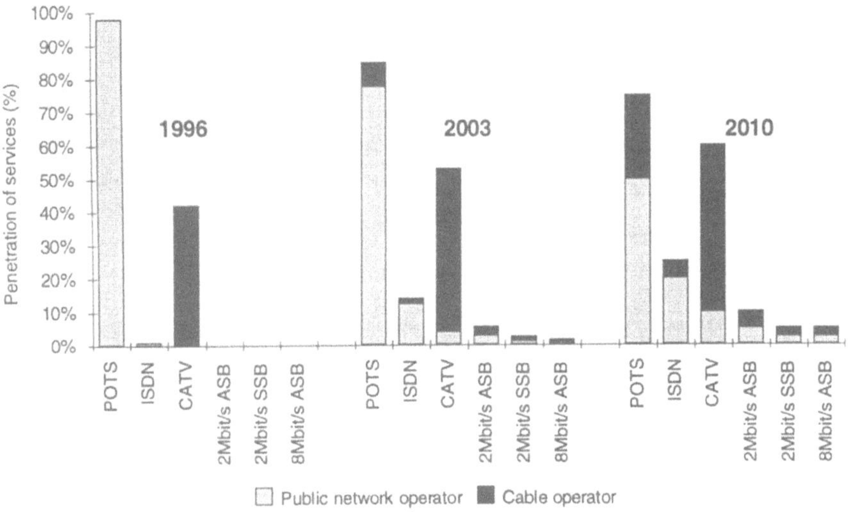

Fig 14.1 Penetration of services and market shares.

14.4 BROADBAND UPGRADE OPTIONS

In upgrading the physical layer of the very cost sensitive access network, the network operators face great strategic challenges, which encompass selection of transmission medium and system technology. The relevant wireline transmission media for the future interactive broadband networks are optical fibre cable, coaxial cable and twisted copper pairs. Alternative system technologies include enhanced copper options like asymmetric digital subscriber line (ADSL), very high-speed digital subscriber line (VDSL) and high bit-rate digital subscriber line (HDSL), the ATM and fibre based broadband passive optical network (ATM PON) and hybrid fibre coax (HFC) for coaxial cable network upgrades [11].

Two different access network evolutions, or evolutionary paths, have been examined for each of the two operators. The network evolutionary paths examined consist of network architectures, upgraded in three steps during the 15 year study period. They are classified as moderate or aggressive evolutionary paths with respect to technology choice and degree of service integration in the upgraded network.

The four PNO and cable operator network evolutions are depicted in Fig 14.2. The two initial networks and corresponding baskets of services are described in the two rectangular boxes at the left side of the figure. The lower box describes the starting situation of the cable operator, the upper one the corresponding network and services provided by the PNO. The moderate and the aggressive evolutionary paths of the operators are depicted as solid lines originating from these two boxes. For each of the two operators the upper line represents the moderate evolutionary path. Network upgrade and changes in the provision of services occur during the three periods indicated on the horizontal time axis in the middle of the figure. The changes are indicated along the solid evolutionary path lines. The rectangular boxes show the status of the network infrastructure and services provided after the respective upgrade periods.

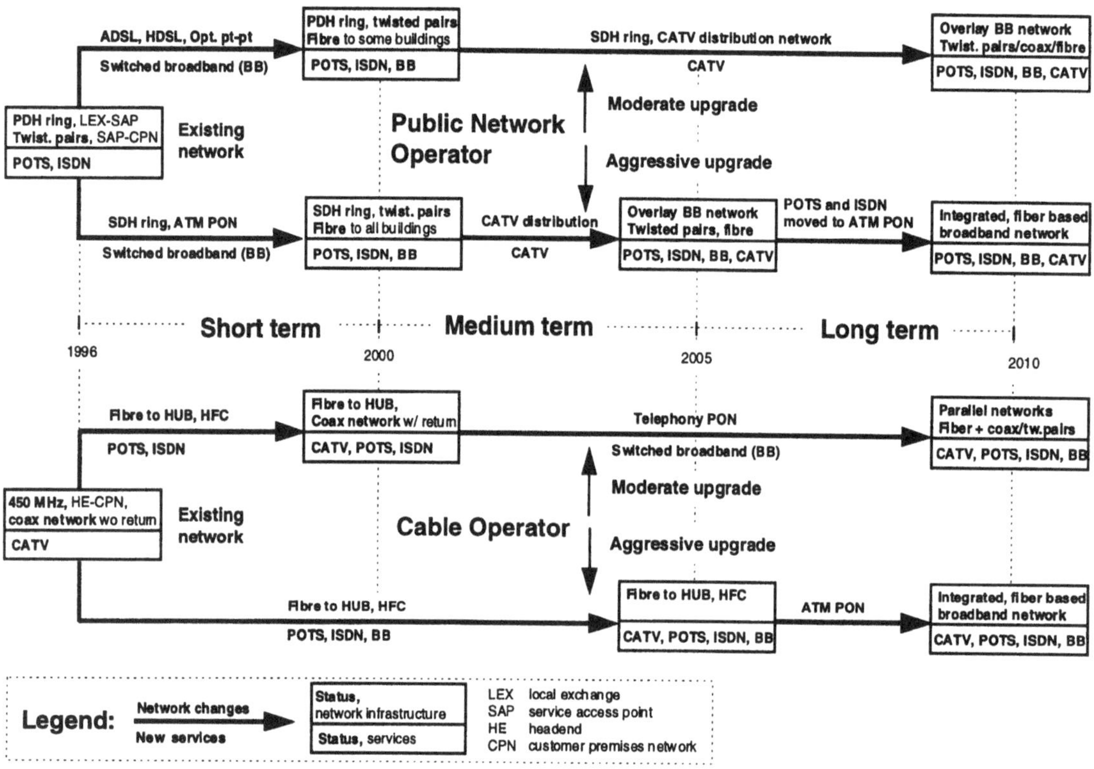

Fig 14.2 The four examined PNO and cable operator network evolutionary paths.

14.5 PNO EVOLUTIONARY PATHS

The evolutionary paths considered highlight the economic and technological implications for moderate operators which aim to utilise the existing twisted pair copper cables as the basic transmission medium, and aggressive operators which extensively upgrade the network with broadband fibre technology. The former represents a limited degree of service integration, whilst the aggressive upgrade enables an introduction of a full service fibre network. Detailed sketches of the final of the architectures for the PNO alternatives are shown in Fig 14.3 and Fig 14.4.

14.5.1 Moderate evolutionary path, PNO

The plesiochronous digital hierarchy (PDH) ring between the local exchange and the service access point (SAP) are kept until 2001. The service access point refers to the network localisation of the concentrator, as used in several European countries with modern access network infrastructures. From 2001 to 2010 an synchronous digital hierarchy (SDH) ring is installed. In the distribution network ADSL and HDSL equipment is installed to provide new services like 2 Mbit/s ASB, i.e. service on demand (SOD), and 2 Mbit/s SSB. 8 Mbit/s SSB customers are connected through a point-to-point fibre/coax link. In addition, a CATV distribution network is installed using the existing fibre base for 8 Mbit/s customers.

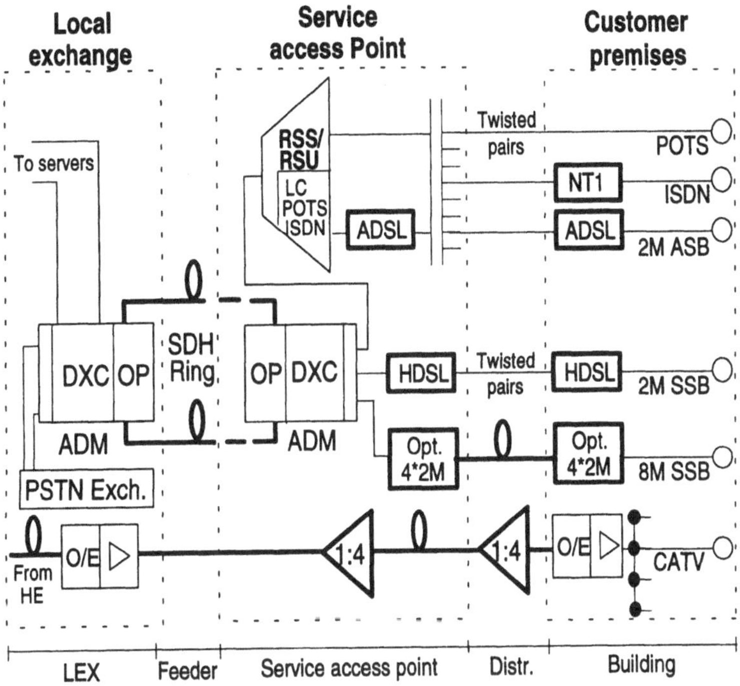

Fig 14.3 The final of three consecutive architectures in the moderate PNO upgrade option: an enhanced copper architecture with optical fibre for the highest demand and a parallel distribution network for CATV.

14.5.2 Aggressive evolutionary path, PNO

New technology is introduced during the period from 1996 to 2000. The PDH point to point connection to the hub is replaced by an SDH based ring structure. The deployment of an ATM PON is started in 1996 in order to provide new services like SOD and 2 Mbit/s SSB. A combined fibre-coax network for CATV distribution is installed in 2001. After 2006 the ATM PON is used to deliver all services. The resulting network is passive between the local exchange and the building where the optical network unit (ONU) is situated.

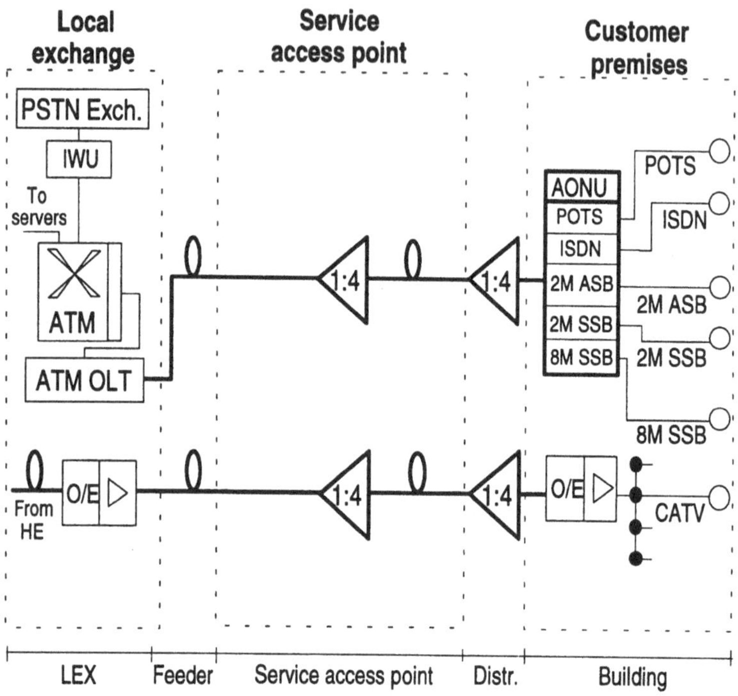

Fig 14.4 The final of three consecutive architectures in the aggressive PNO upgrade option: an integrated fibre optical network (FTTB) based on an ATM broadband passive optical network for all services and a parallel distribution network for CATV.

14.6 CABLE OPERATOR EVOLUTIONARY PATHS

The cable operator evolutionary paths represent moderate operators which only partially integrate their network by sharing duct layout, and aggressive operators which fully integrate the network by providing the service set over the same duct layout, transmission medium and network termination units as well. Detailed sketches of the final of the three successive architectures for the cable operator alternatives are shown in Fig 14.5 and Fig 14.6.

14.6.1 Moderate evolutionary path, cable operator

Only the CATV service is provided from year 1996 to 1998. In 1998 the CATV network is upgraded to bi-directional network with return path. POTS and N-ISDN services are then offered on the integrated coaxial cable network. In 2001 the existing coaxial network is upgraded to an integrated network with building located ONUs for CATV and SOD. In addition a fibre to the building (FTTB) overlay network is deployed. The overlay is a narrowband FTTB PON for POTS, N-ISDN and 2 Mbit/s symmetric switched broadband services. POTS and N-ISDN customers share ONUs, whilst 2 Mbit/s customers have their own dedicated ONUs. The ducts are common for the two overlaid networks, but fibre cables, electronics and drop cables are separate. Twisted pair drop cables are installed inside the buildings in 2001 for this network. The 8 Mbit/s SSB service is not offered at all by the moderate operator.

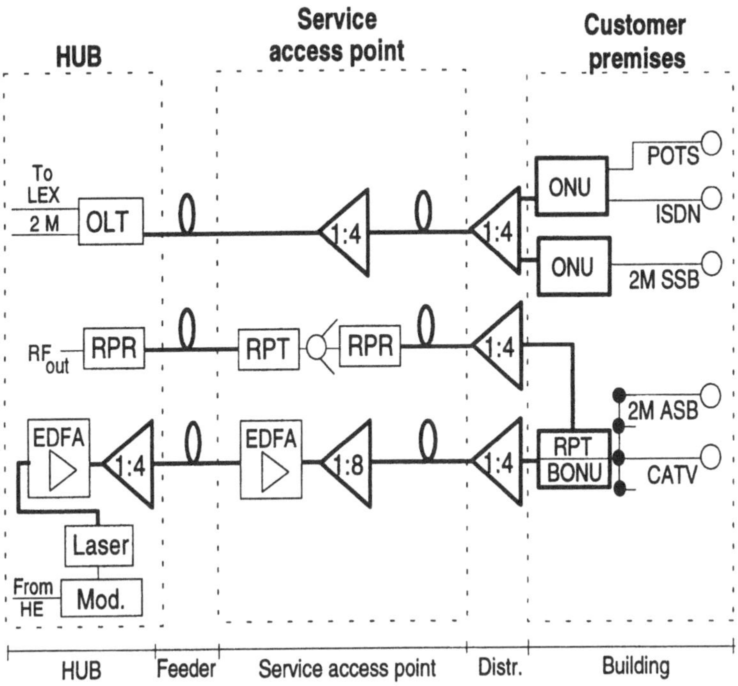

Fig 14.5 The final of three consecutive architectures in the moderate cable operator upgrade option: an optical network (FTTB) based on a TPON (telecommunication over passive optical network) and a parallel hybrid fibre coax (HFC) network for CATV and SOD.

14.6.2 Aggressive evolutionary path, cable operator

Fibre is deployed to the buildings in 1998. This yields a partly integrated network with common ONUs, but separate fibres between the SAP and the buildings and separate drop cables. Twisted pair cables are installed in the drop for POTS, N-ISDN, 2 Mbit/s SSB services. The existing coaxial drops are used for CATV and SOD. In 2005 an ATM PON is introduced in a FTTB configuration. All services except CATV are moved to the ATM-PON. The 8 Mbit/s service is offered with dedicated ATM PON ONUs.

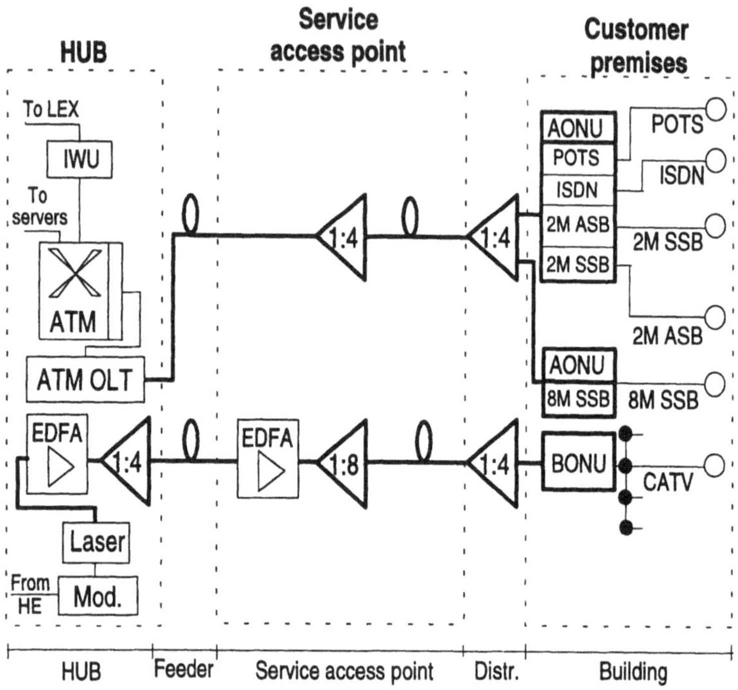

Fig 14.6 The final of three consecutive architectures in the aggressive cable operator upgrade option: an integrated fibre optic network (FTTB) based on a ATM PON for all services and a parallel distribution network for CATV.

14.7 RESULTS AND DISCUSSION

The methodology developed by the RACE 2087/TITAN project has been applied to evaluate the broadband upgrade case [7].

14.7.1 The overall upgrade costs

Fig 14.7 shows the discounted installed first costs (IFC) per connection for the alternatives examined. All new connections are counted as one connection irrespective of the differences among the service connection types in capacity provided. The 1996 costs per connection of the existing networks of the operators are included at the left side of Fig 14.7 for reference. All the figures refer to pure network costs, and do not include equipment at the customer premises. For POTS in particular the costs of the customer premises equipment might be more expensive for the cable operator.

The costs per connection in the existing networks are very similar for the PNO and the cable operators. This is partly due to much lower penetration of subscribers relative to homes passed in the cable network compared to the PNO network. The cost per connection for the cable network has a potential to be reduced to around half of the present value if almost all of the homes passed are connected, i.e. the difference between the two costs per connection reflects the fraction of homes passed.

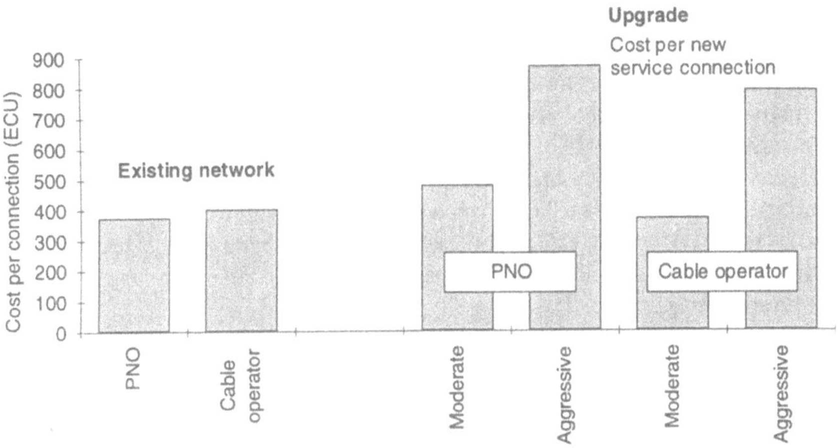

Fig 14.7 The cost per new service connection of the two operator's existing networks in 1996 and the cost per new service connection of the two operator's upgrade alternatives (upgrade period: 1996-2010) @ 250 metre outdoor loop length. 1 ECU ≈ 1,3 US$.

The results indicate that for all upgrading alternatives the PNO must face higher costs than the cable operator. Even for the moderate upgrade, the upgrade costs per new service connection are in the same order of magnitude as the ones for the existing network. The costs of an aggressive upgrade approach are approximately twice higher than the cost per connection in the existing network. The cable operator has significantly lower upgrade costs per new connection than the PNO. This difference is mainly caused by the higher number of new narrowband connections (POTS and ISDN) in the cable operator network as compared to the PNO network.

The main reason for the high upgrade costs per new service connection of the fully integrated PNO upgrade alternative is that it reflects the total replacement of the network with a single broadband PON for all non-distributive services, with costs shared only by the new customers. Extensive fibre deployment and introduction of fibreoptic technology in the residential access network, might be motivated by prospects of future additional revenues or operation and maintenance cost reductions. Nevertheless, such upgrade strategies turn out to be very costly. The cost is likely to become significantly lower if the system introduction is delayed enough to benefit from component price reductions.

These results correspond to low civil works costs, due to high duct availability and short loop lengths in this particular study. In general, when loop lengths are longer, fewer ducts are available for new cables and/or the optical network termination will have to be shared by fewer subscribers, the fibre optic upgrade alternatives are significantly more costly. The influence of the access network loop length on system costs is discussed in [7].

The upfront costs are associated with establishing the required common infrastructure, such as shared cables, network termination units and civil works. Service take rate dependant costs refers to customer and service specific network elements like drop cables and line cards. A closer look at the upfront costs and the take rate dependant costs confirms that even for an area with good existing duct layout, the broadband PON upgrades have higher upfront costs than the other alternatives. However, the service take rate dependant costs are comparable to and even lower than the costs of the other options studied. Hence, broadband PON upgrades will rely on a high service take rate to alleviate a less expensive broadband migration than the enhanced copper alternatives. There will probably be a level of service take rate beyond which the broadband PON solutions can be justified in pure economic terms. The aspects of service penetration sensitivity are further addressed in [4].

The results illustrate that the line costs for broadband upgrades are high both for cable operators and PNOs. The required investment levels are comparable to the cost levels of the existing access networks.

14.7.2 The upgrade project value

The net present value (NPV) divided by the IFC is shown in Fig 14.8 as a measure of the efficiency of the upgrade projects. The NPV as used here is the discounted accumulated revenues minus the discounted accumulated costs (IFC and running costs). The residual network value is not included in the NPV. The upgrade projects are compared to present-day narrowband and distributive access network projects of the PNOs and cable operators respectively. The figure depicts at the left side the ratio of NPV to IFC, for installation of a complete new network in 1996, the PNO offering POTS and ISDN only and the cable operator offering CATV only during the 15 year period. In the middle is shown the upgrade of the existing network: The PNO offers POTS, ISDN, broadband switched services and CATV, and the cable operator offers CATV, POTS, ISDN and broadband switched services, during the study period. At the right side is illustrated the upgrade of the existing network, with only new services taken into account: The PNO offers broadband switched services and CATV, the cable operator offers POTS, ISDN and broadband switched services.

The NPV should be three times the IFC during this period, in order to achieve a reasonable internal rate of return. The establishment of the present access network infrastructure, both for the PNO and the cable operator have typical payback periods in the range of ten years. The benefit when upgrading of having an already established infrastructure with high revenues from the old services, is clearly shown by the high NPV of the total upgraded networks, both compared to the NPV of a greenfield deployment of present day networks (left, Fig 14.8), and the NPV of the upgrade (right, Fig 14.8). However, the upgrade itself (Fig 14.8, right) has a relatively low internal rate of return. The moderate PNO upgrade has a reasonable internal rate of return due to the high revenues of old services. The cable operator moderate and aggressive upgrade options, as well as the PNO aggressive alternative, are all less profitable projects.

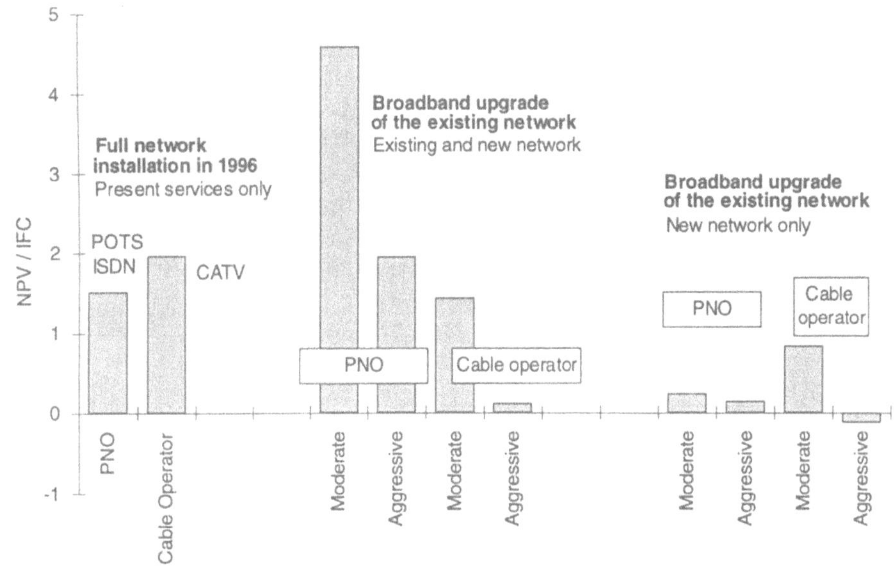

Fig 14.8 Net present value divided by investment.

The evolution of the annual cash flows indicates the long term effect of the broadband upgrade, looking beyond the upgrade period itself. An analysis of the annual cash flows after the upgrade reveals the dominant role of the high revenue narrowband switched services and hence the difference in strengths between the two operators in the original services and the transition to a more balanced situation through the broadband upgrade. In absolute values the cable operator upgrade cash flow is almost twice as high as for the PNO.

Compared with present PNO access network projects the upgrade projects are seemingly quite similar with respect to expected payback period and internal rate of return.

14.7.3 Service take rate and competition

The study illustrates the effect of the competition and the evolution of the service take rate, for a case where the cable operator gains 25% of the initial PNO subscribers, and the new broadband market is shared equally between the two operators. The evolution of the overall revenues of the two operators for the given annual connection fees illustrate the importance of the take rate of the switched broadband services. Very high tariffs will be required to obtain a significant contribution to the overall revenues, with the given penetrations of the switched broadband services and market shares. The revenue gain by the upgrade is very high for the cable operator. For the PNO the upgrade represents an increase in revenue by 50 %. The PNO revenue shifts from being a POTS dominated revenue to a situation in which one third of the revenue comes from POTS, one third from N-ISDN and the remaining one third from the new broadband services. For the cable operator all the added revenue come from new services.

The effect of the size of the total broadband market and the market shares have been studied using a sensitivity analysis, in which the narrowband services market remains unchanged compared to the original service scenario. Fig 14.9 shows the annual cash flow after the upgrade as a function of total broadband market take rate and cable operator market share. The results show that the relative strength between the two operators is more affected by the market share than by the total market take rate. This to a large extent reflects the tariff elasticity used. In our study only the total market take rate influences the tariffs. Nevertheless, the cash flows are not very sensitive to market shares. This underlines the dominant contribution of the narrowband service revenues.

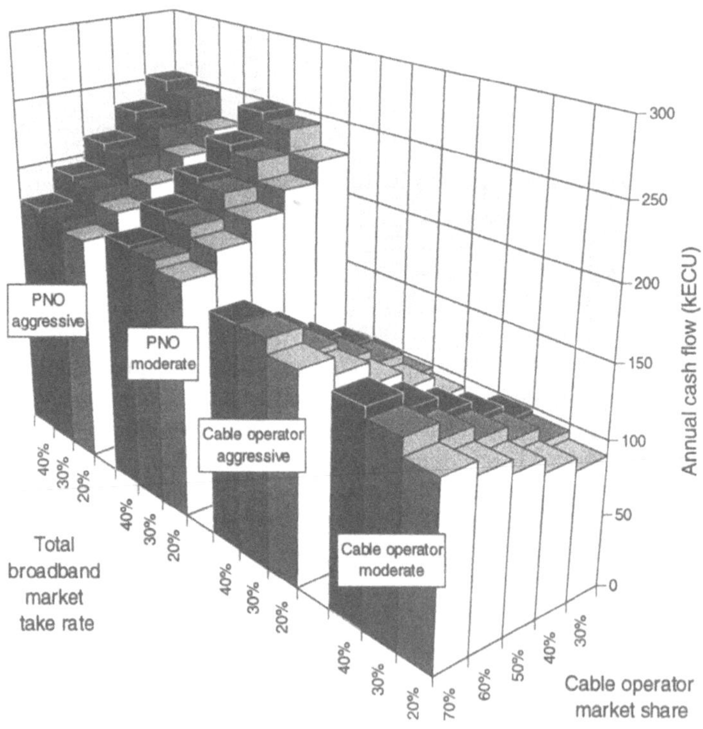

Fig 14.9 The annual cash flow after the upgrade as a function of total broadband market take rate and cable operator market share.

Fig 14.10 shows the payback periods of the projects as a function of total broadband market take rate and cable operator market share. In general, the payback periods for the two operators are higher for the integrated fibre upgrades than for the moderate alternatives. The payback periods are more sensitive to variations in the market shares than fluctuations in the total broadband market. This emphasises the importance of gaining a significant market share of the broadband market, independent of the overall market take rate.

Hence, broadband upgrades will require operators with both significant market shares of new broadband services and significant market shares of the existing narrowband services.

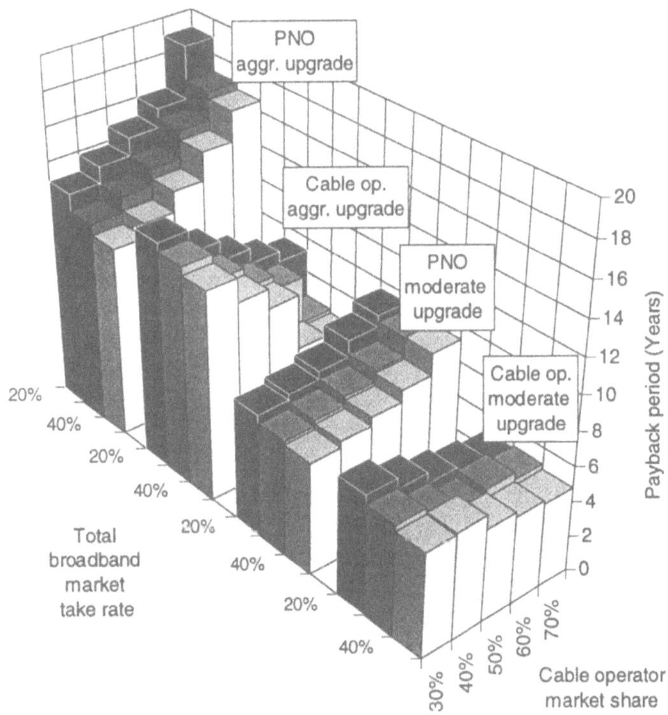

Fig 14.10 The payback period of the upgrade as a function of total broadband market take rate and cable operator market share.

14.8 SUMMARY

For both cable operators and PNOs the required investment levels are comparable to the cost levels of the existing access networks, independently of the upgrade technology choice.

High upfront costs of extensive fibre introduction for broadband requires relatively high service take rates for an investment to be cost competitive compared to other upgrade alternatives.

The expected payback periods and internal rates of return of broadband upgrades are comparable to traditional PNO access network projects.

Broadband upgrades can only be justified by take up of new broadband services and by big shares of the existing narrowband services.

The analysis reveals the dominant role of the high revenue narrowband switched services. Through the upgrade there is transition from a significant difference in strengths between the two operators in the original services to a more balanced situation. The cash flow and revenue breakdown results suggest that for cable operators the narrowband market exhibits higher profitability prospects than the emerging broadband market.

REFERENCES

[1] Ims, L A, Olsen, B T, Stordahl, K, Myhre, D, Zaganiaris, A. Upgrading the PNO and Cable Operator access networks to advanced broadband services: technology alternatives and their economic implications in a competitive environment. *In: Proc. Telecom America 96, Technology Summit, Rio de Janeiro, Brasil, June 10-15, 1996, 243-249.*

[2] Miki, T. Toward the Service-Rich Era. *IEEE Communications Magazine*, 32 (2), 1994, 34-39.

[3] LeBel, P, Oliver, R. Residential Broadband Architectures Strategies and Economics. *In: Proc. The 7th International IEEE workshop on Optical Access Networks (OAN '95), Nuremberg, Germany, Sep. 24-27, 1995, 4.4-1 - 4.4-10.*

[4] Ims, L A, Olsen, B T, D. Myhre, D, Mononen, J, Lähteenoja, M, Ferrero, U, Zaganiaris, A. Multiservice Access Network Upgrading in Europe: a Techno-economic Analysis. *IEEE Communications Magazine,* 34, (12), 1996, 124-134.

[5] Olsen, B T, Ims, L A, Øverli, T, Stordahl, K, Myhre, D, Drieskens, M, Kraushaar, J, Tahkokorpi, M, Ravera, M, De Bortoli, M, Zaganiaris, A, Markatos, S, Lähteenoja, M, Mononen, J, Fleuren, F. PNO and Cable Operator broadband upgrade technology alternatives: a techno-economic analysis. *In: Proc. Optical Fiber Conference 1996 (OFC '96), San Jose, USA, Feb. 25 - March 1, 1996, 2, 140-141.*

[6] Olsen, B T, Ims, L A, Øverli, T, Stordahl, K, Myhre, D, Drieskens, M, Kraushaar, J, Zaganiaris, A. Techno-economic analysis of PNO and CATV operator strategies for broadband upgrade of the access network. *In: Proc. International Teletraffic Seminar 1995, Bangkok, Thailand, Nov. 28 - Dec. 1, 1995, 54-1 - 54-12.*

[7] Olsen, B T, Zaganiaris, A, Stordahl, K, Ims, L A, Myhre, D, Øverli, T, Tahkokorpi, M, Welling, I, Drieskens, M, Kraushaar, J, Mononen, J, Lähteenoja, M, Markatos, S, De Bortoli, M, Ferrero, U, Ravera, M, Balzaretti, S, Fleuren, F, Gieschen, N, De Oliveira Duarte, M, de Castro, E. Techno-economic evaluation of narrowband and broadband access network alternatives and evolution scenario assessment. *IEEE Journal on Selected Areas in Communications,* 14 (6), 1996, 1184-1203.

[8] Luck, D P. Broadband To The Home: Evolution Scenarios for Australia. *In: Proc. 13th Annual Conference European Fibre Optic Communications and Networks (EFOC&N '95), Brighton, UK, June 27-30, 1995, 1, 173-176.*

[9] Stordahl, K, Murphy, E. Methods for forecasting long term demand for wide and broadband services in the residential market. *IEEE Communications magazine*, 13, (2), 1995, 44-49.

[10] Stordahl, K, Olsen, B T. Demand Elasticity and Forecasts of Wide and Broadband Services in the Residential Market Based on Results from an International Delphi Survey. *1995 International Communications Forecasting Conference, Toronto, Canada, June 13-16, 1995.*

[11] Pugh, W, Boyer, G. Broadband Access: Comparing Alternatives, *IEEE Communications Magazine*, 33 (8), 34-45, 1995.

Analysis of video on demand and Internet access

15

Ilari Welling, Markku Tahkokorpi

This chapter examines the profitability of different multimedia services and network technologies used to provide the services from the point of view of network operators. In particular, the profitability of video on demand, Internet access and the combination of these services have been studied [1,2].

15.1 INTRODUCTION

Particularly during the last decade, but also earlier, visionaries have proposed several new communications services. For a long time video telephony has been seen as the successor of ordinary telephony. More advanced services, such as movies on demand, interactive network games and shopping via a network have also been proposed. According to market surveys there is also a customer demand for these services [3,4]. New services, which earlier did not exist in any form, offer a whole new and potentially voluminous revenue source for network operators who provide them. This makes new services very attractive and the target of a lot of discussion. There is, however, one problem in the way. The network infrastructure of today is not capable of providing these new services, and the changes required are not small. Operators are required to rebuild their networks considerably before new services can be offered. This raises a question. Even if the expected revenues are large, are they large enough to justify these investments? This study concentrates on this question. The objective of this study is to find answers to whether given communication technologies are economically viable for providing certain new services or not.

This sort of question cannot be analytically answered and instead models for different scenarios will be built. Five different technologies, which are described in chapter 15.4, will be modelled. Four of these technologies are suitable for the typical networks of telecommunication operators. One of the architectures is suitable for the network of a cable operator. These technologies are applied for two different services, video on demand (VOD) and Internet access.

By nature, this sort of evaluation cannot be precise. It contains speculations both about the future and human behaviour. Several assumptions regarding the modelled environment and services must be made, all of which cannot be thoroughly justified, in order to get any results. Although the results from the study are presented as hard figures, the reader must bear in mind that these figures are the result of the assumptions made, and hence open to interpretation. In order to see how the main results are affected by key assumptions, such as service tariffs and usage rate, a sensitivity analysis will be carried out in chapter 15.7.

15.2 LIMITATIONS

For a new operator the option to build the whole network from scratch is feasible. An existing operator starts with an operational telephony network or a cable television network, with the option of upgrading this network. In this study the latter has been assumed, i.e. an existing operator is updating his network for new services. There is a large variation in the condition and standard of existing networks from operator to operator and from country to country. Some have future proof networks with a large base of installed fibre cable, while others still rely on copper cable. In order to make the results of this study generally applicable, investment calculations solely concentrate on new network equipment and not on possible new cabling. It has been assumed that the operator either has cabling in adequate condition or repairs it anyway regardless of this project.

Conventional cable operators only deliver broadcast programs. Recently, after legislation concerning telephony has been relaxed in many countries, cable operators have started to compete with telephone operators. They have either installed copper pairs in their existing ducts or they have invested in integrated delivery systems, which allow voice telephony to be transmitted over coaxial cable. The latter is often called cable telephony, and in this study it has been assumed that cable operators who upgrade their network for new services already have invested in cable telephony in order to provide telephone services. The business of providing new services is, hence, not encumbered by investments in cable phone, even if they use the basic delivery system provided by cable phone.

It is also worth emphasising that the focus of this study is clearly on the economics of new multimedia networks, not the technology. Network descriptions point out the essential cost components in the network, but they do not intend to be detailed technical descriptions of any particular technology.

15.3 SERVICES

The purpose of new digital, high bandwidth access networks is to open opportunities for several new services. Many candidates for new services have been proposed. These include: videophony, home shopping, home office, interactive games etc. Fast transparent digital subscriber lines offer an excellent transport platform for developing such new services.

Out of the many proposed services two basic services have been selected for evaluation in this study. These are:

- interactive video retrieval service
- interactive information access service.

The first service is better known as video on demand (VOD) and the latter nowadays often means access to the Internet. Interactive information access does not have to mean access to the Internet as we know it today, but access to any interactive information network.

Both services are modelled with various access capacities. VOD has two access capacity options and information access has three. The lowest access capacity is referred to as *narrowband* and means bit rates less than or equal to 64 kbit/s. The next access capacity level is called *wideband* with bit rates up to 2 Mbit/s. Services with bit rates above 2 Mbit/s are called broadband services. In this study the broadband networks are modelled using parameters relevant to the 6 Mbit/s rate.

15.3.1 Tariff assumptions

The tariffs used in this study represent the part of the tariffs received by the network operator. Third parties, such as content providers, will in some cases increase the actual fee paid by the customer. In the case of Internet access, however, the content is mostly free and even when Internet content providers collect fees it is, at least so far, independent from network provider billing. Tariffing of services has been divided into three categories:

1. Connection fee
2. Monthly rental fee
3. Usage tariff (per minute or per recording watched).

For those services which are available today, existing tariffs have been used. Most of the services analysed in this study are not available today, which means that real price lists do not exist. Estimates about how tariffs of new services relate to tariffs of existing services have thus been made. Some of these assumptions are not necessarily clear and hence sensitivity analysis has been carried out both for Internet access tariffs and for VOD recording price, as documented in chapter 15.7. In this study the starting point for VOD tariff assumptions has been competitiveness with video rental and pay TV. Table 15.1. lists applied VOD tariffs.

Table 15.1 Video on demand tariffs. Tariffs are based on existing video rental and pay TV tariffs [8-10].

	Wideband (2 Mbit/s) VOD	*Broadband (6 Mbit/s) VOD*
Connection fee	60 ECU	60 ECU
Rental per month	11 ECU	16 ECU
Usage fee per recording	1.3 ECU	1.9 ECU

Narrowband Internet access is a reality today and hence there is no need to make assumptions regarding its tariffs. Hundreds of access providers around the world compete for customers, and their tariffs are well published. For wideband services tariffs twice as high as integrated services digital network (ISDN) and 64 kbit/s integrated delivery system (IDS) have been assumed. Table 15.2 tabulates Internet access tariffs used in the modelling. It should be noted that wideband and broadband Internet access services do not use POTS.

Table 15.2 Internet access tariffs. Narrowband tariffs are taken from the market and wideband- and broadband tariffs are assumed to be twice that of a narrowband service.

	PSTN	*ISDN and 64 kbit/s IDS*	*Wideband and broadband*
Connection fee	63 ECU	63 ECU	120 ECU
Rental per month	10 ECU	13 ECU	25 ECU
Usage fee per minute	0.05 ECU	0.06 ECU	0.14 ECU
Local call per minute	0.01 ECU	0.01 ECU	not included

As for the price of network components, price erosion has also been modelled for service tariffs. Fig 15.1 in the next chapter shows the shape of the price erosion curve in the same chart as the service penetration. The erosion trend is basically the same for all tariffs, only the reference point changes.

15.3.2 Penetration and service usage

Service tariffs alone are not enough to define revenues in economic models. They must be multiplied by the usage of the service. As with tariffs, service usage rates include assumptions about currently non-existent services. Hence, sensitivity analysis will be performed also for service usage.

Basically the same penetration has been assumed for each service, so that the different cases analysed would be as comparable to each other as possible. The point is to reveal the viability of different technologies and how they compare to each other, rather than to calculate the absolute amount of profits operators may expect from the business. The actual penetration that would take place if networks were built also varies from country to country due to different income levels and consumption habits. This can be seen from the video rental and pay TV statistics referred to before.

Since most of the suggested services are either completely new or still in their infancy, penetration has been assigned to start from below 1.0 %. While the start penetration is easy to set, the end penetration after ten years is not. Comprehensive market surveys have been carried out to forecast probable demand for new services. As part of the project RACE (research for advanced communications in Europe) 2087/TITAN (tool for introduction strategies and techno-economic evaluation of access network), Telenor, the national telecom operator in Norway, conducted an extensive Delphi survey on the subject [4]. This survey covered several countries from southern, central, and northern parts of Europe. The results show potential demand of 10 % to 20 % for residential VOD in the year 2005. Based on these results and further discussions with specialists an end penetration of 15 % has been used in this study. The penetration evolution follows an S-curve, as shown in Fig 15.1.

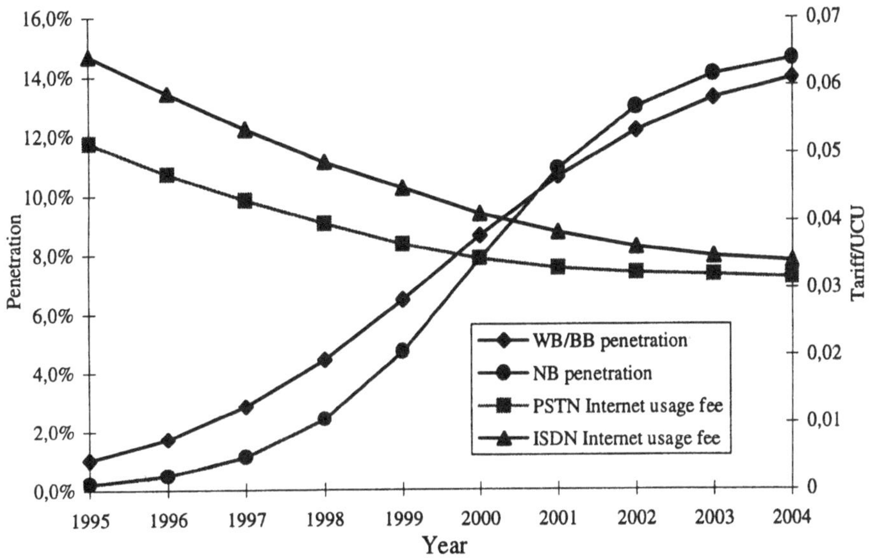

Fig 15.1 Service penetration and narrowband Internet usage fee. The penetration evolution is based on a market survey [4], whereas the Internet usage fee comes from existing commercial service tariffs.

Video on demand

A European home with a video cassette recorder (VCR) rents or purchases on average fourteen recordings per year [8]. It is reasonable to assume that a VOD subscriber is more interested in movies than the average household with VCR. Also the easy availability of recordings will probably contribute to higher usage of VOD recordings than ordinary rental. Here it has been assumed that the VOD subscriber annually orders 30 recordings which is roughly twice the amount of current rentals. This yields 2.5 recordings per month. If it is assumed that on average a recording lasts for 1.5 hours, then consumers will statistically spend 7.5 minutes every day watching VOD recordings. This can be considered to be a quite moderate figure. Effects of values between one and ten for the average number of recordings ordered per month will be investigated in chapter 15.7.2.

Internet access

Even if we already have experience with commercial Internet usage, it is difficult to estimate how much time users will spend on the Internet once the market truly reaches the man in the street. Today many Internet users see the Internet as a hobby, and they can spend several hours every day on the Internet [12]. However, most of these very active users use the service from schools and universities or from other organisations where, from the individual's point of view, the connection is free. Here it has been assumed that the bulk of the residential users are sensitive to the per minute charges, and they will limit their usage to essentials. Based on this reasoning four hours per month average usage has been assumed. Within this time many value adding information retrievals can be performed, as well as electronic mail and home transactions carried out.

Since this four hours per month figure cannot be properly justified, it has been selected as one of the variables for sensitivity analysis. Results from this analysis are presented in chapter 15.7.4.

15.4 NETWORK MODELS

For delivering the set of services described in the previous chapter, a corresponding set of network models has been created. The model area has been selected to represent either a small town or a larger city district. The size of the utilised model area is 38,400 homes passed (h.p.). The exact number comes from the way TITAN builds geometric models. This is also a suitable model area for feasible implementation of the chosen technologies. It is assumed that all investments are made within this area, i.e. no trunk network is included. Models can be divided into three categories:

1. Narrowband networks utilising existing technology
2. Wideband/broadband networks in a telecommunication operator environment
3. Wideband/broadband networks in a cable television operator environment.

One important assumption is that in all cases service will be delivered using existing cables and fibres. Either cables and fibres are in adequate condition, or they will be repaired by the operator, in any case independently from the project under study. It is noteworthy that none of the analysed networks has the fibre to the home (FTTH) architecture.

It might be argued that the most fibre rich solutions of this study are not feasible with the network infrastructure of today. However, the state of the infrastructure varies greatly from operator to operator and from country to country. Since the purpose of this study is to assess the viability of new services as universally as possible, it was decided to exclude possible new copper and fibre cabling altogether. Each reader utilising the results of this study can assess these components if required, with more knowledge of the exact situation in the field. Furthermore, in these sorts of networks the cost of electronics tend to dominate.

Another study [13] carried out by Eurescom (European institute for research and strategic studies in telecommunications), where somewhat similar networks were examined, suggests that the costs due to fibre cable are in any case less than ten percent of the overall investment. In most cases it is much less than ten percent. A potentially greater expense is the installation of these fibres. If new ducts must be constructed, a majority of the fibre related costs will be due to the installation. Again, this expense is heavily dependent on the situation of the particular operator in question. If the availability of existing ducts is good, the expense will not be very high, but if a significant amount of new ducting must be constructed, the price goes up. As with the fibres themselves, this expense is excluded in order to maintain the general applicability of the results. In each case the network can be divided into four parts:

1. Server
2. Backbone network
3. Access network
4. Network terminal.

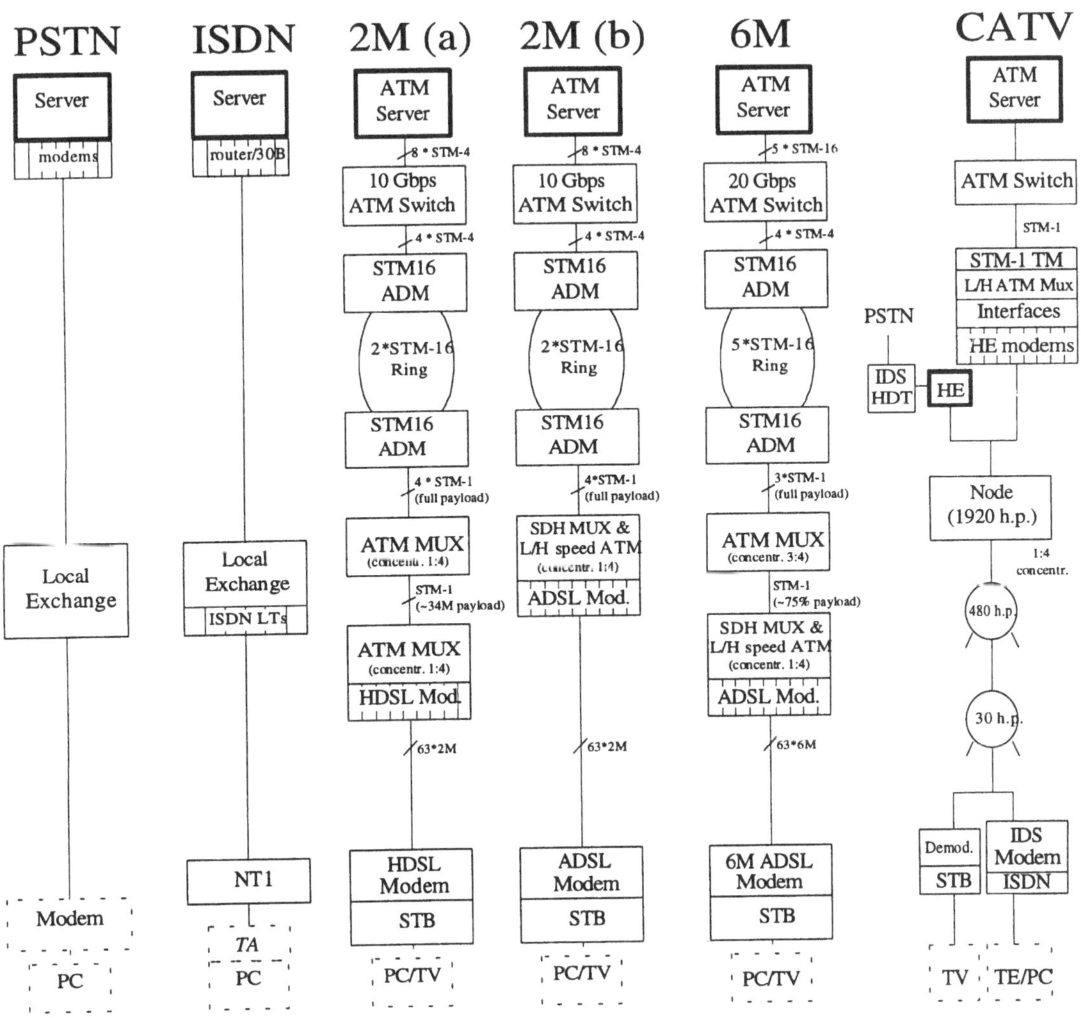

Fig 15.2 The overall structure of the analysed networks.

General assumptions include the backbone network, i.e. synchronous digital hierarchy (SDH) rings, asynchronous transfer mode (ATM) switches and multiplexers etc. being part of the initial investment, while servers, access network elements and network terminals are invested in as customers are connected to the network. This provides full coverage from day one and easy upgradeability of the network as new customers subscribe to the service.

For the network dimensioning a 1:4 concentration ratio is assumed. Traditionally 1:8 has been used for telephone network dimensioning, but since the average session length of the new services is considerably longer than the average phone call, lower concentration must be used. 1:4 concentration has been suggested by specialists. Fig 15.2 presents the general structure of all model networks studied.

15.5 COST COMPONENTS

In this study model networks are dynamically dimensioned over the study period according to real demand. This means that components particularly in the access network and customer premises are invested in as needed during the study period when new customers subscribe to the service. These parts of the network are major contributors to the overall investment per user, because these costs are divided only by a small number of customers, or ultimately only by one as in the case of the set-top box (STB). This will be shown in chapter 15.6.4. Therefore it is very important to take into account the price evolution of these components over time.

In this analysis all cost components have been divided into four classes as follows:

1. The first class contains all terminals, modems, set-top boxes and similar devices. In this class prices erode rapidly and the total price erosion is high.
2. The second class contains new infrastructure. Here the new infrastructure mainly means ATM devices, such as ATM switches. In this class price erosion after ten years is almost as high as in the first class, but the rate of price erosion in the beginning of the study period is slower.
3. The third class contains only servers for VOD and Internet services.
4. The fourth class consists of mature infrastructure. Included here is infrastructure using SDH technology, which is considered to be very much more mature than the ATM technology.

Fig 15.3 graphically represents these component classes as calculated by TITAN. One must note that for servers and mature infrastructure, i.e. for classes three and four the reference year of prices is 1994, hence 1995 prices have already eroded. Another thing to remember is the basic requirement for the price erosion to come true. In order for prices to erode, a mass market must be successful. If different technologies are not accepted by the markets and hence not sold in large quantities, prices will not erode.

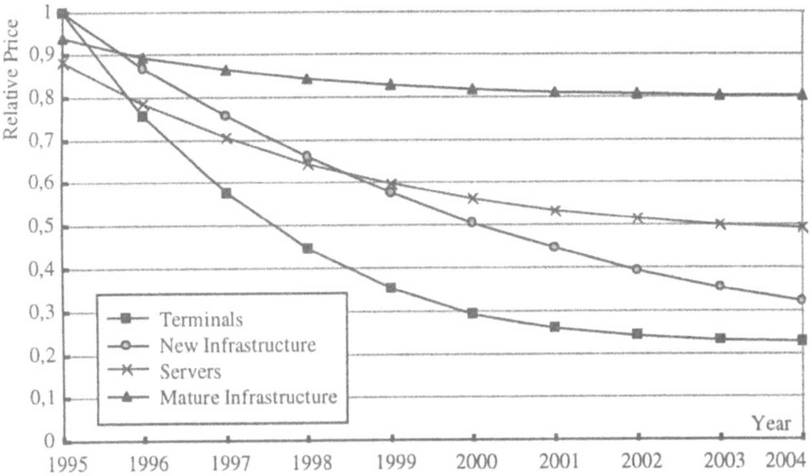

Fig 15.3 Different component classes used in modelling and their relative price evolution over time.

Most of the reference prices for components in years 1994 or 1995 have been taken from the TITAN cost database. This database has been gathered by the TITAN consortium and it does not represent any specific vendor, but the general price level in Europe. For the case of ATM equipment and some other components, prices have been taken from corporate service or price lists of other vendors. Due to the sensitive nature of pricing information, the complete price list cannot be printed here.

15.6 RESULTS

This chapter presents the basic results from the study. Since the TITAN tool that has been used to calculate the results is a product of a European Union project, the monetary unit used is the ECU. The following table lists exchange rates used to convert currencies for calculations.

Table 15.3 Exchange rates applied in calculations. The values correspond to the situation in autumn 1995.

ECU	*USD*	*Fim*
1.00	1.34	5.70

Another important economic parameter used in calculations is the discount rate. For the discount rate 7.5 % has been used, which is the default value in TITAN. It is considered to be the average effective discount rate in the EU by the TITAN consortium.

15.6.1 IRR, NPV and payback period

The three most important economic figures available from the TITAN tool are internal rate of return (IRR), net present value (NPV) and payback period. Of these three NPV is often considered to be the best indicator of whether investment should be made or not [14]. The NPV tells the value of the project in today's money. The IRR is closely related to NPV. It is defined as the discount rate which brings the NPV of the project to zero. If the IRR is higher than the opportunity cost of capital required for the project, the project is viable. The interpretation of IRR is this simple as long as the cash flow of the project behaves monotonically. However, if the cash flow goes from negative to positive or vice versa more than once during the project's life, the IRR is not any more unambiguous. This can lead to wrong conclusions and hence one must be careful when interpreting IRR figures. The NPV does not have this same unpleasant property, which is one of the reasons why it is preferred over IRR. The payback period has the strength of being very esoteric. It simply tells when the cash balance of the project turns positive. Once the earnings are larger than the investments, the payback period has been reached.

Table 15.4 tabulates all the above indices for all analysed network models and service combinations. The public network operator (PNO) results are presented at the top of the table and cable operator results at the bottom. The results for VOD service only, Internet access only or combined service are presented, each where relevant. Internet access service in the public switched telephone network (PSTN) network generates positive cash flow every year within the study period, hence IRR is not applicable for this service.

This table contains all of the basic results of this study at a glance and is therefore worth more detailed investigation. The payback period calculation in TITAN starts from the end of the first year, i.e. when the first closing of accounts is made, not from the calendar start of the project. This means that if study period is ten years, as it is in this study, the payback period TITAN reports can not exceed nine years.

From the results it can be clearly seen that narrowband services belong to a different category than other services when it comes to profitability. PSTN Internet access breaks even already during the first year. This is because required investments for starting the service are minimal. Integrated services digital network (ISDN) and IDS Internet access services als‘ show very good results. Out of these two, the IDS Internet is slightly better. The advantage of IDS is mainly due to the assumptions made. Both technologies offer 64 kbit/s access, but in the case of a cable operator it was assumed that basic IDS is already available while for the case of the ISDN the subscriber line will be updated from PSTN to ISDN. For ISDN a terminal adapter for each subscriber was also an assumed investment to support the customer's old analogue telephone.

For wideband technology asymmetric digital subscriber line (ADSL) seems to be the best when both VOD and Internet access services are offered. High speed digital subscriber line (HDSL) is slightly worse due to higher initial investments and hybrid fibre coax (HFC) comes in the third place. It is notable that these three are the only high bit rate technology/service combinations which break even during the study period. If either a faster technology is exploited or only one of the services is offered, the payback period exceeds the study period. The order of preference changes for these technologies if only the VOD service is considered. In that case HFC changes position from third to number one. The main reason for this is that PTO network technologies, i.e. ADSL and HDSL, offered wideband Internet access whereas the community antenna television (CATV) network offers only narrowband IDS Internet access. When revenues from Internet access are missing, PNO cases suffer relatively more than cable operator cases.

Table 15.4 IRR, NPV and payback period figures for all analysed cases as calculated by the TITAN tool. TITAN starts payback period calculation from the end of the first year, hence maximum payback period reported is study period minus one, i.e. in this case nine years.

Service	*IRR*	*NPV (kECU)*	*Payback period (years)*
PSTN Internet	N/A	2,060	< 1.0
ISDN Internet	45 %	1,490	6.2
2 Mbit/s VOD and Internet (HDSL)	12 %	1,490	7.5
2 Mbit/s VOD (HDSL)	-18 %	-7,130	> 9
2 Mbit/s Internet (HDSL)	5 %	-8,20	> 9
2 Mbit/s VOD and Internet (ADSL)	15 %	2,230	7.3
2 Mbit/s VOD (ADSL)	-17 %	-6,300	> 9
2 Mbit/s Internet (ADSL)	7 %	-40	> 9
6 Mbit/s VOD and Internet (ADSL)	2 %	-3,000	> 9
6 Mbit/s VOD (ADSL)	-21 %	-11,670	> 9
2 Mbit/s CATV VOD (HFC) and 64 kbit/s IDS Internet	10 %	370	8.5
2 Mbit/s CATV VOD (HFC)	-4 %	-1,720	> 9
64 kbit/s IDS Internet (CATV network)	106 %	1,970	1.8
6 Mbit/s CATV VOD (HFC) and 64 kbit/s IDS Internet	4 %	-930	> 9
6 Mbit/s CATV VOD (HFC)	-7 %	-3170	> 9

Both broadband technologies are unprofitable with all service combinations. If both services are offered their joint result is not very unprofitable, but if only VOD is in use ADSL gives the worst result for all analysed cases. Plain VOD service with broadband HFC technology is also unprofitable, but it is recognisably better than ADSL. This is explained by the high investment cost of broadband ADSL technology.

15.6.2 Cash balances

The economic figures presented in the previous chapter showed the overall project profitability in very concise form. To get more information about how a project behaves over its lifetime something more is needed.

Cash balance is a commonly used figure in economics. Cash balance can be presented as a function of time and thus it gives information about different phases of the project. One reason for the wide use of cash balance is probably its intelligibility. With a little simplification it can be said that the cash balance is the amount of money on the bank account of the project. When the cash balance turns positive the project has paid off, i.e. the payback period of the project has been reached.

The cash balance also tells another very important fact about a project. The minimum value of cash balance is the amount of financing the project needs. This is often a critical factor when projects are planned to be run with venture capital. If the investments in the project can be spread so that the minimum value of cash balance becomes smaller, less money is needed from financial institutions yielding lower payable interest and easier negotiations to get the loan.

The following three diagrams depict cash balances for all analysed wideband and broadband cases. They have been grouped according to the service combinations. In Fig 15.4 are shown the cash balances for combined VOD and Internet access services. Fig 15.5 shows the situation when only VOD is provided and Fig 15.6 has curves for the Internet only case.

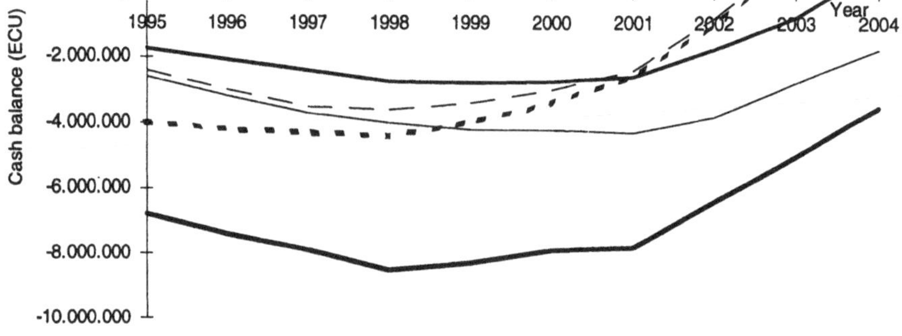

Fig 15.4 Cash balances for combined VOD and Internet access services.

From the previous figure a clear division between wideband and broadband services can be seen. Wideband services do break even, while broadband services do not. Particularly the broadband service in a PNO network suffers from high initial investments. An important observation is that the minimum value of cash balance for broadband ADSL is approximately twice as much as for the next highest. This implies high stakes. If the project is a failure, the loss will be large.

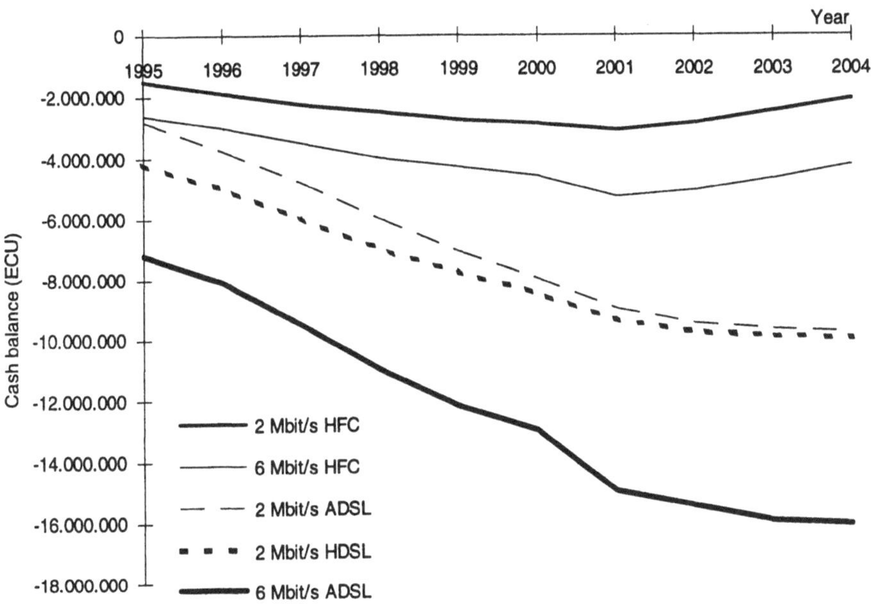

Fig 15.5 Cash balances for different technologies when only the VOD service is subscribed.

The cash balance diagram for the VOD service only shows a detail which quite well describes the results of the whole study when talking about VOD. In the chart the positive value axis is not needed for the cash balance. None of the different technical solution is able to yield positive cash balance within the ten year study period. The broadband ADSL network can not even turn the cash balance curve upward, while other technologies start to head upward around year seven or eight.

The diagram of cash balances for cases where only Internet access service is provided looks very different from the other cash balance diagrams. The distinction is also clear between narrowband and wideband cases. Narrowband services hardly bring about negative cash balance, while wideband services do not break even. Wideband services are, however, just about to break even and it can be said that for both ADSL and HDSL technology in a PNO network the Internet access payback period is approximately ten years.

Internet access in a PSTN network is a good example of how new services in the telephone network can work. Requiring practically no investments at all a new service, i.e. a new product, for the PNO is created. This new product is capable of creating a new revenue source, which did not exist earlier. Hence, a telecom operator can increase revenues from the infrastructure that has been built for other purposes without actual new investments and hence without real risk.

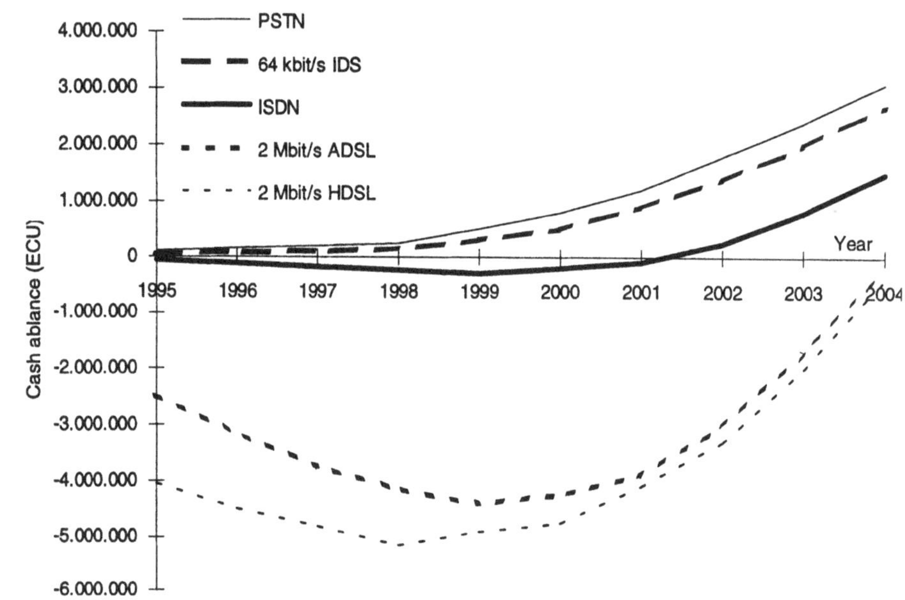

Fig 15.6 Cash balances for different technologies when only Internet access service is subscribed.

15.6.3 Overall investments

In addition to the overall profitability of constructing a network for new services, operators are also interested in how large the investment is in absolute terms. The size of the investment is an indicator of the possible maximum loss in the worst case scenario. Table 15.5 tabulates overall investment, investment per user and running costs in all analysed cases as calculated by TITAN. All figures are discounted values over the ten year study period.

One of the most interesting figures in Table 15.5 is the investment per user. Before drawing any conclusion from these figures, it must be clarified how TITAN calculates it. In TITAN the investment per user is derived from the cumulative discounted investment. Accumulated discounted investment over the study period is divided by the number of connected users at the end of the project. This means that in the investment per user figure both discounting and the price erosion of components over time is taken into account.

Table 15.5 Investments and running costs as calculated by TITAN. Expressed in kECUs, except investment per user which are in ECUs.

Service	*Overall investment*	*Investment per user*	*Running Costs*
PSTN Internet	310	50	80
ISDN Internet	1,850	310	400
2 Mbit/s VOD and Internet (HDSL)	9,060	1,500	2,690
2 Mbit/s VOD (HDSL)	9,060	1,500	2,690
2 Mbit/s Internet (HDSL)	9,060	1,500	2,690
2 Mbit/s VOD and Internet (ADSL)	8,730	1,450	2,460
2 Mbit/s VOD (ADSL)	8,730	1,450	2,460
2 Mbit/s Internet (ADSL)	8,730	1,450	2,460
6 Mbit/s VOD and Internet (ADSL)	13,820	2,280	4,090
6 Mbit/s VOD (ADSL)	13,820	2,280	4,090
2 Mbit/s CATV VOD and 64 kbit/s IDS Internet	5,130	850	1,400
2 Mbit/s CATV VOD	4,430	730	1,250
64 kbit/s IDS Internet (CATV network)	700	120	140
6 Mbit/s CATV VOD and 64 kbit/s IDS Internet	7,280	1,200	1,990
6 Mbit/s CATV VOD	6,580	1,090	1,840

When looking at the investment per user of Internet via PSTN, it becomes very clear why it gives such good economic results. Investments are negligible. For Internet via ISDN investments per user are already significantly higher. This is mainly due to the upgrade of PSTN access to ISDN access. In the CATV network narrowband Internet means less investments than in the ISDN case, even if the level of service is similar. The explanation for this is that the cable phone basic delivery system is assumed to be used for free, since the operator already has invested in it for telephony service. Also the investment of a terminal adapter for analogue telephony is not required with IDS technology.

On the other hand investments per user in wideband and particularly in broadband networks are heavy. They range from 1,000 ECUs to over 2,000 ECUs. One must also remember that these figures do not incorporate any cabling. If operators are required to build new cabling and particularly new ducting, the investment per user becomes even larger. An often quoted figure in telecommunication journals for maximum investment per user for a viable VOD system is around 1,000 USD, excluding the STB. This is in line with the fact that the VOD systems of this study do not pay off.

15.6.4 Investment breakdown

From Table 15.5 it becomes clear that the investments in wideband and broadband networks are too heavy to produce viable business cases. The question is, where is the capital located in the network? If the maximum investment per user can be around 1,000 USD, where is it exceeded? To answer these questions a cost breakdown is made.

The following tables represent cost breakdown for two situations. One must note that in these tables investment per user is calculated on a different basis than in Table 15.5. Here price evolution of components is not taken into account. Instead, these figures represent the investment per user connected to the network in the first year. For those who join later, i.e. most of the customers, the investment is smaller because prices have depreciated. Regardless of this fact these tables show where the bulk of investment per user is located.

Table 15.6 Cost breakdown of 2 Mbit/s HDSL architecture.

Investment	*Cost*	*Sharing*	*Cost/customer*
Server/gateway	450 kECU	6,000	75 ECU
ATM switch	90 kECU	6,000	15 ECU
STM-16 ring equipment	110 kECU	6,000	18 ECU
STM-1 access	10 kECU	250	40 ECU
ATM access MUX	9 kECU	250	36 ECU
HDSL modem pair	2 kECU	1	2,000 ECU
STB	0.8 kECU	1	800 ECU
	Total investment per customer		*2,984 ECU*

Table 15.7 Cost breakdown of 6 Mbit/s ADSL architecture.

Investment	*Cost*	*Sharing*	*Cost/customer*
Server/gateway	900 kECU	6,000	150 ECU
ATM switch	135 kECU	6,000	23 ECU
STM-16 ring equipment	1,500 kECU	6,000	250 ECU
ATM MUX	14 kECU	250	56 ECU
STM-1 access	20 kECU	63	317 ECU
ATM access MUX	10 kECU	63	159 ECU
ADSL modem pair	4,000 ECU	1	4,000 ECU
STB	800 ECU	1	800 ECU
	Total investment per customer		*5,754 ECU*

As can be seen, the investment per user lies in the access technology and the STB. Even if the servers and the synchronous digital hierarchy (SDH) technology in the upper parts of the network are expensive, their contribution to investment per user is small. For the cable network the situation is very similar. Cable modems plus STB dominate investments.

Hence, if the investment per user must be lowered, it is the access technology which one must look at. Together they represent almost 90 % of the investment. The price erosion for these components is assumed to be high, as shown in Fig 15.3, but even so they are the most costly part of the network also at the end of the study period.

15.7 SENSITIVITY ANALYSIS

Making forecasts of the viability of new communications technologies and new applications includes a considerable number of assumptions. Not all of these assumptions can be thoroughly justified. Hence it is necessary to perform sensitivity analysis to see how key results of the study change as a function of key assumptions made.

Assumptions include the new technologies, which are themselves uncertain. Even more, the assumptions include expectations about the behaviour of people. We cannot say for sure that the price of an ATM switch is going to come down, say by 15 % per year, but it is most likely going to be something in that range. On the other hand, if we claim that people will spend four hours per month on the Internet, we can go wrong proportionally by much more than with the price of an ATM switch. For example the development of integrated circuitry has been predicted amazingly accurate by Moore's law, but who could have foretold the popularity of World Wide Web few years ago, let alone predict a formula for its growth. Gordon Moore, co-founder of Intel Corporation, stated in the early sixties that the amount of information storable on a given amount of silicon roughly doubles every year. This is known as Moore's law and it held until the late seventies, after which the doubling period slowed to one and a half years [15].

Based on this reasoning sensitivity analysis has been performed mostly for the revenue side of the economic models where the human factors lie. Namely five parameters have been studied:

1. Price of a VOD recording
2. Number of recordings ordered per month
3. Wideband Internet usage fee per minute
4. Wideband Internet usage hours per month
5. Delayed project start.

The last one of these parameters is different from the others. It is included to see if it would be advisable for eager operators to wait for two years for technology to mature and hence for prices to come down before investing in new technology.

Sensitivity analysis results are presented as a set of curves with the variable under study on the x-axis and either payback period or net present value on the y-axis. In each figure the nominal value of the variable under study, i.e. the value which was used to calculate basic results, is indicated by a dotted line. In some figures two dotted lines appear, for example if narrowband and wideband services have had different nominal values.

15.7.1 Price of a VOD recording

The basic prices for wideband and broadband recordings were assumed to be 1.3 ECU and 1.9 ECU respectively. Since these tariffs were derived from existing video rental prices in order to make VOD competitive, they can be considered to be rather low for the new service. Hence in sensitivity analysis the price range for VOD has been extended more towards higher prices than towards lower prices. The selected range goes from 1 ECU per recording to 4 ECUs per recording. It is possible that due to the competition with video rental, the price of VOD recording goes below even 1 ECU per recording. Since already 1 ECU per recording yields unacceptable economic results for most technologies, lower values have not been considered.

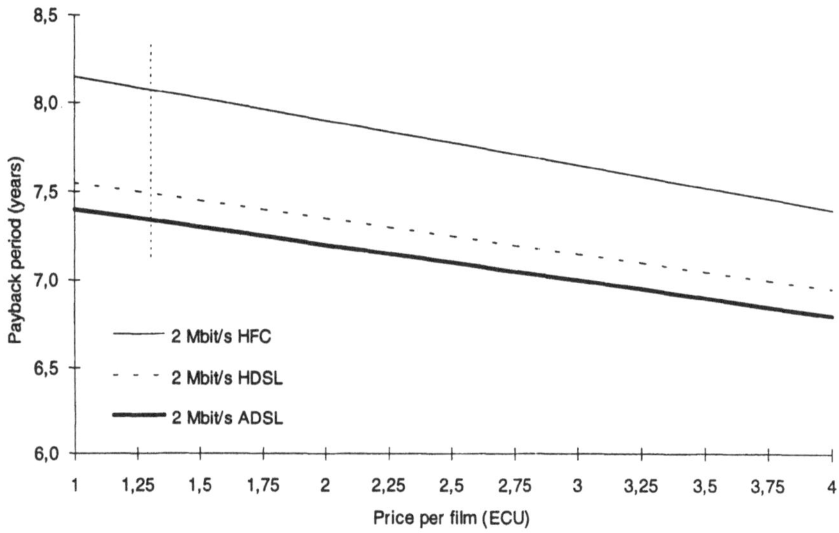

Fig 15.7 Payback period as a function of recording price when both VOD and Internet access services are provided. For broadband services the payback period was greater than the given study period and hence not applicable.

In Fig 15.7 the payback periods for broadband services cannot be seen because even at 4 ECUs per recording the payback period exceeds the study period. This figure, however, tells a lot about the situation. The payback period is almost insensitive to the price of a VOD recording. When the price of a single recording goes up by a factor of four, the payback decreases only by one year, or even less. A good reason can be found for this behaviour.

If we calculate how much revenue the operator collects from transmitting one bit of information for VOD service and for Internet access service and compare these figures, an interesting observation can be made. Let us assume for this purpose that the average recording lasts for one and a half hours. The total number of bits transmitted during one recording is

$$1.5 \cdot 60 \cdot 60 \cdot 2 Mbit/s = 1.13 \cdot 10^{10} \ bits\ per\ recording$$

With the basic price of 1.33 ECU per recording this yields

$$1.33 \ \mathrm{ECU} / 1.13 \cdot 10^{10} = 1.17 \cdot 10^{-10} \ \mathrm{ECU}/bit$$

Comparable revenue per bit for the case of 2 Mbit/s Internet access is

$$\frac{0.14 \ \mathrm{ECU}}{2 \ Mbit/s \cdot 60s} = 1.11 \cdot 10^{-9} \ ECU/bit$$

Now these two figures can be compared.

$$\frac{\text{Internet bit price}}{\text{VOD bit price}} = \frac{1.11 \cdot 10^{-9} \ \mathrm{ECU}}{1.17 \cdot 10^{-10} \ \mathrm{ECU}} \approx 9.5$$

This shows that the operator gets almost ten times more revenue from transmitting one bit for the Internet access service than for the VOD service! Hence it can be concluded that most of the revenues come from Internet access and VOD can be considered to be a supplementary service for the moment. This is the reason why the payback period is almost insensitive to VOD recording price. This same phenomena can be seen in all of the sensitivity analysis figures in this chapter where VOD service is concerned.

15.7.2 Number of VOD recordings ordered per month

The default value for the average number of recordings ordered per month was derived from video rental statistics. In derivation of this value it was already assumed that VOD customers are probably more interested in movies than the average VCR owner. To see what happens if they order significantly more recordings, a range from one recording per month up to ten recordings per month has been studied. Fig 15.8 shows how the NPV changes as a function of the number of recordings ordered per month.

In Fig 15.8 a threshold value for the profitability of broadband HFC is seen. After six recordings per month the net present value becomes positive, which is considered to be the requirement for sensible investment. It also means that at this point the internal rate of return for the project is equal to the discount rate, i.e. 7.50 %. Another interesting development is the behaviour of broadband ADSL. It cannot turn NPV positive and its NPV when ten recordings per month is ordered equals roughly the NPV of broadband HFC with only one recording per month ordered.

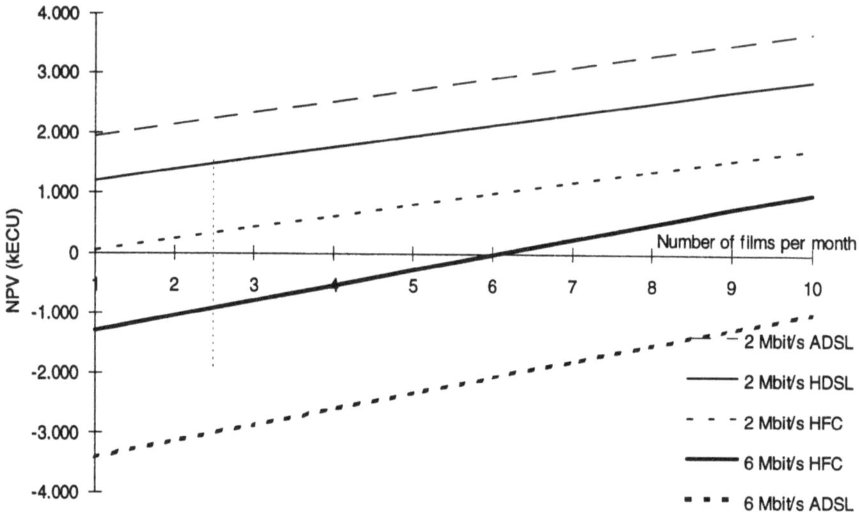

Fig 15.8 Net present value as a function of the number of recordings ordered per month when both VOD and Internet services are provided.

15.7.3 Wideband Internet usage fee per minute

For the wideband and broadband Internet access the usage fee per minute was assumed to be twice that of the ISDN service for which real tariffs already exist. This reasoning gave 0.14 ECU/min for the basic tariff. Doubling of tariffs when the bit rate goes to thirty-fold can be considered a rather moderate increase. For this reason the tariff range for sensitivity analysis has been extended from 0.075 ECU/min up to 0.375 ECU/min. Fig 15.9 shows NPV as a function of per minute fee for cases where wideband or broadband Internet access is provided along with VOD service.

It can be seen that the effect of the Internet usage fee on the overall profitability is on a different scale to the effect of a VOD recording price. When the usage fee per minute doubles, the NPV of the project triples.

With higher Internet access tariffs also the broadband ADSL technology case is, for the first time, able to break even. If the broadband Internet access per minute charge is 0.25 ECU/min, i.e. approximately four times higher than the current ISDN Internet tariff, broadband ADSL yields zero NPV. In the basic calculations it was assumed that wideband and broadband Internet access services bring the same revenue. It can be concluded that if broadband Internet access brings higher revenue, it indeed has a big effect on the project's profitability. Once more accurate information about wideband and broadband Internet access fees becomes available, calculations need to be re-performed with new initial values.

Wideband HDSL and ADSL technologies are firmly above the zero NPV level with the basic per minute tariff. For them the zero crossing threshold happens when per minute tariffs come into the range of current ISDN Internet tariffs.

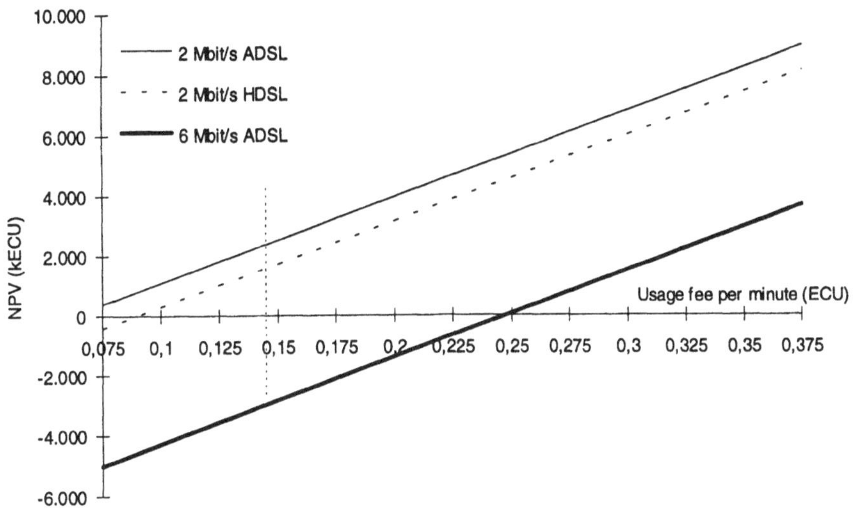

Fig 15.9 Net present value as a function of wideband Internet per minute fee when both VOD and Internet services are provided.

15.7.4 Internet usage hours per month

The average number of Internet usage hours per month is probably the parameter that involves the most uncertainty in this study. A recently performed user survey [12] in Finland suggests that those who use the Internet today, spend significantly more time on the Internet than assumed here. Currently average usage may be even up to 20 hours per week. There are, however, several reasons why this figure should not apply to this study. As the same study shows the majority of the current Internet users use the service from schools, universities or from work, which means that they do not themselves pay directly for the service. Once the per minute charge applies to the individual user he or she will more carefully choose what to do on the Internet. Also the target user group in this study are the normal households, while the bulk of the current Internet user are oriented towards electronic data processing (EDP) because of their work or field of study. One must also note that the survey referred to here was conducted in Finland, and Finns are among the most active Internet users in the world.

If consumers are going to use any commercial service, be it a source of information or a barber shop, it has to provide some added value. If the value added compared to the cost is high, the service is likely to be used a lot. Can the Internet provide high added value? The answer depends on the content of the Internet, which itself is a source of uncertainty. Internet is a new medium which today provides certain, not too well defined, content, but the content of tomorrow is a big question mark. Valuable services, like home banking, have already emerged, but something might be in store which can radically change the role of the Internet. The trigger for this progress could be the development of secure anonymous transactions on the Internet.

For these reasons a range of one decade for average Internet usage hours has been studied, extending from two hours per month up to 20 hours per month. Fig 15.10 shows the NPV as function of monthly Internet usage hours.

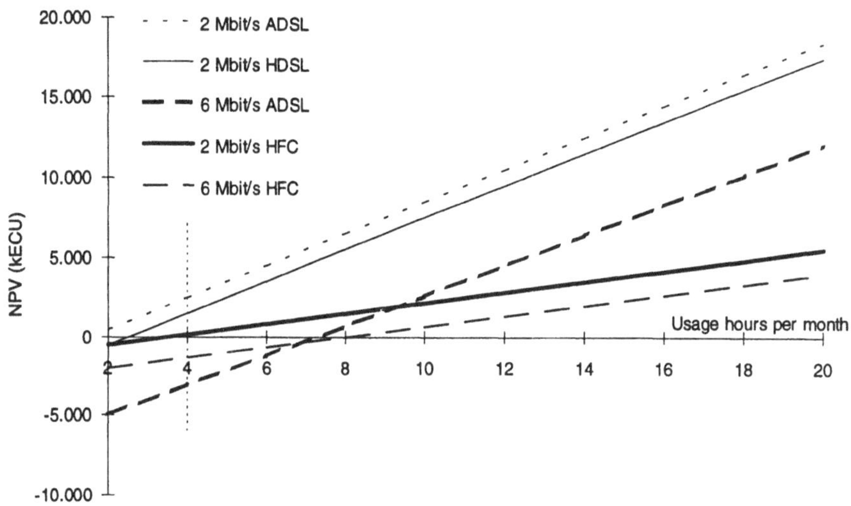

Fig 15.10 Net present value as a function of monthly Internet usage hours when both VOD and Internet access are offered.

For wideband PNO technologies the situation looks very good. They stay above the zero NPV level for almost all studied values, and yield an NPV almost twice as big as the overall investment when usage of twenty hours per month is attained.

For broadband services and wideband HFC the economic situation is very interesting around eight hours of usage per month. Broadband ADSL intersects broadband HFC, and the zero level, at seven hours average usage, and wideband HFC at nine hours average usage per month. This level for average usage is possible and thus causes further uncertainty about preference between technologies.

Fig 15.11 further shows how the profitability of Internet access as the only service is affected by average usage hours per month.

In this figure also the narrowband Internet access options are included. They are all clearly above the zero level, even at the low limit of the sensitivity study range. This tells about the low risk involved with narrowband Internet service from operator's point of view. Even if the utilisation rate is really low, no losses will occur.

Another interesting detail is the intersecting points for wideband and narrowband services. At around seven or eight hours usage per month, wideband technologies get even with narrowband services, and beyond that they yield a better NPV. This is the same level of Internet usage as previously required for broadband ADSL to beat HFC technologies. Hence, it can be concluded that the average Internet usage rate is indeed a critical factor contributing to the viability of different technologies and service scenarios.

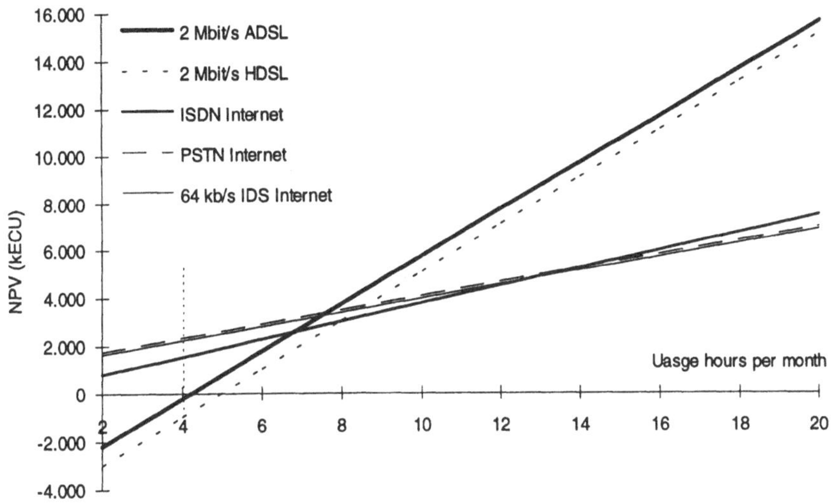

Fig 15.11 Net present value as a function of monthly Internet usage hours when only Internet service is offered.

15.7.5 Delay in project start

A completely different factor effecting a project's profitability from those discussed so far, is the starting year of the project. The technology used to build high bandwidth multimedia networks is very new, and in the beginning of its life cycle. As learned in chapter 15.5, the price of this technology rapidly decreases and this price decrease is modelled in the TITAN tool.

For many operators the question is not whether or not to invest in new network technology, but when to invest. It is obvious that change towards wideband and broadband technologies will be made, but should it be done today or later. To see how key financial parameters are affected by a possible delay in project start, the situation of starting the project in 1997 has been studied. Relative component price erosion during the two year delay can be seen in Fig 15.3. Results for combined VOD and Internet access services, and ISDN Internet access alone are tabulated in Table 15.8. The PSTN Internet case has not been included, because PSTN technology is mature and investments already in year 1995 are negligible. Thus, a delay provides no advantage.

It can be seen that the delay in project start truly improves the profitability of different scenarios. IRR is improved on average by 10 % and the payback period is reduced by more than a year. If the project is started in 1997 all different technologies, including broadband, are able to break even within the ten year study period.

Although it looks like a very tempting option for an operator to wait for two years before starting, there is also another side to the matter. If the operator is not in monopoly position, the customer base is shared with competitors. Since the penetration of any service in the market is the sum of all service provider's penetrations, there is a possibility that those who are first on the market gather most of the potential customers. Even if it is cheaper to build the network two years later, it may not advisable to do so if potential customers are already being served. On the other hand, if everybody waits, there will not be mass market for new technology, which is the basic requirement for price erosion of network components to take place.

Table 15.8 The effect on key economic results of delaying the project start by two years.

Service	*Project start year*	*IRR*	*NPV (kECU)*	*Payback period (years)*
ISDN Internet	1995	46 %	1,490	6,2
	1997	62 %	1,610	5,4
2 Mbit/s VOD and Internet (HDSL)	1995	12 %	1,490	7,5
	1997	20 %	3,370	6,0
2 Mbit/s VOD and Internet (ADSL)	1995	15 %	2,230	7,3
	1997	29 %	4,330	5,4
6 Mbit/s VOD and Internet (ADSL)	1995	2 %	-3,000	>10
	1997	8 %	370	8,2
2 Mbit/s CATV VOD and	1995	10 %	370	8,5
64 kbit/s IDS Internet	1997	19 %	1,520	6,7
6 Mbit/s CATV VOD and	1995	4 %	-930	>10
64 kbit/s IDS Internet	1997	12 %	770	8,1

15.8 RISK ANALYSIS

The previous four subsections studied what kind of effect the VOD recording price, number recordings ordered per month, Internet usage hours per month and Internet per minute fee have on overall project profitability. To see what kind of combined effect they have when changed simultaneously, a ten thousand round Monte Carlo simulation was performed. In a Monte Carlo simulation random numbers are used to measure the effects of uncertainty in input variables on output variables. For each input variable a probability distribution is assigned and during each simulation round a random value is drawn from this distribution. Output values are recorded after each round, and after a large number of rounds statistical behaviour of output variables can be determined. Since this kind of simulation is very time consuming, it has been performed only for one case. According to basic results 2 Mbit/s ADSL was the best technology candidate for new services, and hence it has been selected for the Monte Carlo study case. The simulation was performed by the use of Crystal Ball® in conjunction with TITAN. Probability distributions for different variables have been selected to reflect the basic assumptions presented in section 15.3.2 and the sensitivity analysis assumptions presented in this chapter. Table 15.9 tabulates distributions and their parameters.

Table 15.9 Probability distributions used in Monte Carlo simulation.

Variable	*Parameters*
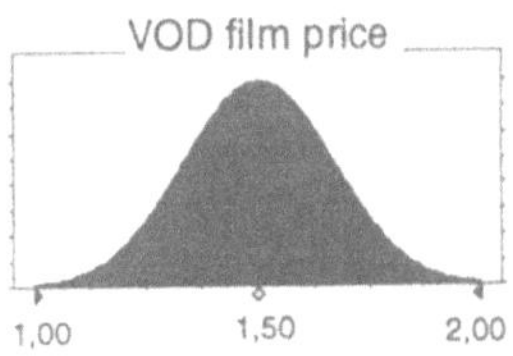	Distribution type: Normal. Mean = 1.5 ECU/recording. Variance = 0.065. 95 % confidence interval = [1.0, 2.0]. Correlation = -0.5 with recordings/month.
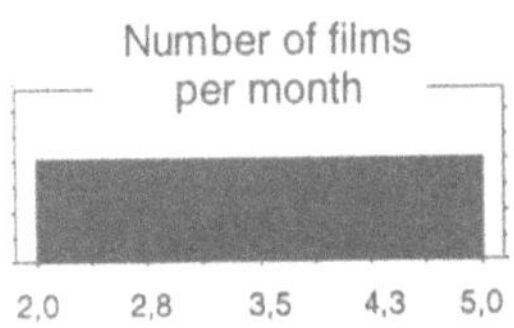	Distribution type: Uniform. Min. = 2. Max. = 5. Correlation = -0.5 with VOD recording price.
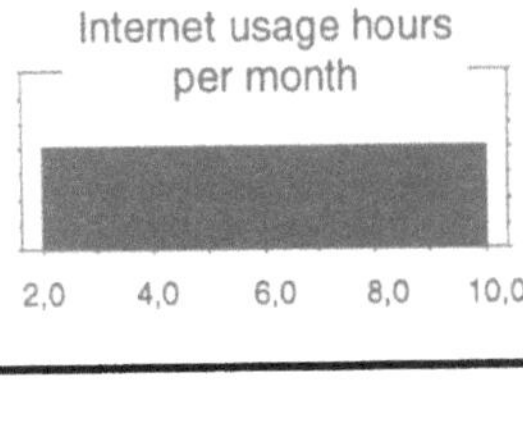	Distribution type: Normal. Mean = 0.14 ECU/min. Variance = 0.0016. 95 % confidence interval = [0.06, 0.22]. Correlation = -0.5 with usage hours/month.
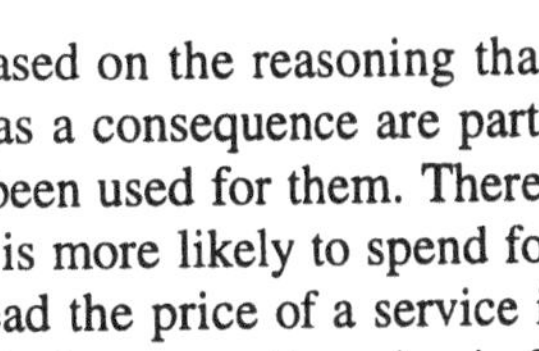	Distribution type: Uniform. Min. = 2. Max. = 10. Correlation = -0.5 with minute fee.

Based on the reasoning that service usage rates are affected by human factors and as a consequence are particularly difficult to forecast, a uniform distribution has been used for them. There is no proper justification to claim that the average user is more likely to spend four hours per month on the Internet than six hours. Instead the price of a service is likely to be derived from competing services and the further away the price is from the price of the competitor, the less likely that price is to be used. Hence, a normal distribution has been used for service prices.

The price of a service and the service usage rate cannot be independent. Basic price elasticity theory says that if service is cheap, it is going to be used more than if it is expensive. There are exceptions to this rule, but there is no reason to believe that VOD or Internet access service are one of the exceptions. Hence a correlation of -0.5 has been assumed between service usage rate and tariffs. A correlation of ±0.5 is generally considered to be mediocre, but not strong. When investigating the results of the simulation one must bear in mind that these distributions and their characteristics have been selected according to best understanding, but there is no concrete justification for them.

The probability distribution of the NPV for the wideband ADSL case after Monte Carlo simulation is presented in Fig 15.12. Table 15.10 lists statistical parameters for the distribution.

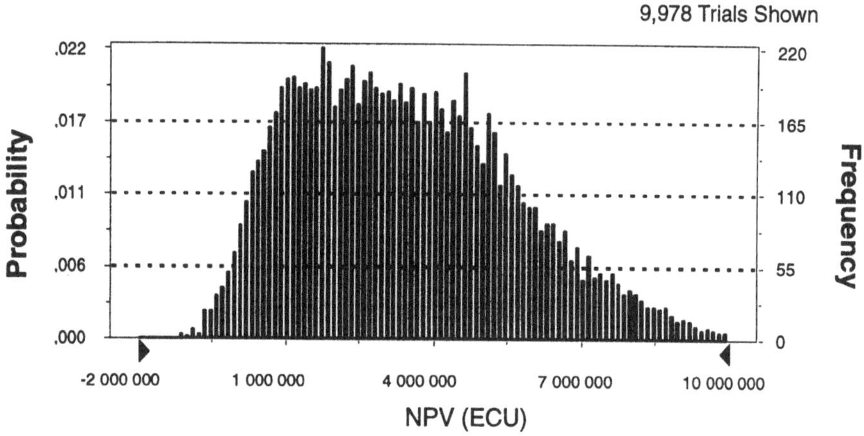

Fig 15.12 Probability distribution of the NPV after a 10, 000 round Monte Carlo simulation for wideband ADSL technology when both VOD and Internet access services are offered.

Table 15.10 Statistical parameters of the NPV probability distribution which is presented in Fig 15.12.

Mean	*3,467,000 ECU*
Median	3,252,000 ECU
Standard deviation	2,206,000 ECU
Variance	4.87^{12} ECU^2
Range minimum	-1,170,000 ECU
Range maximum	11,981,000 ECU

The evident feature of the NPV distribution in Fig 15.12 is its wide spread. The difference between minimum and maximum value is over 13 MECUs and standard deviation is over 2.2 MECUs. This implies that the business incorporates high uncertainty. Under basic assumptions the NPV showed a value of 2.23 MECU, but these results suggest that the actual realisation of NPV can be rather far away from the nominal value. Furthermore it must be remembered that this simulation takes into consideration only four of all the possible variables affecting the result. In this distribution the 50 % confidence interval for NPV extends from 1.66 MECU to 4.97 MECU. An important observation is however the fact that probability of getting a positive NPV under these assumptions is 97.7 %, i.e. a very high probability.

Another interesting result is the respective contribution of each variable to the uncertainty of the NPV. For this purpose Crystal Ball© calculates the contribution of each variable to the variance of the distribution. Table 15.11 shows the results.

Table 15.11 Proportional contributions of different variables to the variance of the NPV distribution.

Variable	*Contribution to variance*
Internet usage hours per month	92.2 %
Internet usage fee per minute	6.9 %
Number of recordings ordered per month	0.8 %
VOD recording price	0.1 %

The importance of the VOD service as a whole is even smaller than one might expect from the comparative revenue calculated previously in chapter 15.7.1. As expected, the average Internet usage hours dominate in this respect The uncertainty of the monthly Internet usage yields a wide distribution for the possible input values, and since the weight of Internet service on overall profitability is high, it has a major impact on the results. From here it can be concluded that in order to produce more precise forecasts about the profitability of combined Internet access and VOD service, one should pay special attention to the estimation of how much Internet will be used. If the probability distribution for Internet usage can be squeezed, e.g. after a comprehensive market survey, the final results will be considerably more accurate.

15.9 SUMMARY

The basic question asked in this study was whether or not new services are viable with the given network technologies. Unfortunately the answer is elusive. There are so many variables and assumptions contributing to the analysis, that an unambiguous answer cannot be given. Each party considering different options must go through the results and underlying assumptions to see if they fit their particular situation. For accurate results a case-by-case analysis incorporating specific circumstances is required.

Even so, certain general guidelines can be seen. The study also points out important factors to be considered when making decisions, and it serves as a way to gain further understanding of the fundamental questions involved. The framework of the questions was set to see how VOD and Internet access work out as single services and how their combination succeeds.

Perhaps the clearest result of the study applies to the VOD service. VOD, as the only service, was not profitable in any of the analysed cases within the study period. Both wideband and broadband scenarios examined by the study in either telecommunication or cable operator networks cannot achieve satisfactory economic results. Initial investments are too high, and even more important, the revenue from the VOD service is very low. If VOD is trying to compete with video rental by means of price, it cannot generate enough revenue to pay for the investments. HFC technology in a CATV environment gets closer to the break even point, while PNO technologies are highly unprofitable.

As the cost breakdown in chapter 15.6.4 shows, the vast majority of the per-user investment is tied to equipment in user premises and modems in the operator's end of the subscriber loop. To significantly lower the overall investment, the cost of this equipment must be reduced. Either a new cheaper technology must be developed, or the cost of current technologies, i.e. ADSL, HDSL, and cable modems must be substantially cut. Even a significant reduction of 50 % of these costs is not enough to turn plain VOD service to profit because revenues are so weak. It has been shown in chapter 15.7.1, that operators collect almost ten times higher revenue from transmitting one bit for Internet access service than for VOD service. To be profitable VOD needs revenues from other sources, possibly from advertisers. Alternatively, plain VOD can be realised only in specialised areas of well off customers, who do not care about the cost as long as the service is available. In other words the results of this study suggest that VOD is not feasible as the only service in the short term.

Internet access, instead, seems to have considerable potential. This service in a PSTN network pays off immediately, thanks to negligible required investments. ISDN and IDS options are not quite as good, but still very profitable. To further improve their profitability an integrated, and hence cheaper, adapter combining network termination, computer adapter, and adapter for analogue telephony would be useful.

Wideband Internet access does not quite pay off under the basic assumptions, but if Internet usage exceeds five hours per month or the per minute tariff is approximately 2.5 times the current ISDN tariff, the service breaks even. The situation is quite similar for both ADSL and HDSL technology, ADSL being slightly better. The potential of the Internet access service is based on its relatively high tariff as compared to VOD.

The profitability of the combined case, when both VOD and Internet access are provided, very much depends on Internet access. If Internet access sells well, the whole business does well. Under basic assumptions, all analysed wideband technologies, i.e. ADSL, HDSL and HFC break even. Broadband technologies that are suitable for PNO networks still remain unprofitable. The break even threshold for broadband ADSL is around eight hours average Internet usage per month, or alternatively, the Internet per minute tariff has to be four times the current ISDN tariff. In general the combination of Internet access and VOD service looks very promising, especially at wideband speed. The best technology candidate, according to this study, is wideband ADSL. HFC technology suffers from inability to support wideband Internet access due to lack of reverse channel capacity.

Other studies carried out on the subject, show rather inconsistent results [11,16,17]. Depending on who is behind the study, different technologies are reputed to be more cost effective and robust than the others. For example a BroadBand Technologies spokesman claims that FTTC is cheaper than HFC if actual service penetration is 15 % or higher, while in the same article Pacific Bell presents results that show HFC to be at least 30 % cheaper than FTTC [17]. Robert Olshansky in the Telephony magazine suggests a third technology called BDSL to be more cost effective than either of the main rivals, whereas Dr. Negroponte from MIT calls everybody short-sighted who does not build fibre all the way to the home regardless of the cost [18]. In general the only fact different parties debating the subject agree upon, is that the problem is obscure.

Vern Mackall from Northern Business Information states that: "In the longer term, the requisite core competency in broadband services will be one of marketing. Questions of technology and network architecture will be of little relevance" [16]. This can be very true for what was found in this study. If operators are capable of marketing Internet access service well enough, any of the analysed technologies is viable. Probably operators will start to market many other services in addition to VOD and Internet access to gain return from their investment. In this way, the utilisation rate of new, possibly expensive access technology, and at the same time the overall profitability, can be improved.

REFERENCES

[1] Saijonmaa, J, Tahkokorpi, M, Welling, I. Cost of Investment and Revenue Modelling and Analysis of Various Networked Multimedia Services in PTO and CATV operator Environments. *In: Proc TELECOM '95, Technology Summit, Geneva, Switzerland, October 3-11, 1995, 2, 629-633.*

[2] Welling, I. Economics of VOD and Internet access service upgrade. *In: Proc. Eurescom Workshop on Optical Multiservice Access Networks (WOMAN), Turin, Italy, September 23-24, 1996.*

[3] Schmoe, C, Spiegel, K, Steis, B. Interactive interest. *Telecom report international*, 18, (2), 1995, 6-9.

[4] Stordahl, K, Murphy, E. Methods for forecasting long term demand for wide and broadband services in the residential market. *IEEE Communications magazine*, 13, (2), 1995, 44-49.

[5] Zaganiaris, A, Gieschen, N, Olsen, B T, Stordahl, K, Tahkokorpi, M, Drieskens, M, Ajibulu, A. A methodology for achieving life-cycle costs of optical access networks-from RACE 2087/TITAN. *In: Proc. 11th Annual Conference European Fibre Optic Communications and Networks (EFOC '93), the Hague, the Netherlands, June 30-July 2, 1993, 2, 136-141.*

[6] Olsen, B T, Stordahl, K. Forecast of price development of network components, based on an extension of the learning curve model. *Telektronikk,* 90, (1), 166-172, 1994.
[7] RACE 2087/TITAN: final project report. Editor: Zaganiaris, A. Brussels, Belgium, 1996. (R2087/CNE/SCL/DS/L/020/b1).
[8] Fisher, D. (editor). European video software: Rentals stabilise as sales grow. *Screen Digest*, 25, (5), 1995, 105-112.
[9] Fisher, D (editor). Pay TV movie channels, 1: Western and Eastern Europe. *Screen Digest*, 25, (11), 1995, 249-252.
[10] Fisher, D (editor). Worldwide video software market: Patchy development. *Screen Digest,* 25, (11), 1995, 253-264.
[11] Olshansky, R. Broadband digital subscriber line: A full service network for the copper plant. *Telephony*, 24, (228), 1995, , 52-60.
[12] Sokala, H. Tällainen minä olen. *Helsingin Sanomat*, February 16, 1996, part D, page 1.
[13] Ferrero, U, de Bortoli, M, Myhre, D, Ims, L A, Olsen, B T, Mononen, J, Lähteenoja, M, Zaganiaris, A, Pousa, M, Szomolányi, K, Orfanos, I. Access Network Upgrade Strategies: mapping the European Diversity. *In: Proc. 7th International IEEE workshop on Optical Access Networks, Nuremberg, Germany, September 24-28, 1995, 4.2-1 - 4.2-10.*
[14] Brealey, R A, Myers, S C. Principles of corporate finance. McGraw-Hill, International edition, 4th edition, 1991, 11-125.
[15] Pucknell, D A, Eshragian, K. Basic VLSI design. Prentice Hall, International edition, 2nd edition, 1988, 2-3.
[16] Mackall, V. The Broadband Payback. *Telecom Perspectives*, 10, 1995, 109-117.
[17] Ellis, L. HFC, SDV cost differential still muddy. *CED*, 21, (10), 1995, 48-50.
[18] Negroponte, N. Special report: Germany moves ahead with fiber deployment. *Fiber Optics News*, 7, (1), 1996, 4.

Mixed residential and business area upgrades

16

Leif Aarthun Ims, Borgar Tørre Olsen, Dagfinn Myhre, Kjell Stordahl

The chapter addresses some of the economic implications of broadband upgrade strategies for the access network. The major contribution of the paper is a techno-economic evaluation of broadband access network upgrade alternatives for the mixed residential and business market. An aggressive asynchronous transfer mode *(ATM) based passive optical network upgrade solution is compared to a more conservative upgrade solution with enhanced copper technology combined with point to point fibre connections for the highest capacity demand. The network life cycle costs and revenues from the bearer services provided during a ten year upgrade period are taken into account. In addition the effect of competition on the overall budget is studied [1,2].*

16.1 INTRODUCTION

An increasingly more diverse market demand, both with respect to service types, terminal mobility and bandwidth requirements, is expected in the coming years [3-6]. Thus, the traditional wireline access networks of the operators are likely to be upgraded to future-proof fixed broadband access networks accommodating a wide range of demographic diversity. As such, the development of the existing narrowband copper based access network infrastructure to a broadband network represents one of the major challenges of the telecommunications network operators. Cost effective migration paths will be crucial in the highly competitive environment now emerging [2,7-9].

This chapter presents techno-economic analyses of some of the key aspects of access network upgrading for the mixed residential and business market, like technology choice, service penetrations and the effect of competition. These issues have been highlighted through a study of access network evolution alternatives for two different demographic areas over a ten year study period from 1995 to 2005. The emphasis is on the technological possibilities and their economic implications [10-12].

16.2 DESCRIPTION OF SELECTED ACCESS NETWORK AREAS

Areas representing quite different existing network situations have been selected in order to highlight some of the key issues of upgrade technology choices and strategies. Two different mixed demographic areas have been studied, one downtown, business customer dominated area and one suburban, residential customer dominated area. The two main differences between the areas are the housing structure and the distribution of the customer groups.

Downtown area: The first area under consideration is a downtown area, with apartment and business blocks. The area represents a typical central part of a city, and is a business dominated area, with 90 % business customers (small, medium or large businesses) and 10 % residential customers. The business customers are categorised according to the typical service connection of today: Small business: POTS; medium business: POTS and 64 kbit/s leased line; large business: 2 Mbit/s leased line. The small business and residential customers are co-located in apartment- and business blocks with between 20 and 30 potential customers. The medium and large business customers are co-located in blocks housing on average two businesses per block.

Suburban area: The second area is a mixed suburban area, with single house dwellings and offices. This is a residential customer dominated area, with 90 % residential and 10 % business. Both the business and the residential customers are located in individual houses.

In both areas one fourth of the business customers are assumed to be medium to large businesses. The spread of customer categories in the downtown and the suburban areas is shown in Fig 16.1 and Fig 16.2 respectively.

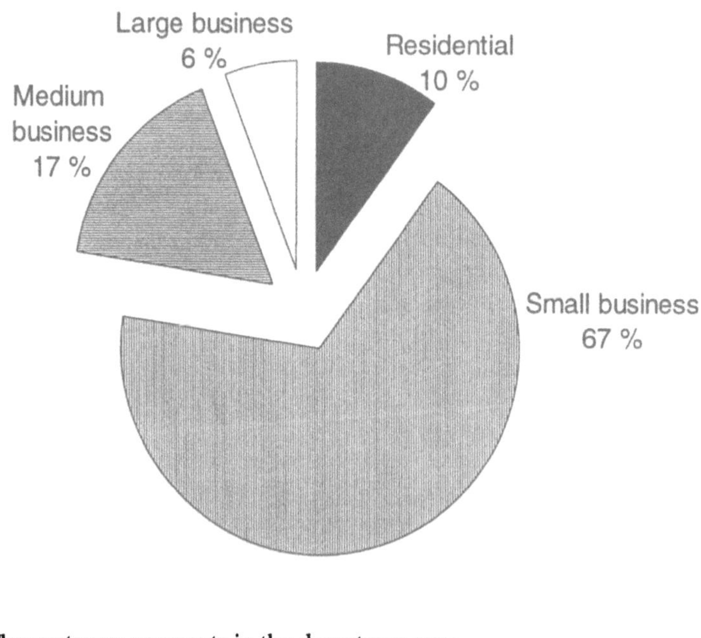

Fig 16.1 The customer segments in the downtown area.

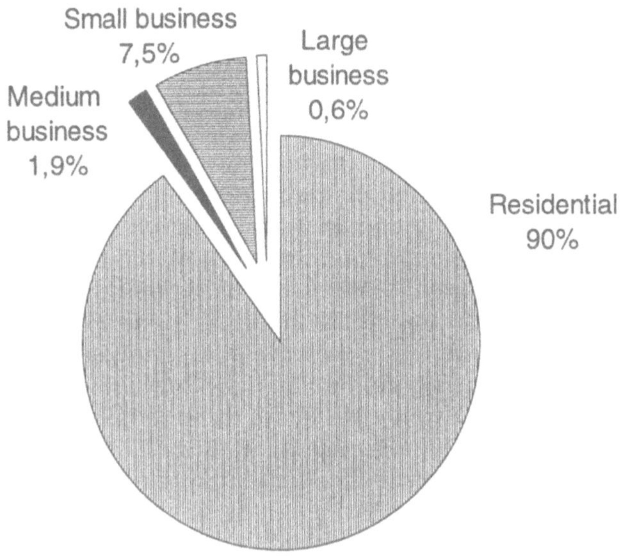

Fig 16.2 The customer segments in the suburban area.

In the downtown area one building on average occupies ~6 acres. The single house suburban dwellings on average occupies approximately 1 acre. The average loop lengths from the main distribution frame to the customer premises are ~500 m and ~550 m for the downtown and suburban areas respectively. The areas under consideration are approximately 1 km^2 in size. In the paper all results shown are for a subscriber density of 1000/km^2 unless otherwise stated.

Considered here is only the portion of the network between the local exchange (LEX) and the customer premises. In the feeder section of both areas an synchronous digital hierarchy (SDH) ring is assumed to already exist, connecting four so called service access points (SAPs) in a fibre ring structure. The SAP refers to the network localisation of the concentrator, as used in several European countries with modern access network infrastructures. The SAP as used in this study has 1,024 potential broadband customers connected. The concentrator in the SAP and the line cards are assumed to have been installed prior to the study period. For both areas the distribution part of the network (connected to the SAP) is assumed to have an existing modern, twisted pair based double-star copper infrastructure and a good quality duct layout.

16.3 EVOLUTION OF BROADBAND SERVICES

The provision of a set of bearer services in a mixed residential and business market has been the focus of this work. Only bearer services are addressed, and as such this relies on a transformation of higher level services into bandwidth requirements and respective bearer services. It is assumed that plain old telephone service (POTS), narrowband integrated services digital network (N-ISDN), 64 kbit/s and 2 Mbit/s leased line type bearer services are already offered over the public network operator's (PNO's) existing, modern copper based network.

For the residential and small business market POTS and N-ISDN are expected to remain as a basic set of services for the foreseeable future. The residential and small business market service set has been selected based on a comprehensive international Delphi survey, carried out within the project RACE (research for advanced communications in Europe) 2087/TITAN (tool for introduction strategies and techno-economic evaluation of access network) [13,14]. According to the survey, the relevant services are: N-ISDN, wideband (capacity ≤ 2 Mbit/s.) and broadband (capacity > 2 Mbit/s.). The forecasted penetrations of the services as a percentage of the residential and small business market are shown in Fig 16.3. Advanced asymmetric switched 2 Mbit/s services like service on demand (SOD) or video on demand (VOD) will be required by some customers. In addition, symmetric 2 Mbit/s or 8 Mbit/s bearer services will be provided to some residential and small-business customers.

For the medium to large business market, representative figures for the ranges of demand are selected from available statistical material. Three alternative bearer services have been considered: n·64 kbit/s, 2 Mbit/s and 8 Mbit/s, with POTS/ISDN included in the two last services. Estimated penetrations of the services for the study period, as a percentage of the total number of medium to large business customers, are given in Fig 16.4.

The effect of competition has been modelled by an independent linear reduction during the study period of the service penetrations of the business and residential market segments [12]. The loss of subscribers is given by the percentage lost for the two market segments at the end of the study period.

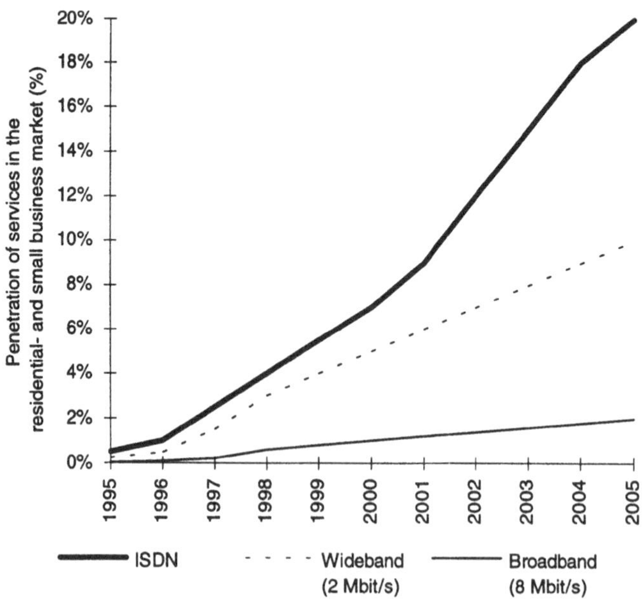

Fig 16.3 Penetration of services in the residential and small business market in the nominal scenario.

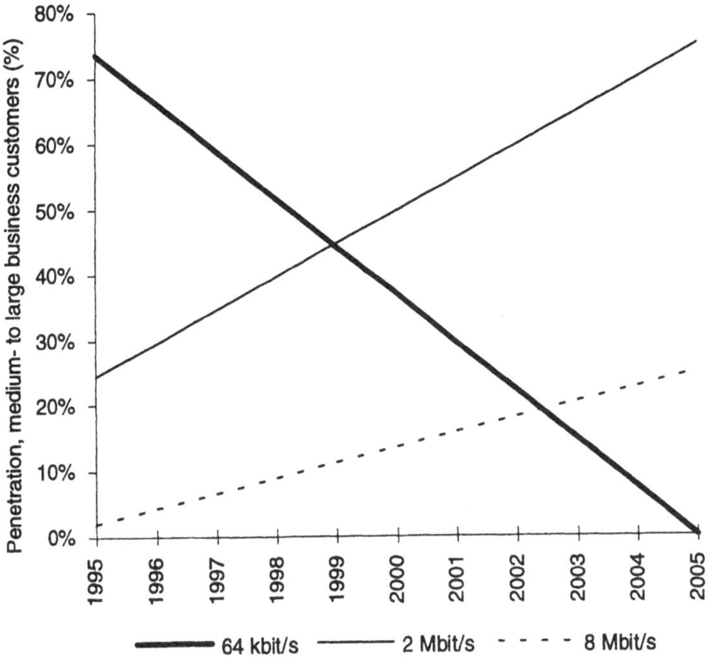

Fig 16.4 Penetration of services in the medium- to large business market, nominal scenario.

Bearer service tariffs have been included in the calculations, in addition to the service penetrations, in order to produce upgrade project revenues. The residential and small business market tariffs are based on the Delphi survey and represent what are considered to be European averages. For the medium- to large-business market, representative figures for the tariffs are selected from available statistical material.

The tariff structure implemented includes an evolution of the tariffs during the study period. The tariffs for all bearer services considered in the initial (1995) and final year (2005) of the study period are shown in Fig 16.5.

The tariffs used are the yearly subscription tariffs, and do not include traffic revenue. For the residential and small business market, the tariffs as a function of demand estimated from the Delphi survey for wide- and broadband connections have been derived [15,16]. The medium and large business customer tariffs have a stronger elasticity than the residential tariffs, due to the expected strong competition in this market segment: an annual decrease in tariffs by 10 % for n·64 kbit/s, 2 Mbit/s and 8 Mbit/s is assumed.

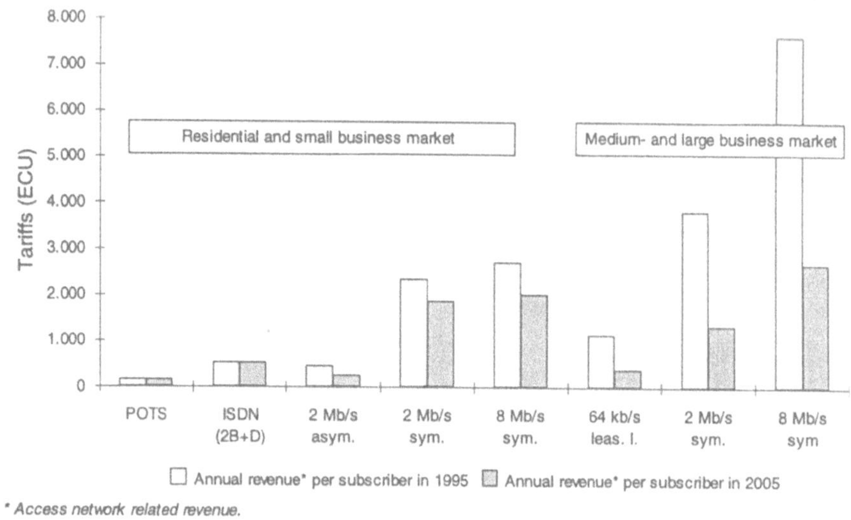

Fig 16.5 The undiscounted tariffs in 1995 and 2005 for the services considered.

16.4 DESCRIPTION OF TECHNOLOGY ALTERNATIVES

In the study two different architectures based on existing and emerging technologies have been considered, one enhanced copper architecture and one optical fibre architecture. The main characteristics of the two upgrade options are:

16.4.1 Enhanced copper architecture

This architecture represents the likely approach taken by a technologically conservative operator. It illustrates what is expected to be a minimum investment strategy in which as much as possible of the current infrastructure is kept during the upgrade period. POTS/N-ISDN and asymmetric switched broadband services are offered over the existing copper network. asymmetric digital subscriber line (ADSL) modems are installed for provision of 2 Mbit/s asymmetric switched services, and high bit rate digital subscriber line (HDSL) modems are used for providing the 2 Mbit/s symmetric services. Dedicated fibre connections are offered to provide the 8 Mbit/s symmetric services. Hence, only provision of the latter requires installation/changes in the outdoor cable plant. The enhanced copper architectures for the downtown and suburban areas are shown in Fig 16.6 and Fig 16.7 respectively. The main difference between the application of the conservative upgrade technologies in the downtown area and the suburban area is the much higher civil works cost of the fibre connections incurred by the individual housing structure of the suburban area.

16.4.2 Optical fibre architecture

This network architecture describes one probable choice of an aggressive operator with respect to new technology. It represents an extensive changeout of the existing network infrastructure. The network is upgraded to an integrated fibre based network offering all services except POTS and N-ISDN, which are kept on the copper network. Customers demanding broadband services are served via an ATM PON. The optical fibre architectures for the downtown and suburban areas are shown in Fig 16.8 and Fig 16.9 respectively. The ATM PON used has an optical split of 1:16, with a 622 Mbit/s downstream capacity and a 155 Mbit/s upstream capacity. Optical fibre is installed in the distribution part in either fibre to the building (FTTB) configurations (downtown) or fibre to the curb (FTTC) configurations (suburban). The optical network units (ONUs) are shared between customers.

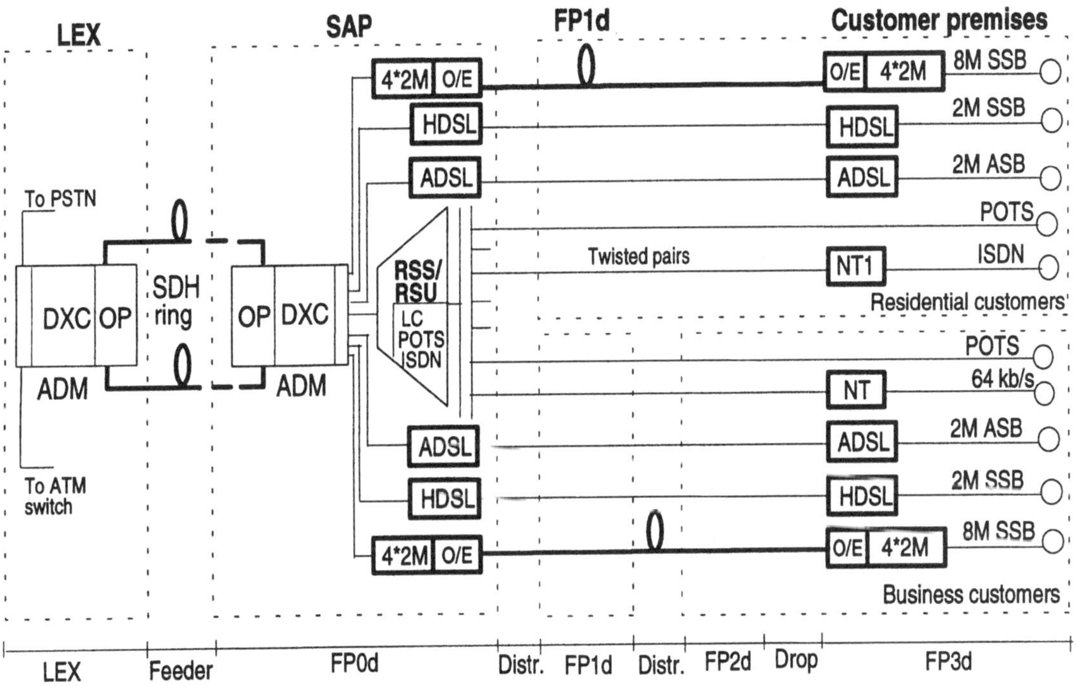

Fig 16.6 Downtown area, enhanced copper architecture.

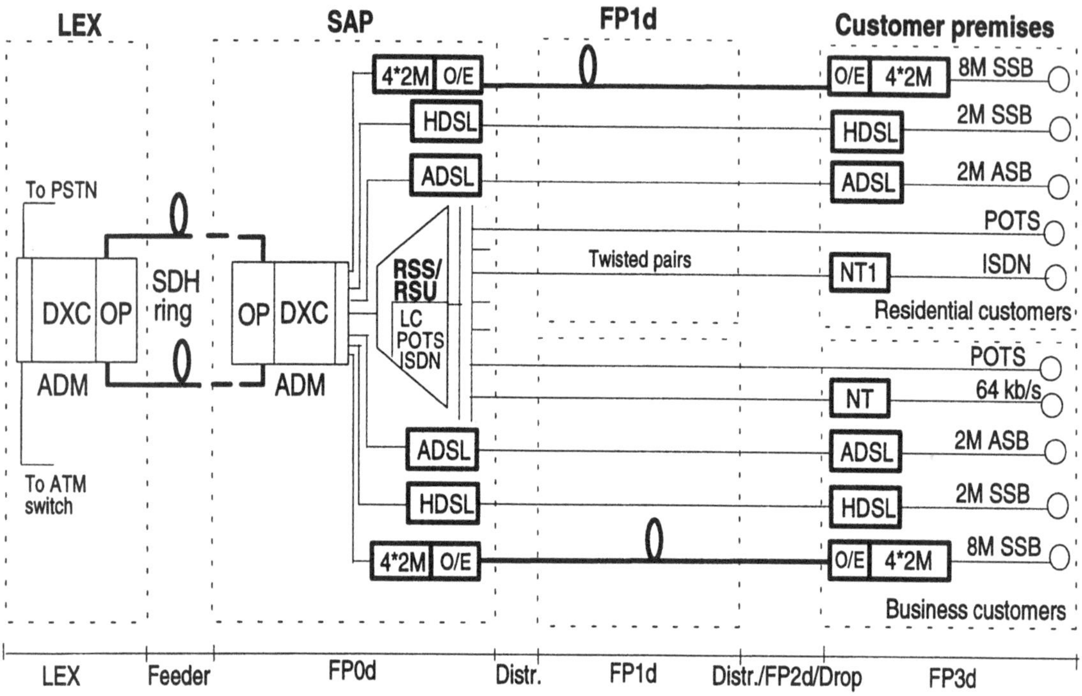

Fig 16.7 Suburban area, enhanced copper architecture.

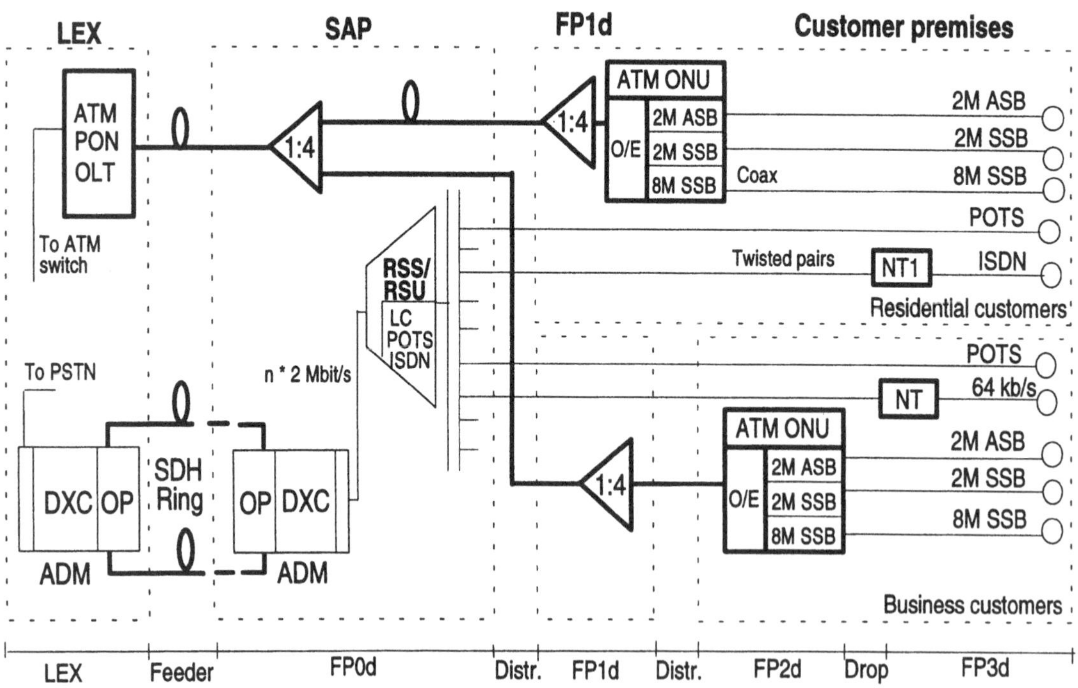

Fig 16.8 Downtown area, optical fibre architecture.

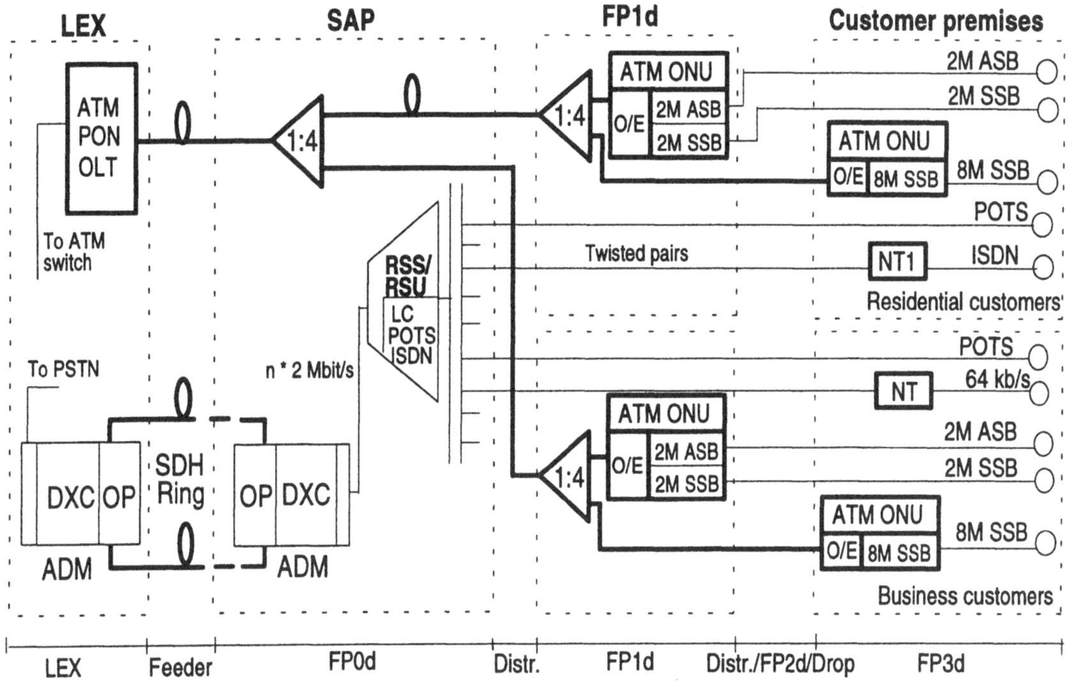

Fig 16.9 Suburban area, optical fibre architecture.

The main differences between the application of the optical fibre architecture in the downtown area and the suburban area is the outdoor installation of ONUs and the much higher civil works costs enforced by customers being located in individual houses in the suburban area.

16.5 ASSUMPTIONS

The analysis presented in this paper is based on several general assumptions, some of which are described in brief here. The study mainly addresses upgrades of two non-greenfield areas in which there is an existing modern copper based infrastructure in the distribution part. For reference, in addition greenfield deployment of the same upgraded architectures in the two demographic areas selected have been analysed.

The duct availability in the upper part of the distribution network is high (50 % for copper and 90 % for fibre), whilst there are no ducts available in the last drop. All new optical fibre cables in the upper level of the distribution network are installed at the start of the project period. However, new drop-cables such as coax and fibre and service specific equipment like modems, are installed during the study period at the time of connection of new service subscribers. Civil works costs used in the distribution part are 25 ECU/m. An aerial drop cost of 2 ECU/m is used for the final stretch.

16.6 RESULTS AND DISCUSSION

The TITAN methodology and the tool have been applied in the analysis [17-19]. In the context of this work some of the key issues of access network upgrading have been highlighted. Each of these aspects will be discussed in greater detail in the following sections.

Fig 16.10 shows the discounted installed first costs (IFC) and discounted accumulated revenues of the optical fibre architecture and the enhanced copper architecture for both the downtown (left) and the suburban (right) area access networks. Shown in white and grey are, the IFC and revenues for the existing network and existing services and the new network and new services, respectively.

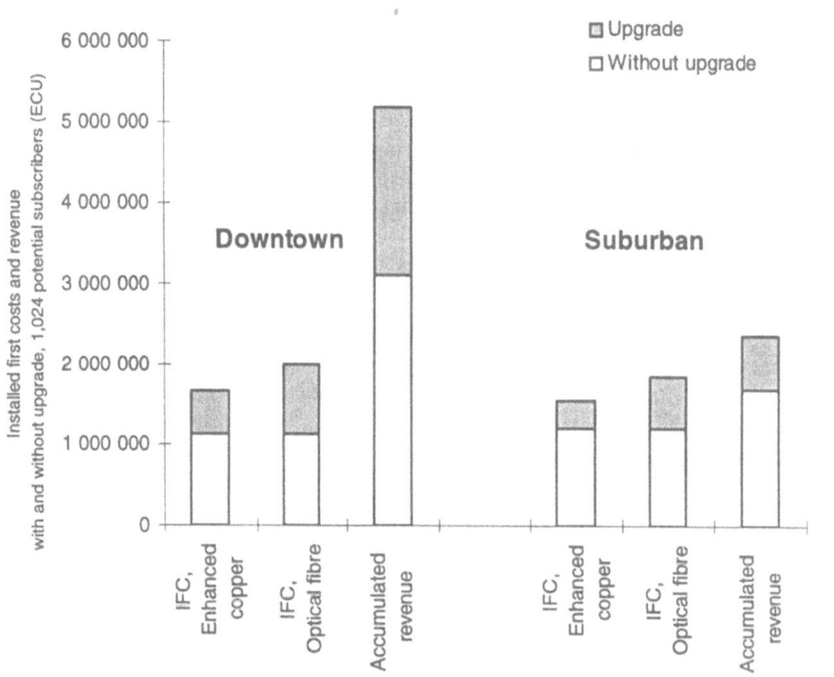

Fig 16.10 Discounted (IFC) and accumulated revenues, 1,024 potential subscribers.

The IFC without upgrade is greenfield costs including the costs of establishing the whole of the existing twisted pair copper base and equipment. The civil work costs and duct availabilities used are in accordance with the assumptions stated in the previous section. The similar levels of IFC of the suburban and downtown networks before upgrade are due to the fact that the higher civil works costs in the second distribution link level of the single dwelling suburban area are being compensated by higher 64 kbit/s modem and 2 Mbit/s line terminal costs in the downtown area.

The revenues "without upgrade" include accumulated study period revenues from all service connections existing at the start of the study period, i.e. POTS and N-ISDN for small businesses and residential customers and 2 Mbit/s, 64 kbit/s leased lines and POTS for medium and large businesses. The upgrade revenues comprise additional revenues from the take up of new service connections during the upgrade period from 1995 to 2005, i.e. 2 Mbit/s asymmetric switched broadband and 2 Mbit/s and 8 Mbit/s symmetric switched broadband for both market segments.

16.6.1 Overall upgrade costs

Both in the suburban and the downtown areas utilising the existing copper seems to enable a migration path to broadband service provision which not only is less expensive than the fibre alternative considered, but also may constitute less than 50 % of the investments laid down in the present access infrastructure. However, a fibre upgrade in addition to the upgrade of the terminal equipment includes a complete upgrade of the network infrastructure from twisted pairs to fibre. The additional cost of this infrastructure upgrade of a block dominated area seems to be obtainable with marginal costs of 50 % compared to the upgrade costs of the pure terminal equipment. In a suburban area with single house dwellings the fibre upgrade marginal costs can be closer to 100 % higher compared to the enhanced copper alternative.

In general, a fibre upgrade of the existing copper network may, largely dependent on the duct availability, involve investments of the same order as the ones once required to establish the current narrowband infrastructure.

16.6.2 Upfront costs and penetration dependant costs

The fact that fibre upgrades requires an initial upgrade of the existing infrastructure as opposed to revitalising the existing copper suggests that the initial, or upfront costs of fibre upgrades are higher than for enhanced copper options. This is illustrated in Fig 16.11 which show the discounted upfront costs and discounted service penetration dependant costs per new broadband customer for the two technology options examined. All 2 Mbit/s asymmetric and symmetric service connections are counted as one broadband connection. Each 8 Mbit/s symmetric connection is taken as four broadband connections. The upfront costs are associated with establishing the required common infrastructure. For the fibre optic alternative typically the optical line terminal (OLT), the feeder and distribution fibre cables, the optical couplers and the ONU, which all are shared by all or several customers, are installed in the first year of the upgrade period and hence included in the upfront costs. Customer specific equipment like ONU interface cards and ONUs with fibre drops for customers with the highest demand, are installed at the time of connection to the network, according to service penetrations, and thus considered as service penetration dependant costs. Typical upfront costs in the enhanced copper architecture are new synchronous digital hierarchy (SDH) add and drop multiplexers (ADMs) for increased ring capacity.

However, the 2 Mbit/s ports of these ADMs together with broadband modems (HDSL, ADSL), twisted pair line terminals and fibre drop cables are considered service penetration dependant costs.

In the downtown and suburban areas ~31 % and ~13 % respectively of the total broadband customer base are connected at the end of the study period, with the given market segment shares (10 %-90 %) and service penetrations.

The upfront and service penetration dependant costs are seemingly influenced by characteristics which vary between the two areas, such as housing structure and the average capacity demanded in the upgraded network. The results in Fig 16.11 show that for both demographic areas the enhanced copper alternative has significantly lower upfront costs than the fibre counterpart.

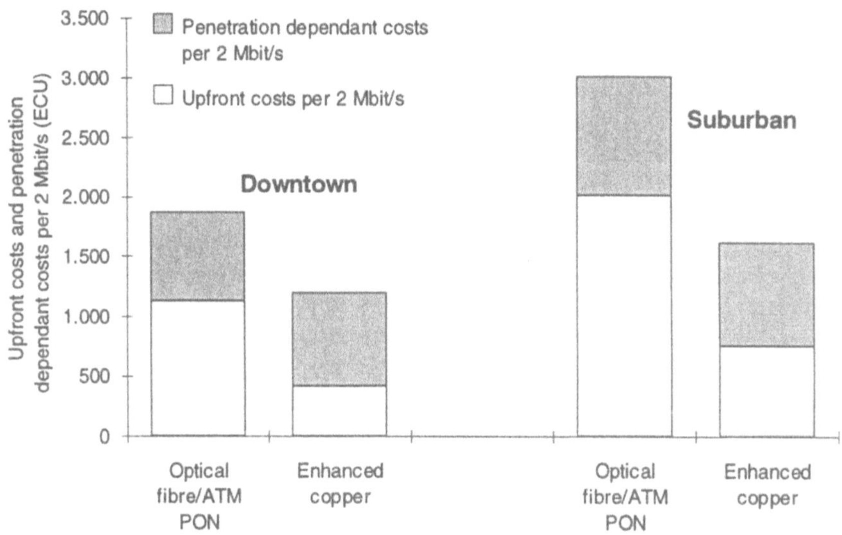

Fig 16.11 Upfront costs and service penetration dependant costs per new 2 Mbit/s connection.

Additionally, the comparison of the downtown and the suburban areas in Fig 16.11 illustrates that the enhanced copper upgrade has upfront costs per new broadband customer almost independent of the variations in housing structure and overall network capacity demand. However, in general the fibre alternatives have service penetration dependant costs similar to or less than the corresponding costs of the enhanced copper alternatives.

These results indicate that probably there is an average capacity demand beyond which the lower service penetration dependant costs of the fibre alternatives in pure economic terms can justify a more aggressive fibre upgrade.

16.6.3 Cost of electronics

Fig 16.12 shows the discounted IFC broken down on material types, installation and civil works. The significant contributions from the electronics to the overall costs for both architectures and both areas are evident. The electronics part of the overall costs varies between approximately 50 % and 80 % of the total IFC. For both areas studied, the service penetration dependant costs of placing the electronics investments in a fibre infrastructure are approximately 50 %, with the reasonably good existing duct layout assumed in this paper.

Since the upgrade to broadband means high electronics investment for any of the upgrade alternatives, the operator is faced with the question of placing these significant electronics investments in the old infrastructure (copper, twisted pair) or new future proof infrastructure (fibre).

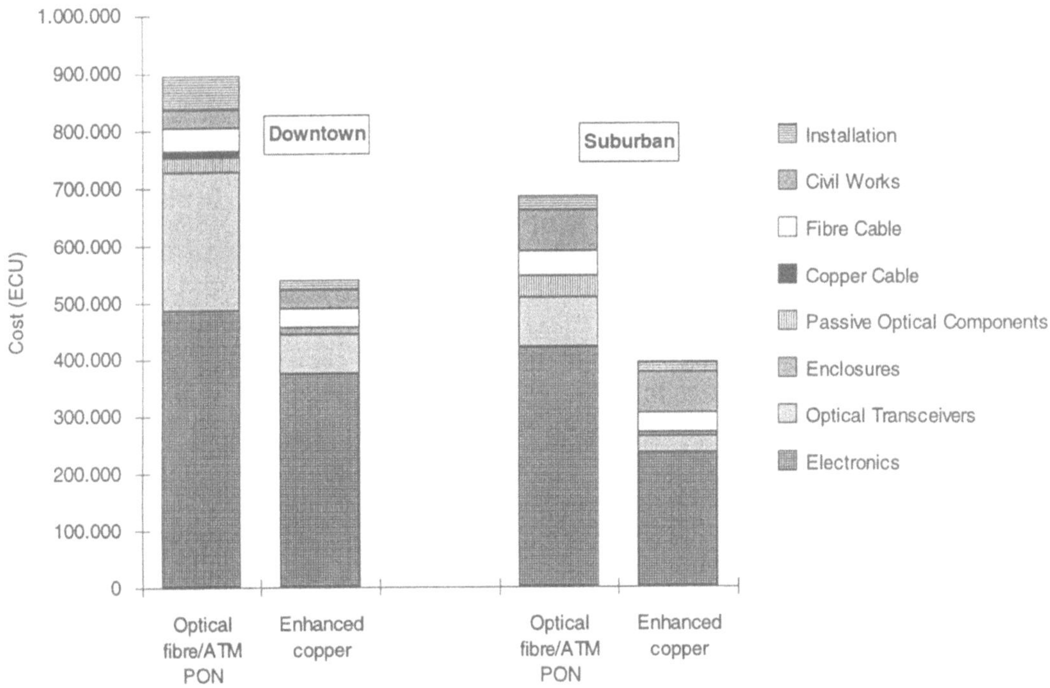

Fig 16.12 Cost breakdown of optical fibre and enhanced copper architectures, downtown and suburban.

16.6.4 The effect of the upgrade on the revenue base

The results in Fig 16.10 illustrate that in a ten year project perspective the business dominated downtown area probably has a potential for a much higher revenue per unit currency spent by the network operator than the residential dominated suburban area. With the given assumptions, a downtown enhanced copper upgrade may result in a gain in revenue of four times the investments, whilst for the more expensive fibre upgrade only twice the upgrade investments are returned as revenue. In the suburban area the fibre upgrade seems even less prosperous with a revenue gain due to the upgrade at the same level as the upgrade costs. However, upgrading the suburban network with enhanced copper technology may result in a revenue return of twice the investments.

Fig 16.13 depicts the ten year period discounted revenue breakdown for the two demographic areas studied. The revenues in the downtown and the suburban areas are dominated by symmetric 2 Mbit/s and POTS respectively. 50% of the revenue in the suburban area is POTS revenue, whilst the 2 Mbit/s service in the downtown area contributes in excess of 40 % of the overall revenue.

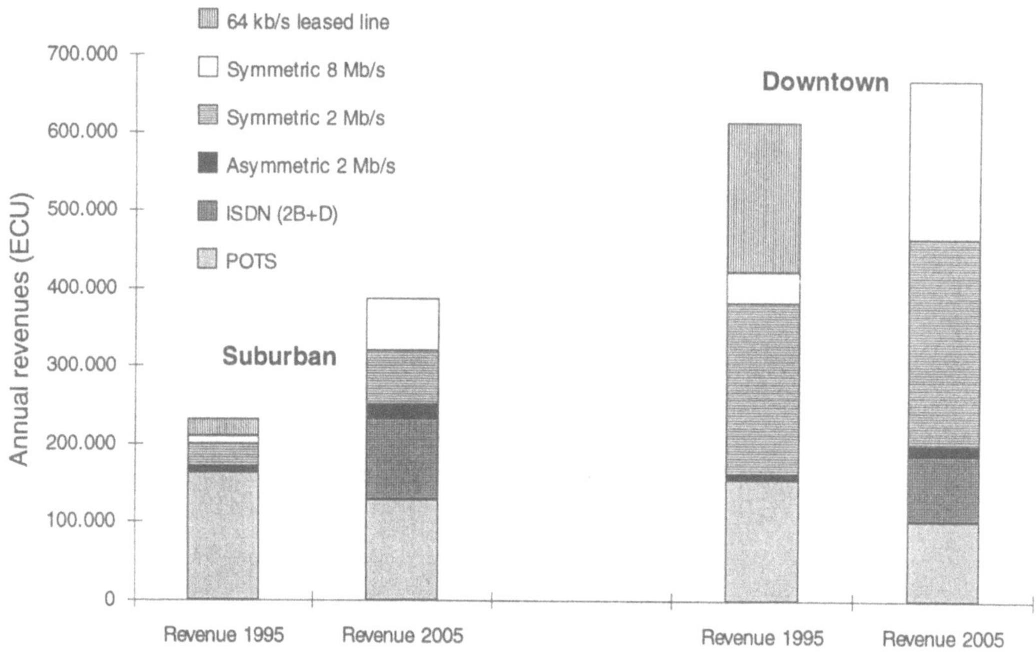

Fig 16.13 The revenue splits in 1995 and 2005 for the downtown and suburban areas.

The revenue splits in the initial (1995) and final (2005) year of the study period are shown in Fig 16.13. The figure illustrates the difference in the potential of the revenue base expansions of the two different areas. In the business dominated downtown area only the bandwidth provided in an already established broadband market is upgraded. In the residential dominated suburban area an existing pure narrowband market is expanded with brand new broadband services through the upgrade. In the latter case, the additional revenues from the upgrade services almost doubles the final year revenue as compared to the revenue prior to the upgrade. With the assumed tariffs and service penetrations, the value of the N-ISDN service at the end of the study period is quite significant: the N-ISDN revenue develops during the period to a level equal or higher than the POTS revenue level. Conversely, for a final year 6,7 % penetration and the given tariff evolution the asymmetric switched 2 Mbit/s service contribution to the overall revenue at the end of the period is negligibly small in both cases.

Hence, with the given tariffs and penetrations the broadband upgrade results in a significant expansion of the revenue base of a market which is narrowband dominated at present. For an already established broadband market, providing more bandwidth mainly has the effect of compensating for the tariff pressure and probably only marginally will expand the overall revenue base.

16.6.5 The long term perspective

The evolutions of the net present value (NPV) of the enhanced copper and the optical fibre architectures are illustrated in Fig 16.14 and Fig 16.15. NPV of the project at one particular point in time as used here is the discounted accumulated revenues minus the discounted accumulated costs (IFC and running costs). The residual value of the network at the end of the project is not included in the NPV. The NPVs are shown for three selected years and are split between the NPV of the existing network and services and the NPV of the upgrade network and services. The zero-crossing point in time of the NPV gives an indication of when the investments are returned, and hence the payback period.

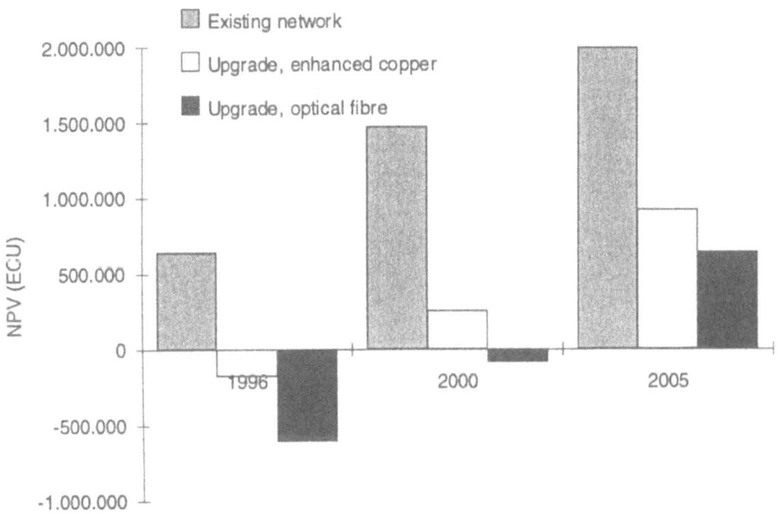

Fig 16.14 NPV, downtown area, enhanced copper and optical fibre architectures.

The longer payback periods of the suburban upgrade projects are indicated by the figures as are also the longer periods for the return of the investments of the fibre upgrades compared to the enhanced copper upgrades. The contribution of the upgrades to the overall NPV in the suburban case may seem negligible, whilst the downtown upgrades significantly improve the overall NPV even within a ten year timeframe. Cable upgrade projects represent investments in infrastructure, which has a longer technical life time than equipment, and as such usually will require payback periods in excess of ten years. This is also reflected in this study by the longer time constants of the fibre upgrades.

Fig 16.15 shows the NPV divided by IFC for the two architectures and areas under consideration. The ratio NPV to IFC gives an indication of the level of the projects yearly internal rate of return (IRR). A ten year linear project with a yearly IRR of 10 % will have a NPV/IFC ratio of ~2,6. The results are split between "existing and new services" and "upgrade", with the former including all service connections including the ones existing at the start of the study period, whilst the latter only includes the new network elements and services resulting from the respective upgrades. The figure illustrates quite clearly that the upgrade project benefits from the high revenues on the existing narrowband network. The NPV/IFC ratios for the overall projects ("existing and new services") are within a range in which a reasonable IRR may be achieved. However, this is not so for the upgrades themselves. Especially in the suburban area, an access network upgrade investment seems to be a very long term project.

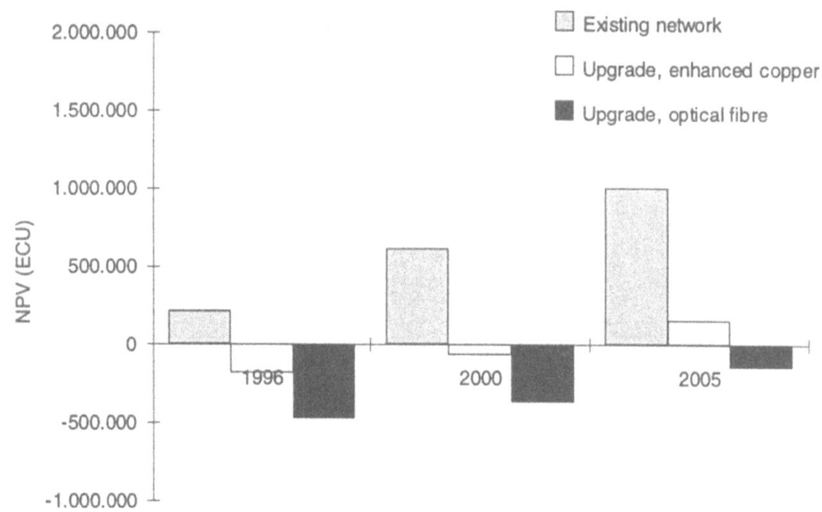

Fig 16.15 NPV, suburban area, enhanced copper and optical fibre architectures.

Fig 16.17 shows the upgrade costs of the previous figure together with, for reference, the NPV/IFC ratio for a greenfield installation of a copper network including all the service connections existing prior to the broadband upgrade, i.e. upgrade broadband services and network elements included. The duct availability in the distribution part of the greenfield reference is 50 % for copper and 90 % for fibre. The optical fibre upgrades, even though a not very profitable project within a ten year timeframe, have IRRs in the same order as typical present day access network greenfield projects in the respective areas. The enhanced copper upgrades as such are in fact more profitable than the reference greenfield projects.

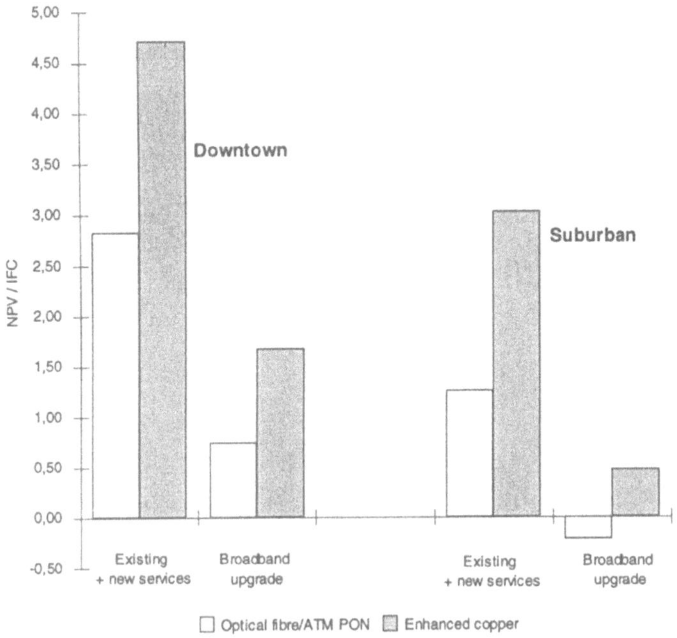

Fig 16.16 NPV divided by IFC.

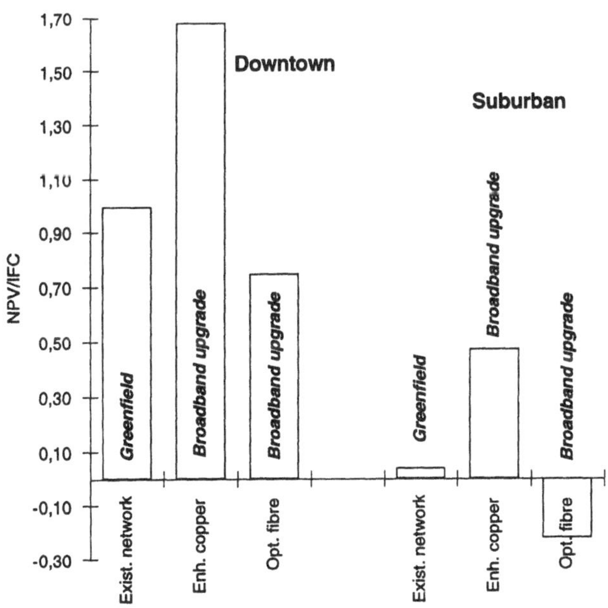

Fig 16.17 NPV divided by IFC for upgrades and greenfield installation.

However, due to the general long term nature of access network upgrading the revenues beyond the initial ten year period should probably be just as important as the efficiency of the upgrade period itself. The cash flow trend is useful as a long term predictor, looking at the prospects beyond the upgrade period.

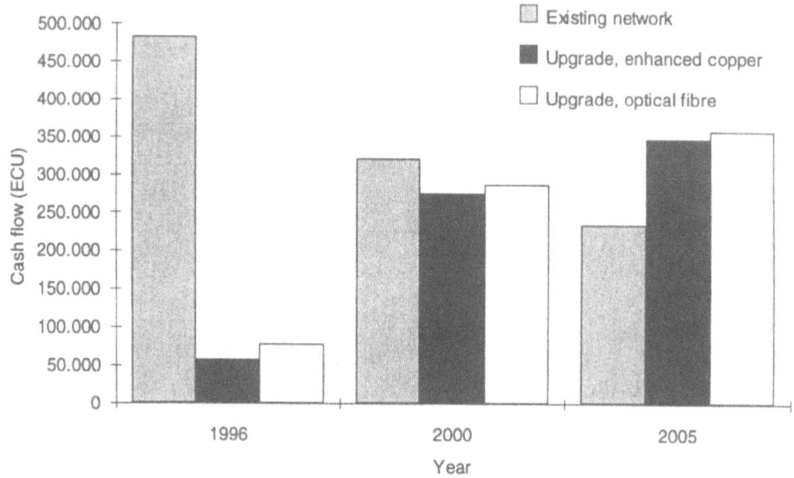

Fig 16.18 Cash flow, downtown area, enhanced copper and optical fibre architectures.

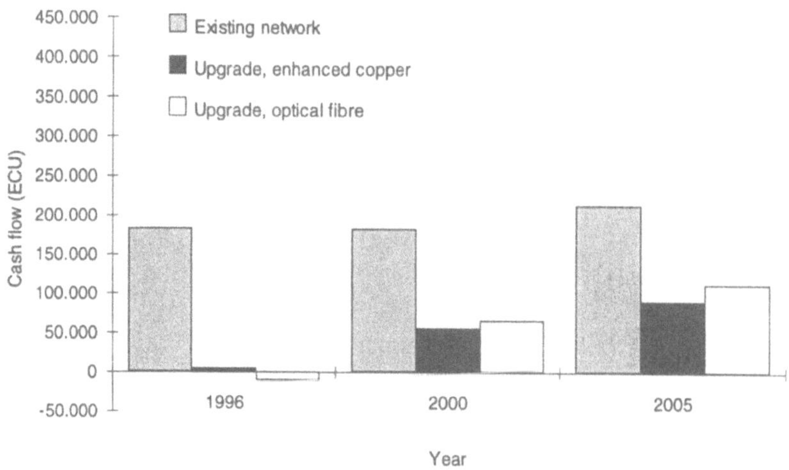

Fig 16.19 Cash flow, suburban area, enhanced copper and optical fibre architectures.

The evolutions of the cash flows of the enhanced copper and the optical fibre architectures are illustrated in Fig 16.18 and Fig 16.19. The cash flows are shown for three selected years and are split between the cash flow of the existing network and services and the cash flow of the upgrade network and new services. The significant contribution of the upgrade to the overall cash flow for both architecture options in both areas is obvious. For the downtown upgrade projects, even if the overall cash flow in the final year does not improve compared with the cash flow in the beginning of the upgrade period, the upgrade at least compensates for the 50 % reduction in the cash flow of the existing network and services. In the suburban area project the existing network and existing high penetration narrowband services remain the dominant contributor to the overall cash flow. Nevertheless, the broadband upgrade results in a 50 % overall cash flow increase from the beginning to the end of the upgrade period.

The higher cash flow of the optical fibre in the downtown case indicates that the service penetration dependant costs of connecting new broadband customers are lower for the fibre alternative than for a continuous use and upgrade of the existing copper. The influence of the case characteristics (average subscriber capacity demand and housing structure) on this relationship is clearly illustrated in Fig 16.19 which shows that a suburban residential and single dwelling dominated area in contrast to the downtown block dominated business area has higher service penetration dependant costs associated with a fibre upgrade than with an enhanced copper upgrade.

In conclusion, the results illustrate that a broadband upgrade probably will need several years before the investment costs are returned, even in the case where the already installed copper base is utilised. However, the yearly cash flows in this study are improved due to the upgrade. This suggests that securing long term revenues in the access network most likely will imply long term investments.

16.6.6 The effect of competition

The effect of competition has been modelled through appropriate reductions of the service penetrations. For the business dominated downtown area only losses in the business market segment are modelled, whilst in the residential dominated suburban area only residential market losses are modelled. The cash flow and the NPV of the upgrade projects with different degrees of competition are seen in Fig 16.20- Fig 16.23. Percentage of business and residential market losses are indicated. The results show a clear difference in the degree of sensitivity to market losses between the two area types.

Fig 16.20 shows that in the downtown area even a 75 % loss of business customers results only in a marginal reduction (~25%) in the NPV in the final year. Fig 16.21 illustrates that a 75 % market loss with a marginal impact on the final year NPV severely influences the final year cash flow, and reduces it to ~50% of the final year cash flow without market loss and ~50 % of the cash flow prior to the upgrade.

In the suburban area the situation is quite different from the downtown upgrade illustrated in the previous figures. In this case a 75 % loss of the dominating market segment, now the residential, approximately halves the final year NPV as compared to the no-loss NPV in year 2005 (Fig 16.22). However, the impact on the final year cash flow is more severe relatively in this case as compared to the above business dominated area (Fig 16.23). Here, a 75 % market reduction reduces the final year cash flow by 70 % as compared to the no-loss cash flow in year 2005. Nevertheless, when compared to the cash flow prior to the upgrade, the operator may loose 50 % of the residential segment and still maintain the initial year cash flow.

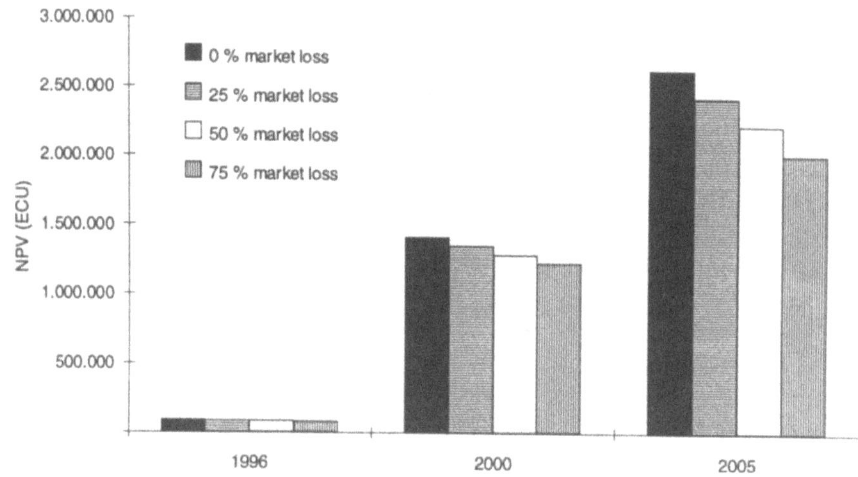

Fig 16.20 NPV with market loss, optical fibre architecture, downtown.

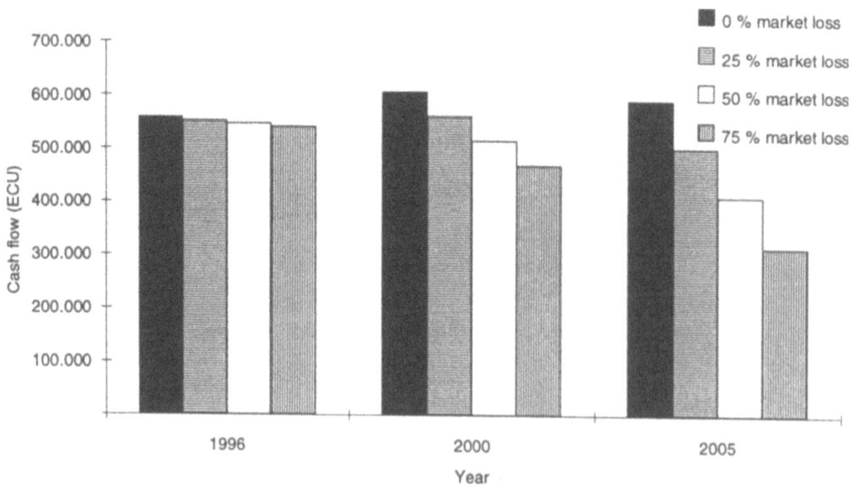

Fig 16.21 Cash flow with market loss, optical fibre architecture, downtown.

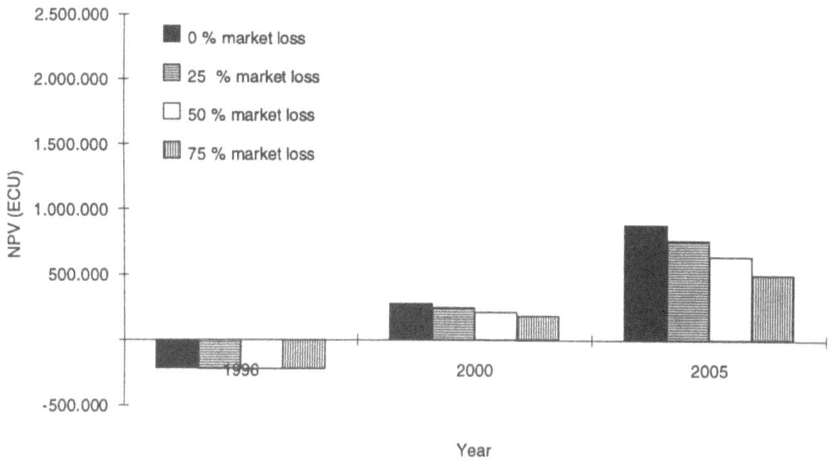

Fig 16.22 NPV with market loss, optical fibre architecture, suburban.

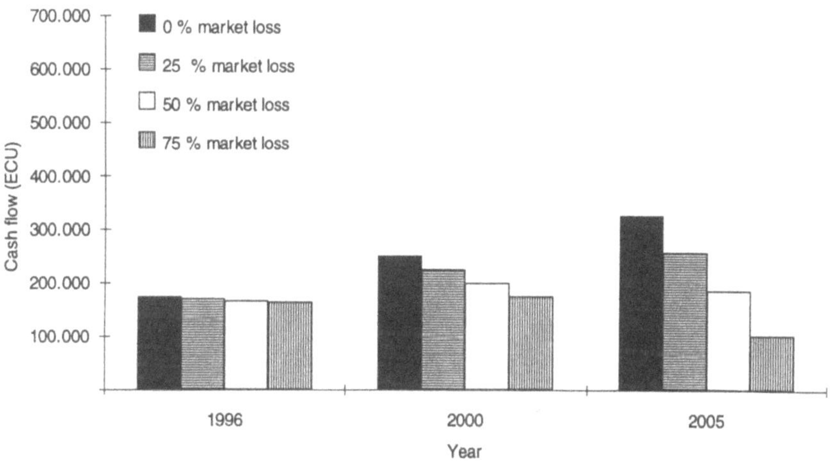

Fig 16.23 Cash flow with market loss, optical fibre architecture, suburban.

16.7 SUMMARY

The comparisons of the two technical solutions for broadband services in the business dominated area presented in this paper indicate that the cost difference between very modest access network upgrading strategies and considerably more aggressive solutions can be marginal compared to the overall investments associated with any upgrading alternatives.

Broadband upgrades to the access network are likely to turn out to be long term projects with payback periods and internal rate of returns in the same order as the present day establishment of narrowband, twisted pair based infrastructures.

The costs of electronics dominates the overall costs independent of upgrade technology selected. Hence, upgrading to broadband turns out to be a choice between significant electronics investment in old (copper) or new (fibre) infrastructure. The operator is thus faced with the strategic choice of upgrading the cable infrastructure or not.

The effect of competition does not necessarily have a severe impact on the overall economic results of the upgrade projects but may significantly affect future revenues and prospects. The dramatic reduction in the cash flow at the end of the study period with increasing competition shows that even reasonably profitable access network projects in terms of NPV and payback period may be severely affected by revenue losses when looking beyond the period under study.

REFERENCES

[1] Ims, L A, Olsen, B T, Myhre, D, Stordahl, K. Migration paths to broadband access for the mixed residential and business market. *Accepted for publication in Telecommunications Systems.*

[2] Ims, L A, Myhre, D, Olsen, B T, Stordahl, K. Paths towards a versatile broadband Access Network. *In Proc. International Conference on Telecommunications 95 (ICT '95), Nusa Dua, Indonesia, April 3-5, 1995, 215-219.*

[3] Pugh, W, Boyer, G. Broadband Access: Comparing Alternatives. *IEEE Communications Magazine*, 33, (8), 1995, 34-45.

[4] Miki, T. Toward the Service-Rich Era. *IEEE Communications Magazine*, 32, (2), 1994, 34-39.

[5] Reed, D P. Residential Fiber Optic Networks - An Engineering and Economic Analysis. Artech House, Norwood, MA, USA, 1992.

[6] K.W. Lu, K W, Eiger, M I, Lemberg, H L. System and Cost Analysis of Broadband Fiber Loop Architectures. *IEEE Journal of Selected Areas in Communications*, 8, (6), 1990, 1058-1067.

[7] Cook, A, Stern, J. Optical Fiber Access - Perspectives Towards the 21st Century. *IEEE Communications Magazine*. 32, (2), 1994, 78-86.

[8] Ims, L A, Olsen, B T, Myhre, D. The Economic Implications of Broadband Access Network Upgrading. *In: Proc. Broadband Networks, London, UK, November 8-9, 1995.*

[9] LeBel, P, Oliver, R. Residential Broadband Architectures Strategies and Economics. *In: Proc. The 7th International IEEE workshop on Optical Access Networks (OAN '95), Nuremberg, Germany, September 24-27, 1995, 4.4-1 - 4.4-10.*

[10] Myhre, D, Ims, L A, Olsen, B T, Stordahl, K. Access Network alternatives for broadband services. *In: Proc. ConTEL '95, Zagreb, Croatia, June 7-9, 1995, 17-24.*

[11] Ims, L A, Olsen, B T, Myhre, D, Stordahl, K. Economic Implications of Broadband Access Network Upgrade. *In: Proc. Broadband Islands 1995 (BRIS '95), Dublin, Ireland, September 4-5, 1995, 33-43.*

[12] Olsen, B T, Ims, L A, Myhre, D, Stordahl, K, Theivindrein, R. Techno-economic Evaluation of Optical Broadband Access Network Scenarios for the residential and business market. *In: Proc. 21st European Conference on Optical Communications (ECOC '95), Brussels, Belgium September 14-17, 1995.*

[13] Stordahl, K, Murphy, E. Methods for forecasting long term demand for wide and broadband services in the residential market. *IEEE Communications magazine*, 13, (2), 1995, 44-49.

[14] Stordahl, K, Olsen, B T, Murphy, E, Zaganiaris, A. de Castro, E, Ravera, M. Delphi Survey: Forecasting Demand for Wide- and broadband services. From RACE 2087/TITAN. *In: Proc 6th International IEEE workshop on Optical Access Networks (OAN '94), Kyoto, Japan, Oct. 16-19, 1994, 7.3-1 - 7.3-10.*

[15] Stordahl, K, Olsen, B T. Demand Elasticity and Forecasts of Wide and Broadband Services in the Residential Market Based on Results from an International Delphi Survey. *1995 International Communications Forecasting Conference, Toronto, Canada, June 13-16, 1995.*

[16] Olsen, B T, Ims, L A, Øverli, T, Stordahl, K, Myhre, D, Drieskens, M, Kraushaar, J, Zaganiaris, A. Techno-economic analysis of PNO and CATV operator strategies for broadband upgrade of the access network. *In: Proc. International Teletraffic Seminar 1995, Bangkok, Thailand, Nov. 28 - Dec. 1, 1995, 54-1 - 54-12.*

[17] Olsen, B T, Zaganiaris, A, Stordahl, K, Ims, L A, Myhre, D, Øverli, T, Tahkokorpi, M, Welling, I, Drieskens, M, Kraushaar, J, Mononen, J, Lähteenoja, M, Markatos, S, De Bortoli, M, Ferrero, U, Ravera, M, Balzaretti, S, Fleuren, F, Gieschen, N, De Oliveira Duarte, M, de Castro, E. Techno-economic evaluation of narrowband and broadband access network alternatives and evolution scenario assessment. *IEEE Journal on Selected Areas in Communications,* 14 (6), 1996, 1184-1203.

[18] Olsen, B T, Zaganiaris, A, Gieschen, N, Tahkokorpi, M, Drieskens, M, Markatos, S. RACE 2087/TITAN: Tool for Introduction scenarios and Techno-economic studies for the Access Network. *In: Proc. RACE Open Workshop on Broadband Access, Nijmegen, the Netherlands, June 7-8, 1993.*

[19] Zaganiaris, A, Ferrero, U, Balzaretti, S, Tahkokorpi, M, Kalervo, M, Olsen, B T, Stordahl, K. Methodology for risk assessment and techno-economic evaluation of optical access networks. *In: Proc The 20th European Conference on Optical Communications (ECOC '94), Florence, Italy, September 25-29, 1994, 3, 83-90.*

Broadband upgrades in residential areas

17

Jari Mononen, Markku Lähteenoja

This chapter presents an in depth analysis of the access alternatives for broadband upgrades in residential dominated areas. The following architectures have been assessed: hybrid fibre coax (HFC), advanced fibre to the home passive optical network (FTTH PON), a so-called SuperPON, and asynchronous transfer mode (ATM PON). A wide range of sensitivity analyses have been performed with respect to service take up, housing structure and the effect of the public network operator (PNO) having an existing community antenna television (CATV) network or not. The installed first costs, life cycle costs, net present values, internal rate of return and payback periods for the different scenarios have been calculated. This paper is based on work done in EURESCOM (European institute for research and strategic studies in telecommunications) project P413 "Optical Networking", aimed at studying the introduction of optical technologies in the access network [1,2].

17.1 INTRODUCTION

In the residential dominated area the evaluated strategy for the future broadband multiservice access network deployment is the combination of the two basic strategies suggested in [3] which are a *copper changeout strategy* and an *overlay strategy*. In this combinatory strategy a new broadband access network is built primarily to deliver services to those customers subscribing to the new multimedia broadband services. However, narrowband customers are also moved gradually to the new access network. This combinatory strategy will give the PNO a single managed and integrated broadband access network infrastructure, after a long investment and renewal period (e.g. ten years).

17.2 CASE DESCRIPTION

This study is focused directly on the three different integrated broadband access network architectures which represent the main optical access network technologies currently under intensive development or evaluation around the world. These technologies, HFC, SuperPON and ATM PON are evaluated performing a wide range of sensitivity analyses on the service penetrations and housing structures. In addition, the effect of assuming the existence of a PNO CATV network at the beginning of the study period is analysed.

In all three integrated access network technology options, three different housing structures, two different service penetration scenarios (scenario 1 and scenario 2) and three different PNO CATV network starting situations are assumed and evaluated. In scenario 1 services will be more popular than in scenario 2, exact penetration curves and the associated tariffs are documented in chapter 17.4. Therefore, for each technology 18 separate calculation rounds are performed, resulting in a total of 54 separate calculations. The evaluated 18 separate residential dominated area situations represent a comprehensive coverage of different European PNO situations of residential access areas, and thus give a good possibility to compare these technology options from the perspective of each PNO.

The evaluated area is a residential dominated non-greenfield area of 1,024 potential customers where the twisted copper pair access network is sufficient for existing services. For new cable installations 20 % and 5 % of duct space is available for fibre and copper respectively.

The majority (95 %) of the customers in the area are residential and small business customers. The housing structure in the area consists of single family houses (one customer per building) and block apartment houses (32 customers per building). A few (5 %) of the customers in the area are medium business customers.

The services provided for the residential and small business customers are: the existing plain old telephone service (POTS), narrowband integrated services digital network (N-ISDN), 2 Mbit/s symmetric switched broadband (SSB), asymmetric switched broadband (ASB; 2 Mbit/s downstream, 64 kbit/s upstream) and analogue CATV service. Following the previously described combinatory strategy POTS customers are first served with the existing twisted pair network and those customers are shifted smoothly from the existing twisted pair network to the new integrated access network according to the POTS transition parameter. This transition is done smoothly during the study period, removing 8 % of old POTS customers each year to the new access network.

For the medium business customers a gradual shift from 64 kbit/s leased line and POTS (ten lines for each medium business customer) connections towards a 2 Mbit/s SSB service is assumed.

In the techno-economic study the area is characterised and evaluated by the use of TITAN (tool for introduction scenario and techno-economic evaluation of access network) [4] and TITAN geometric model, which enable customer density and housing structure variations to be modelled. The costs of the network components are calculated using cost data gathered from many European sources. These costs represent typical European averages. The operation, administration and maintenance (OAM) costs are calculated using the TITAN OAM methodology [4]. The techno-economic study period is ten years long and covers the years 1996-2005.

17.3 ARCHITECTURES

The architectures described are the target, or final year (2005) architectures. The common equipment and infrastructure are installed in the first year (1996). Penetration dependent equipment (e.g. line cards) is installed during the period according to service penetration.

17.3.1 Hybrid fibre coax architecture

The analysed hybrid fibre coax (HFC) architecture is shown in Fig 17.1 In the local exchange separate host digital terminals (HDTs) for interactive services (POTS, ISDN, SSB 64 kbit/s, SSB 2 Mbit/s and ASB 2 Mbit/s) are used as interworking units between the optical part of the HFC network and the local exchange (LEX) or an ATM network.

An optical node is located between the coaxial and optical parts of the HFC network. The optical node is connected with the following three linear point to point optical transmission systems to the local exchange:

1. One system for downstream traffic for analogue CATV services and ASB (service on demand, video on demand) services
2. One system for downstream digital POTS, ISDN and SSB services
3. One system for upstream control of the ASB service, digital POTS, ISDN and SSB services.

The optical node serves a tree and branch type coaxial cable network, that has bi-directional transmission capabilities. In the "pessimistic" service scenario 2 the coaxial network is separated in two segments, each segment containing 512 customers. In the "optimistic" service scenario 1 a split of the coaxial cable network in four segments is required, with each segment containing 256 customers.

In the apartment house area the coaxial cable network and coaxial network units (CNUs) are installed in the basement of the buildings in the first year. The final coaxial cable drops from the basement to the customer apartments are installed at the time when the customer orders the service.

In the small house area the coaxial cable network is installed up to the FP2d (serving eight houses, FP = flexibility point) in the first year and final coaxial cable drops from the FP2d are installed at the time when the customer orders the CATV or ASB service. CNUs are installed in the street cabinets (in FP1d) serving 32 houses in the first year. The existing twisted pair copper drops from the street cabinets (FP1d) to customer premises are used for all but CATV and ASB service.

The linecards for interactive narrowband services are installed in the CNUs at the time when the customer orders the service.

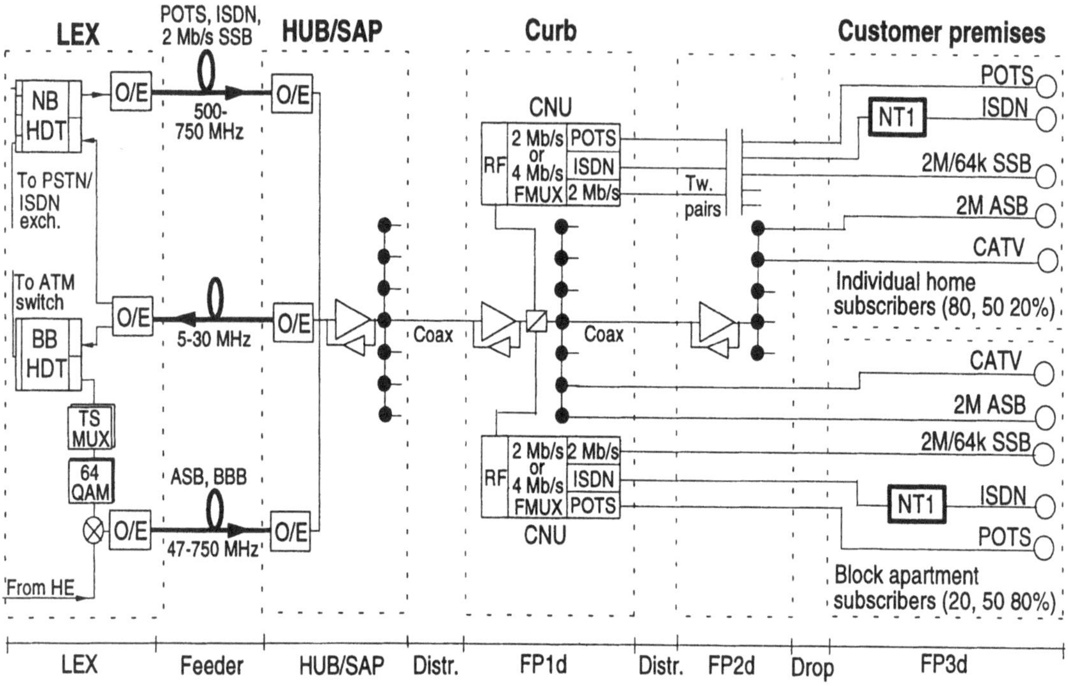

Fig 17.1 Hybrid fibre coax (HFC) architecture.

17.3.2 SuperPON architecture

The analysed SuperPON architecture is shown in Fig 17.2. The SuperPON optical line terminal (OLT) is centralised, located at the core exchange. The splitting ratio of the SuperPON is 2,048. I.e. one OLT serves up to 2,048 optical network units (ONUs). The implemented network structure is a true FTTH network where ONUs are located at the customers premises. One SuperPON network serves two areas of 1,024 customers each. The total capacity upstream and downstream of the SuperPON is 1.2 Gbit/s (enough for both scenarios 1 and 2) In the upstream direction the capacity is partitioned into eight sub-PONs with an active regenerator structure. Each sub-PON carries 155 Mbit/s in the upstream direction.

In the apartment house area the fibre and the last 1:32 splitter are installed in the basement of the buildings in the first year. The final fibre drops from the basement to the customers' apartments are installed at the time when the customer orders the service.

In the small house area fibre cable and the last 1:8 splitter is installed up to the last flexibility point (FP2d) in the first year. The final fibre cable drops from the last flexibility point are installed at the time when the customer orders the service.

New CATV customers are served by the SuperPON system. Old CATV customers are retained in the old CATV network. The CATV service delivered through the SuperPON has been dimensioned to accommodate 80 Mbit/s transmission capacity (20 channels of 4 Mbit/s motion picture expert group (MPEG) 2 signals).

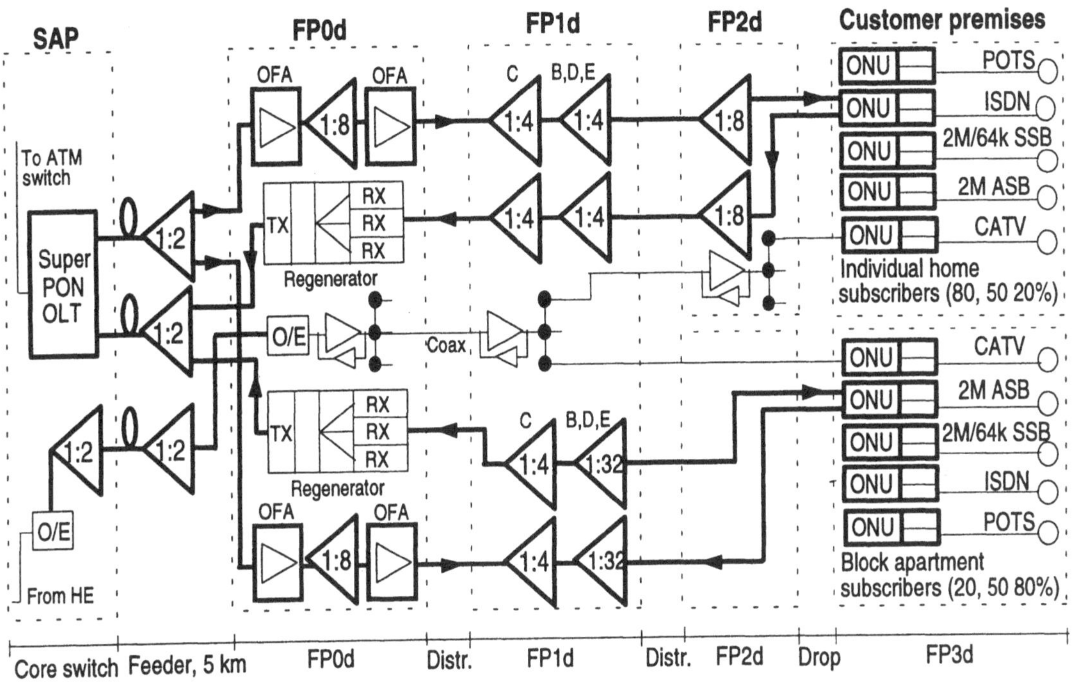

Fig 17.2 SuperPON architecture.

17.3.3 ATM PON architecture

The analysed ATM PON based point-to-multipoint fibre feeder access architecture is shown in Fig 17.3. The ATM OLTs are centralised, located at the local exchange. Each ATM OLT serves two physical PONs. The assumed splitting ratio of one PON is 16. Upstream and downstream transmission is on separate fibres. The assumed ATM PON capacity is 155 Mbit/s downstream and 155 Mbit/s upstream.

In the apartment house area an optical fibre based network is deployed to the basement of the buildings (32 apartments), where also ONUs are installed in the first year. The linecards in the ONUs are installed at the time when the customer orders the service. The existing twisted pair copper drops from the basement to the customers' apartments are used for all but CATV service.

In the small house area an optical fibre based network is built and ONUs are installed in the street cabinets (in FP1d) serving 32 houses in the first year. The existing twisted pair copper drops from the street cabinets to customer premises are used for all but the CATV service.

The CATV service has been assumed to be implemented with a conventional CATV network. The feeder section of the CATV network is an optical AM-transmission system. A coaxial tree and branch network extends from FP0d, serving 1,024 potential customers. An ATM PON access architecture and a

CATV network construction uses the same cable but different fibre in the feeder section. In the coaxial distribution section the CATV network uses the same digging and ducting as the ATM PON access construction. In the small house area the final coaxial drop from FP1d to customer premises has to be installed separately for each new CATV customer.

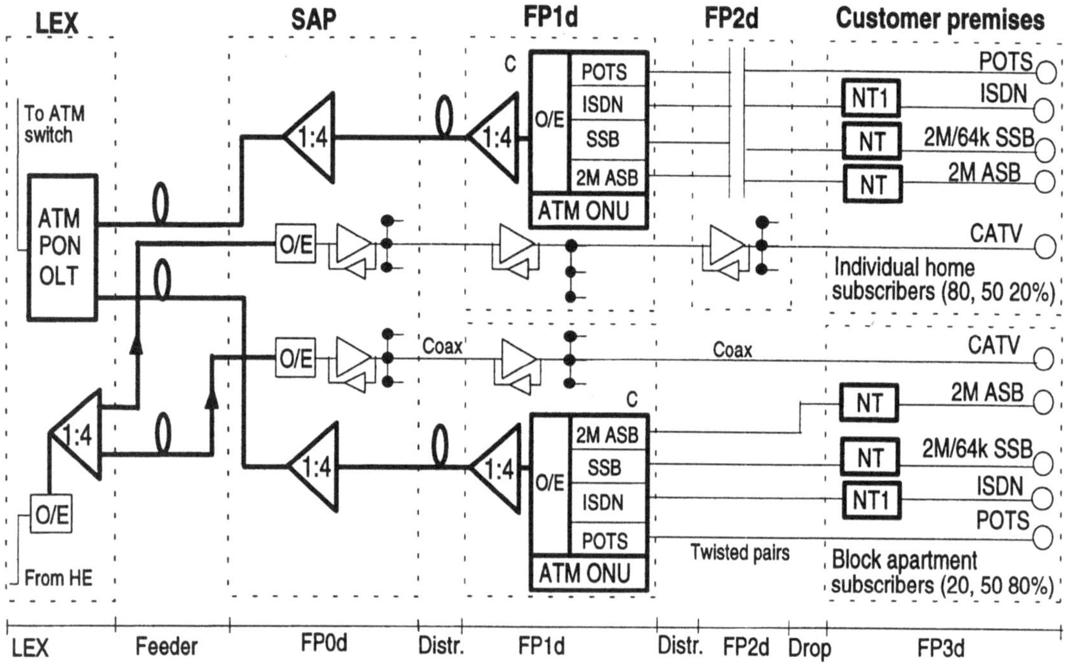

Fig 17.3 ATM PON architecture.

17.4 SENSITIVITY ANALYSIS

Since the residential dominated area is assumed to be more sensitive to the costs than the large and medium business dominated area, a wide range of sensitivity analyses was performed of the service penetration, of the PNO's existing CATV network situation and of the housing structures.

17.4.1 Service penetration

For the service penetration two different service take up scenarios have been analysed:

Scenario 1: New services will be popular and the PNO's market share will increase or will not reduce significantly.

Scenario 2: New services will be less popular and market shares will decrease due to competition and other effects (e.g. mobile services).

The service penetration curves for residential and small business customers in scenarios 1 and 2 are shown in Fig 17.4.

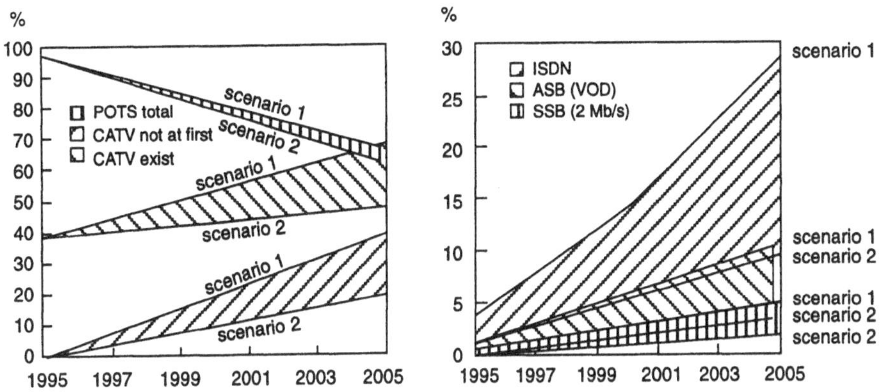

Fig 17.4 Service penetration curves for POTS, CATV, ISDN, ASB and SSB for residential and small business customers.

The service penetration curves for medium business customers (5 % of the customers in the area) are shown in Fig 17.5 Each medium business POTS customer has ten POTS lines on average.

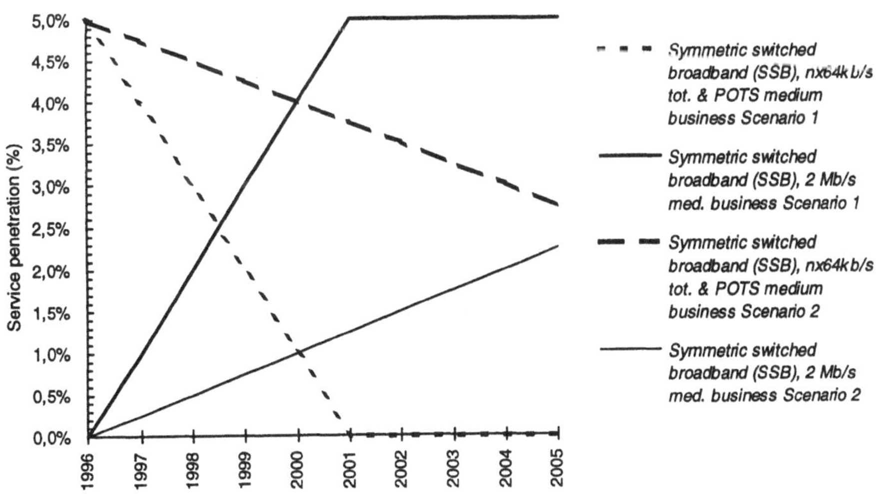

Fig 17.5 Service penetration curves for POTS, SSB 64 kbit/s and SSB 2 Mbit/s for medium business customers.

Service tariffs have been included in the calculations, in order to produce upgrade project revenues. Only the part of the tariff is included, that can be directly associated with the access network. The tariff structure used for the calculations is shown in Fig 17.6. No sensitivity analysis on tariffs was performed.

Table 17.1 The tariff structure used in the calculations.

Service	*Tariff 1995 (ECU)*	*Tariff 2005 (ECU)*	*Annual reduction*
POTS	160	160	0 %
ISDN	320	192	5 %
CATV	150	150	0 %
2 Mbit/s ASB, residential	460	160	10 %
2 Mbit/s SSB, residential	2,300	802	10 %
64 kbit/s leased line	950	569	5 %
2 Mbit/s SSB, business	2,600	907	10 %

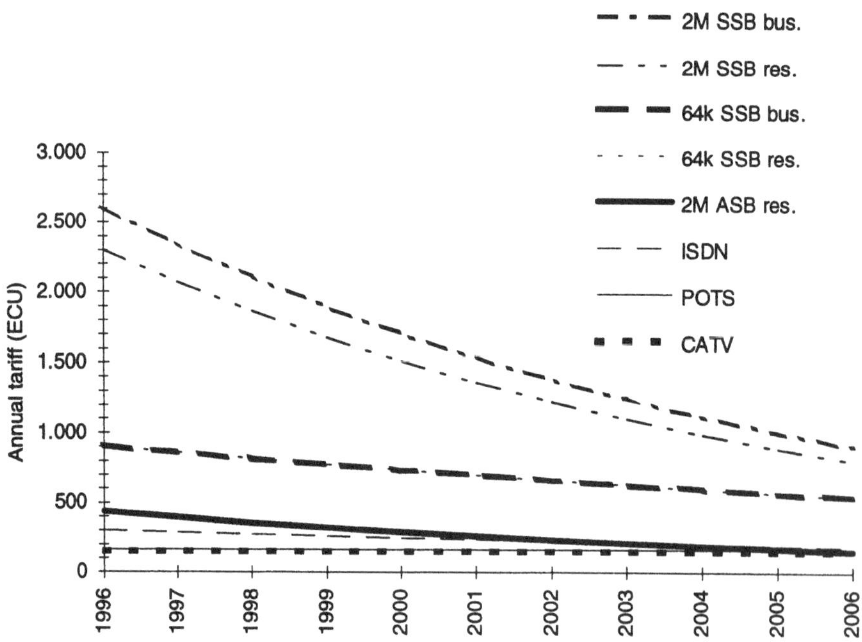

Fig 17.6 The tariff structure used in the calculations.

17.4.2 Existing CATV network situation

Three different existing CATV network situations have been assumed:

1) The PNO has a CATV network at the beginning of the study period
2) The PNO has no CATV network at the beginning, but intends to build one
3) The PNO has no CATV network at the beginning and does not build one.

17.4.3 Housing structure

Three different housing structures have been studied:

1) 80% of customers live in single houses and 20% in apartment buildings
2) 50% of customers live in single houses and 50% in apartment buildings
3) 20% of customers live in single houses and 80% in apartment buildings.

17.5 RESULTS

The installed first costs (IFC) for an area of 1,024 potential customers with customer density of 1,000 potential customers/km^2 of all the analysed situations in the residential dominated area are summarised in the following table (values in million ECUs).

Table 17.2 The IFC for the residential dominated area (in MECUs).

Situation (FTTB= % of apartment houses)	*ATM PON*	*HFC*	*SuperPON*
Scenario 1, CATV not at first, FTTB = 20 %	1.46	2.30	1.50
Scenario 2, CATV not at first, FTTB = 20 %	1.29	1.58	1.42
Scenario 1, CATV not at first, FTTB = 50 %	1.26	2.05	1.30
Scenario 2, CATV not at first, FTTB = 50 %	1.10	1.35	1.22
Scenario 1, CATV not at first, FTTB = 80 %	1.06	1.80	1.11
Scenario 2, CATV not at first, FTTB = 80 %	0.92	1.12	1.03
Scenario 1, CATV not at all, FTTB = 20 %	0.81	2.19	1.46
Scenario 2, CATV not at all, FTTB = 20 %	0.69	1.50	1.40
Scenario 1, CATV not at all, FTTB = 50 %	0.81	1.98	1.26
Scenario 2, CATV not at all, FTTB = 50 %	0.68	1.30	1.20
Scenario 1, CATV not at all, FTTB = 80 %	0.81	1.78	1.06
Scenario 2, CATV not at all, FTTB = 80 %	0.68	1.10	1.01
Scenario 1, CATV exists, FTTB = 20 %	0.73	1.52	1.38
Scenario 2, CATV exists, FTTB = 20 %	0.56	0.80	1.30
Scenario 1, CATV exists, FTTB = 50 %	0.71	1.43	1.18
Scenario 2, CATV exists, FTTB = 50 %	0.55	0.73	1.10
Scenario 1, CATV exists, FTTB = 80 %	0.69	1.35	0.99
Scenario 2, CATV exists, FTTB = 80 %	0.55	0.66	0.91

The NPV of the ten year access network upgrade projects for the three technologies and the 18 scenarios studied are illustrated in Fig 17.7. The NPV used here is the sum of all discounted retained cash flows and the discounted rest value of the investment)

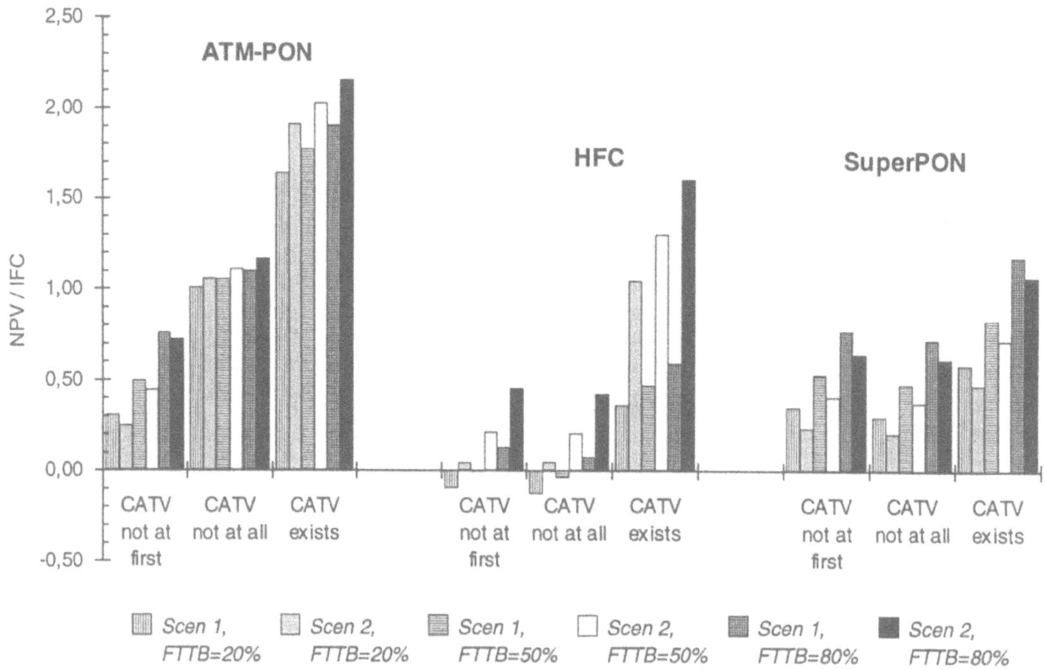

Fig 17.7 NPV divided by IFC for the residential dominated area.

These results show that many of the three "full service network" upgrade projects with the service take up rates and tariffs assumed have NPV/IFC < 1 and represent as such not very efficient projects.

However the upgrade projects with the *ATM PON technology* with the existing CATV network have NPV/IFC around 2 independent of service scenario and housing structure and represent as such quite acceptable upgrade projects. If the CATV service will not be offered at all NPV/IFC will be near 1 and if the CATV network will be built parallel with the ATM PON network, the NPV/IFC will be around 0.5 depending more on housing structure. These two latter projects are not so efficient in economic terms.

The upgrade projects with *HFC technology* are not attractive if the PNO does not own the existing CATV network. Especially in small house dominated areas with optimistic service take up scenario (scenario 1) NPV of the upgrade projects is near zero or even negative. Zero net present value represents return on investments within the project period. However, if the PNO has the existing CATV network and the interactive service take up scenario is more moderate (scenario 2), then the upgrade project with HFC technology is reasonable in apartment building dominated areas.

The upgrade projects with FTTH *SuperPON technology* are still clearly more expensive than with ATM-PON or HFC technologies. The best results with the SuperPON technology are achieved with optimistic service take up scenarios in apartment building dominated areas.

To analyse how the access network upgrade affects the prospects beyond the upgrade period the conclusions are not evident from the NPV or NPV/IFC. Instead the cash flow (revenue minus the sum of running costs and investments during the year under consideration) in the final year is useful as a long term predictor. The cash flow in 2005 for the three technologies and the 18 scenarios studied is illustrated in the following Fig 17.8.

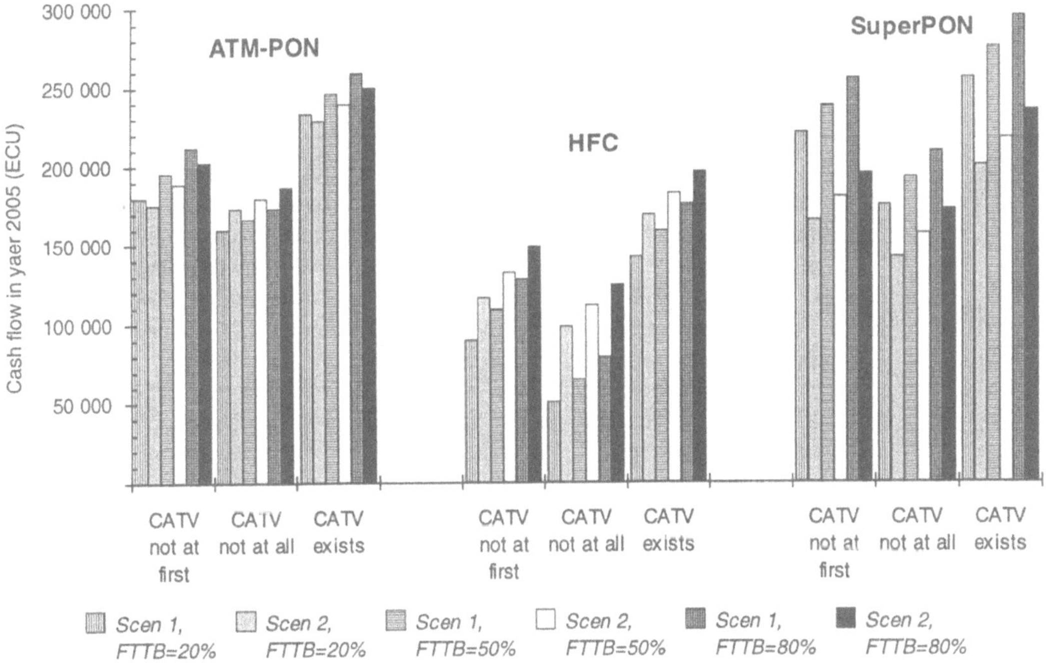

Fig 17.8 Cash flows in the year 2005.

With the tariffs, service take up scenarios and equipment price cost evolution assuming an FTTH based SuperPON solution have gained strength compared to the FTTB/FTTC ATM PON and HFC solutions, and it represent as such a longer term oriented access network upgrade strategy.

17.6 SUMMARY

In all the situations analysed, the ATM PON solution seems to have similar or lower costs than the other solutions. The HFC technology is an economically viable and attractive solution if the PNO has a CATV network already available and broadband interactive service penetrations are moderate (scenario 2). In that case the possibility of a timely upgraded HFC access network architecture improves the economy of the project and prevents losses from too early investments. However, an access network upgrade with HFC technology is not an attractive solution if the PNO does not have a CATV network, is not allowed to offer CATV service or broadband interactive service penetrations are more optimistic (scenario 1).

The cost of the HFC architecture is more sensitive to the service take up than the other architectures. The installed first costs and life cycle costs of HFC architecture are 40-100 % higher in scenario 1 than in scenario 2. For comparison, the similar cost variations of the other architectures ranges from 10 % to 30 %.

The advanced FTTH SuperPON is clearly more expensive than ATM PON technology in an FTTC/FTTB configuration. Based on all the 18 situations analysed, the FTTH PON costs were 52 % higher in average than the ATM PON.

The study shows that the housing structure has a considerable influence on the installed first costs and life cycle costs. Typically, we get 15-40 % higher costs in small house dominated areas (80 % of the customers live in small houses) than in more dense areas (only 20 % of the customers live in small houses). An exception is the ATM PON architecture, which shows negligible cost differences in situations where no new CATV network is built. This is due to the fact that existing copper pairs are used in the drop section.

In this study the PNO has a good existing customer revenue base. Hence, the PNO has the strength needed for upgrading the existing access network for broadband services. The overall profitability of the broadband access network upgrading scenarios changes dramatically if the revenues from existing services are not included in the total upgrade project.

This document is based on results achieved in a EURESCOM project; this does not imply that it reflects the common technical position of all the EURESCOM shareholder/parties.

REFERENCES

[1] Mononen, J, Lähteenoja, M. Broadband access alternatives in the residential dominated area, a techno-economic study. *In: Proc. EURESCOM Workshop on Optical Multiservice Access Networks (WOMAN), Turin, Italy, September 23-24, 1996.*

[2] Myhre, D, Ims, L A, Olsen, B T, Lähteenoja, M, Mononen, J, Ferrero, U, Vangelista, L. Roll out Strategies and Upgrade Access Alternatives for the Mixed Residential and Business Market: A Techno-economic Evaluation. *In: Proc. Networks and Optical Communications '96 (NOC '96), Heidelberg, Germany, June 24-28, 1996, 143-150.*

[3] Stern, J R. Deployment options for broadband networks. *In: Proc. 7th International IEEE workshop on Optical Access Networks, Nuremberg, Germany, September 24-28, 1995, 1.3-1 - 1.3-14.*

[4] Olsen, B T, Zaganiaris, A, Stordahl, K, Ims, L A, Myhre, D, Øverli, T, Tahkokorpi, M, Welling, I, Drieskens, M, Kraushaar, J, Mononen, J, Lähteenoja, M, Markatos, S, De Bortoli, M, Ferrero, U, Ravera, M, Balzaretti, S, Fleuren, F, Gieschen, N, De Oliveira Duarte, M, de Castro, E. Techno-economic evaluation of narrowband and broadband access network alternatives and evolution scenario assessment. *IEEE Journal on Selected Areas in Communications,* 14 (6), 1996, 1184-1203.

Broadband upgrades in business areas

18

Leif Aarthun Ims, Borgar Tørre Olsen, Dagfinn Myhre

This chapter presents a techno-economic assessment of optical access network upgrade technologies for subscribers with high capacity demands [1,2]. The major contribution is an in-depth analysis of broadband upgrade alternatives in the mixed residential and business market. The study focuses on physical layer upgrades, and examines the installed first costs of upgrading the typical one-way, point-to-point fibre connection of current large business customers to higher capacity with a two-way connection for protection, as well as introducing broadband passive optical networks for medium-, small business- and residential customers, with 2 Mbit/s as the basic bearer service for these segments.

18.1 INTRODUCTION

Today the development of the existing leased line/point-to-point business specific network infrastructure to a switched broadband network with introduction of optical fibre access technologies is among the major challenges of the telecommunications network operators. The operators will most likely have to develop the traditional overlay access networks along a cost effective migration path to future proof fixed broadband access networks accommodating a wide range of demographic diversity for an increasingly more diverse market demand, both with respect to customer groups, service types, terminal mobility and bandwidth requirements Appropriate upgrade strategies will be crucial in the highly competitive environment expected at the end of the century.

Previous work in this field has emphasised overall project evaluations of specific mixed market areas, rather than sensitivity analysis of the key factors of influence [3,4]. The latter is targeted in this work. The chapter presents an analysis of the cable infrastructure costs (fibre, civil works and installation only) of different network roll-out strategies for two-way protection in a business dominated access area. In addition, the overall upgrade costs were calculated, taking into account the costs of applying different broadband upgrade technologies to the upgraded cable infrastructure during a ten year project period. Four technical alternatives have been studied for high capacity demand (in excess of 2 Mbit/s), including ring and point-to-point solutions based on asynchronous transfer mode (ATM) or synchronous digital hierarchy (SDH). For 2 Mbit/s subscribers enhanced copper alternatives and broadband passive optical networks (PONs) have been analysed. The work has been carried out within the EURESCOM (European institute for research and strategic studies in telecommunications) project P413 "Optical Networking" [5]. The study is split in three parts:

1. Analysis of a cable infrastructure costs for protection
2. Analysis of different high capacity fibre systems applied to the upgraded cable infrastructure
3. Analysis of different service scenarios for the various fibre upgrade alternatives.

18.2 DEMOGRAPHICS

A typical mixed demographic area has been studied, in which guidelines for network evolution mainly are set by the high capacity demands of large business customers. There are three customer segments in the business dominated access area under study: large business customers, medium business customers and residential- and small business customers. The case specific demographics of the area under study are shown in Table 18.1

Table 18.1 Case demographics.

	Distribution of subscribers	*Number of subscribers*	*Housing structure*	*Number of buildings*
Large business	5 %	51	1	51
Medium business	25 %	256	8	32
Residential and small business	70 %	717	16	45
Total	100 %	1024		128

18.3 SERVICE PORTFOLIO

The provision of a set of bearer services in a mixed residential and business market during the period from 1996 to 2005 is investigated. Only bearer services are addressed. It is assumed that plain old telephone service (POTS), narrowband integrated services digital network (N-ISDN), 64 kbit/s and 2 Mbit/s leased line type bearer services are already offered over the public network operator's (PNO's) existing network.

The residential and small business market service set has been selected based on the Delphi survey, carried out within the project RACE (research for advanced communications in Europe) 2087/TITAN (tool for introduction strategies and techno-economic evaluation of access network) [6]. 2 Mbit/s advanced asymmetric switched broadband services (ASB) like service on demand (SOD) or video on demand (VOD) will be required by some customers. In addition, a 2 Mbit/s symmetric switched broadband (SSB) bearer service will be provided to some residential and small business customers. For the medium to large business market, representative figures for the ranges of demand are selected from available statistic material. Three alternative bearer services have been considered: n·64 kbit/s, 2 Mbit/s and 34 Mbit/s, with POTS/ISDN included in the two last services. Table 18.2 summarises the bearer services defined at the start and at the end of the study period.

Table 18.2 The set of bearer services at the start and at the end of the study period.

	1996	*2005*
Large business	2·2 Mbit/s	34 Mbit/s
Medium business	n·64 kbit/s + POTS	2 Mbit/s
Residential/small business	POTS, N-ISDN, 2 Mbit/s asymmetric and 2 Mbit/s symmetric	POTS, N-ISDN, 2 Mbit/s asymmetric and 2 Mbit/s symmetric

The nominal penetrations of services are shown in Fig 18.1, Fig 18.2 and Fig 18.3. The large business market is described by a linear transition of capacity demand from 2·2 Mbit/s to 34 Mbit/s during the study period and an upgrade to a two-way fibre connection for protection (Fig 18.1).

The medium business customers replace their n·64 kbit/s connection and ten additional POTS lines with 2 Mbit/s (Fig 18.2). The penetration of POTS (ten lines per subscription) and n·64 kbit/s evolves similarly during the period, as shown in Fig 18.2.

In the residential and small business market the number of narrowband lines remains constant, with a transition of 20 % of the POTS lines to N-ISDN (Fig 18.3). The total penetration of 2 Mbit/s services (asymmetric and symmetric) in this market segment grows to around 10 % during the period, in accordance with results from the TITAN Delphi Survey [6].

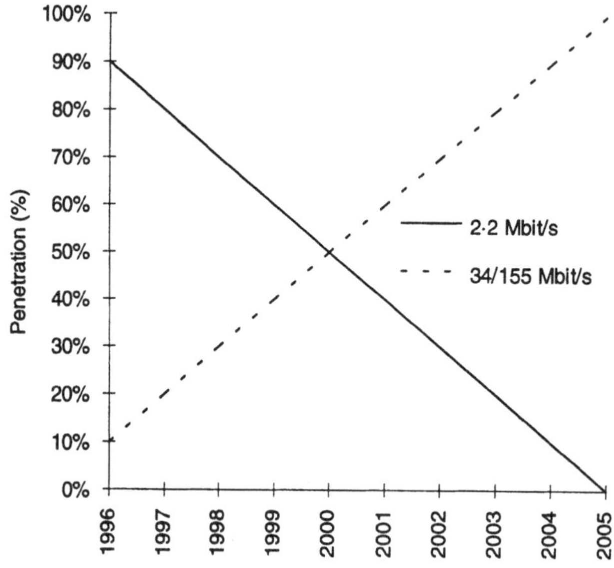

Fig 18.1 Penetration of services in the large business market.

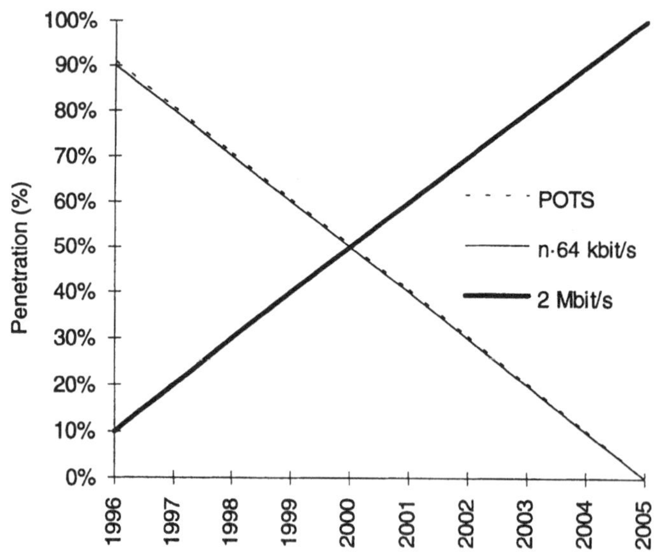

Fig 18.2 Penetration of services in the medium business market.

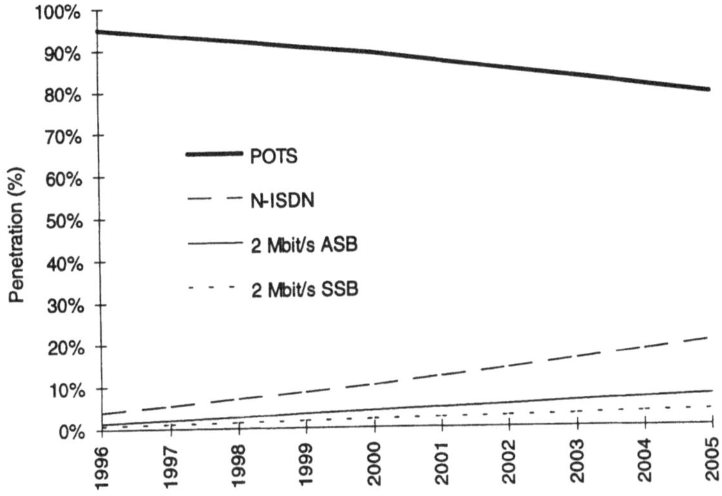

Fig 18.3 Penetration of services in the residential and small business market.

Bearer service tariffs have been included in the calculations, in addition to the service penetrations, in order to produce upgrade project revenues. The tariff evolutions for the three different market segments are shown in Fig 18.4, Fig 18.5 and Fig 18.6. The tariff structure used for the calculations is summarised in Table 18.3.

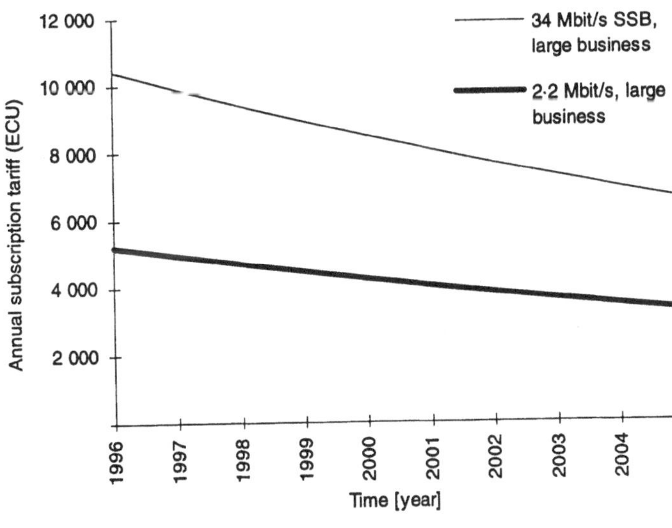

Fig 18.4 Tariff evolution for the bearer services in the large business market.

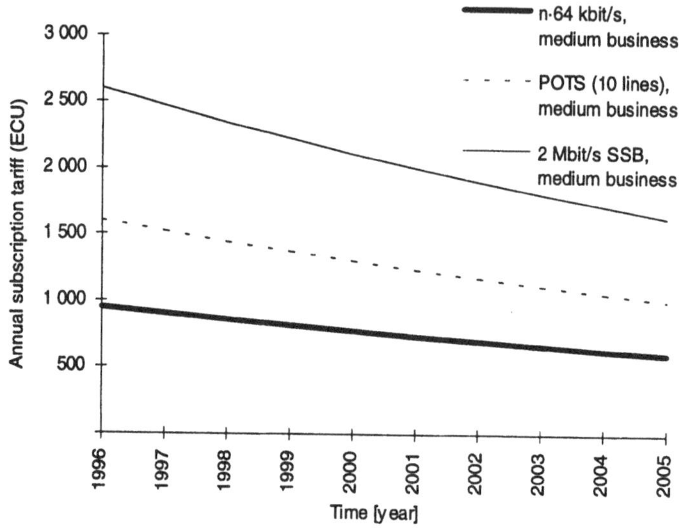

Fig 18.5 Tariff evolution for the bearer services in the medium business market.

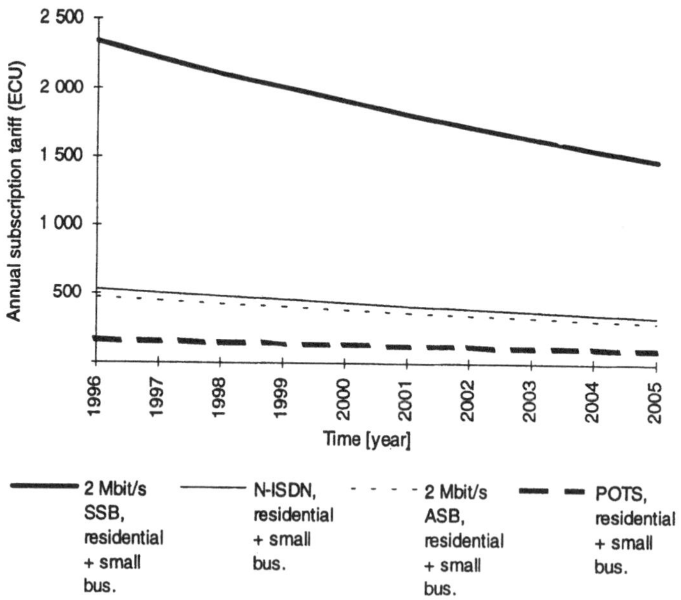

Fig 18.6 Tariff evolution for the bearer services in the residential and small business market.

Table 18.3 The tariff structure used in the calculations.

Service	*Tariff 1996 (ECU)*	*Tariff 2005 (ECU)*	*Annual reduction*
POTS	160	160	0 %
ISDN	304	192	5 %
CATV	100*	100*	0 %
2 Mbit/s ASB, residential	414**	160**	10 %
2 Mbit/s SSB, residential	2,070***	802***	10 %
64 kbit/s leased line	903	569	5 %
2 Mbit/s SSB, business	2,340	907	10 %
34 Mbit/s SSB, business	9,360	2,626	10 %

*: *The average tariff of ~200 ECU is split in a 50 % share to the content provider and 50 % to the network operator.*
**: *Based on the TITAN Delphi survey, tariff elasticity curve, 1 % demand. Only 20 % of the total 2,070 ECU annual tariff (i.e. 414 ECU) in 1996 is allocated to the network operator.*
***: *Based on the TITAN Delphi survey, tariff elasticity curve, 1 % demand.*

18.4 EXISTING NETWORK

In the analysis only the network between the local exchange (LEX) and the customer is considered. The existing network is shown in Fig 18.7.

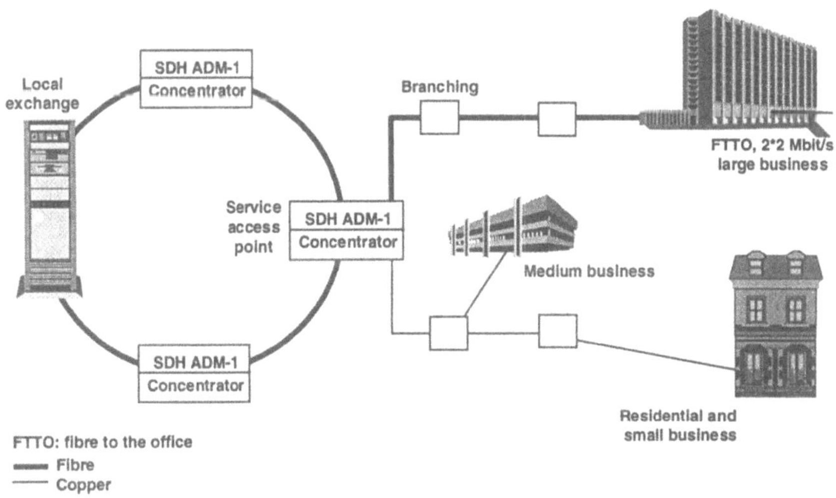

Fig 18.7 The existing network in the area with a one-way fibre connection network for large business customers and a twisted pair copper network for the other customer groups.

All large business customers are connected to the service access point (SAP) with a point-to-point connection, as shown in Fig 18.7. The existing LEX serves typically 4,096 customers, which are uniformly distributed over the area. Each SAP serves 1,024 subscribers, and four SAPs are connected in an SDH-ring structure to the LEX. The SAP is equipped with an SDH add and drop multiplexer (ADM). The average total cost per metre installed fibre depends on the civil work and duct availability. The product of civil works costs per metre and duct unavailability is called total civil work (TCW). The TCW of the distribution network is approximately 12.5 ECU/m.

18.5 CABLE INFRASTRUCTURE UPGRADE FOR PROTECTION

The cable infrastructure costs of instantaneous fibre roll-outs for various civil work costs have been analysed and compared to the total upgrade cost, which also includes upgrade to higher capacity per node. The cable infrastructure is upgraded from a one way connection network with one fibre pair per node (Fig 18.7) to a protection network with two-way fibre connections in separate network paths between the customer premises and the SAP, as illustrated in Fig 18.8.

Both the average loop length (subscriber density) in the area and the number of large business locations are input parameters which impact the costs of such cable infrastructure upgrades. Thus, we have examined areas characterised by different combinations of average loop length and large business locations.

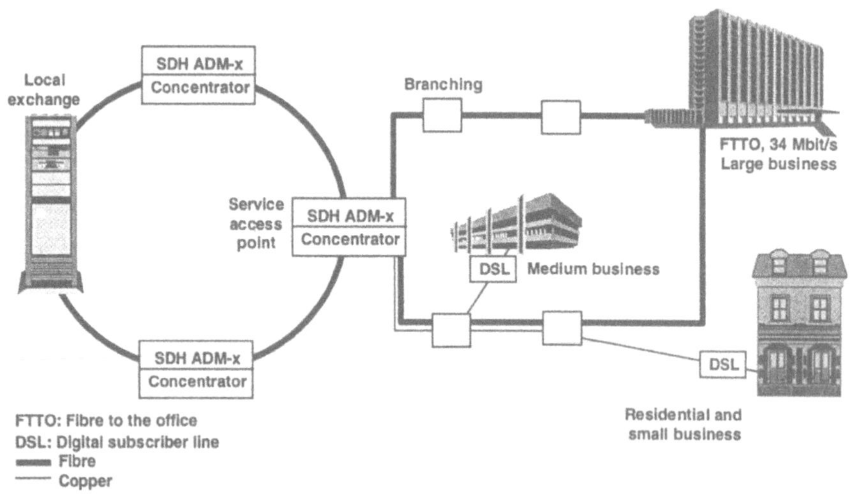

Fig 18.8 The upgraded network is a protection network with two way fibre connections in separate network paths between the large business customer premises and the SAP.

Three different subscriber densities have been analysed: 300, 1,000 and 3,000 subscribers/km^2. The corresponding radius of the network area for these three TITAN subscriber densities are 1,100 m, 600 m and 350 m respectively. The existing cable infrastructure is assumed to be a full star network from the SAP to the respective nodes, with two fibres to each node. The additional cabling provides an additional fibre pair to each node along a different physical path.

The upgrade cost for a two way connection have been calculated for four different numbers of large business customers in the area, namely 32, 48, 64 and 80 large business nodes. There is one large business customer per large business node or building, located in the FP1 and/or FP2 in the modified SYNTHESYS geometric model, which is the model most commonly used in published techno-economic studies in which TITAN has been applied [7].

- 32 nodes: large business customers located in 32 FP1 points;
- 48 nodes: large business customers located in 32 FP1 and in 16 FP2 points;
- 64 nodes: large business customers located in 32 FP1 and in 32 FP2 points;
- 80 nodes: 80 large business customers located in 32 FP1 and in 48 FP2 points.

The study analysed different configurations of the nodes in the geometric model. As illustrated in Fig 18.9, additional crosslines were added to the geometric double star model in order to establish the two way physical ring structure for path protection.

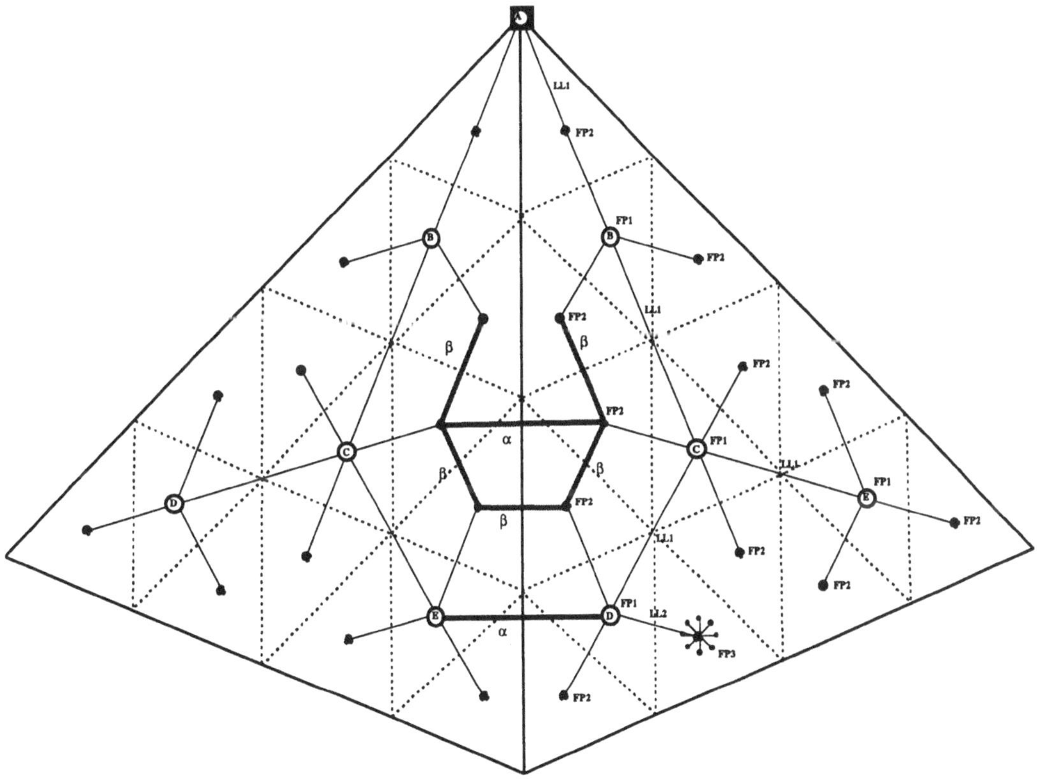

Fig 18.9 Additional connections in the geometric model. Two sectors of the model are shown.

Only one additional length in addition to the TITAN-length has been introduced, named α, which is calculated to be approximately 2πAB/8. All the other lengths, called β, are set equal to LL2. Fig 18.9 shows two sectors of the network model.

The non-greenfield assumption certainly implies that the upgrade project will not require installation of new ducts everywhere in the existing distribution network. Hence, in the upgrade project duct availabilities of 0.5 and 0 were assumed for fibre cable in the existing part of the network and the new crosslines respectively. The average distance (loop length) from the SAP to the large business customer premises is 420 m.

The different large business locations for the four situations are defined in a way such that the average loop length of the new points in the model are close to the average for all 128 possible FP1 and FP2 points.

18.6 COST ANALYSIS RESULTS OF CABLE UPGRADES

Fig 18.10 shows the upgrade costs and the existing network costs as a function of the number of nodes, with a duct availability 50 % and civil work costs of 25 ECU/m for both projects.

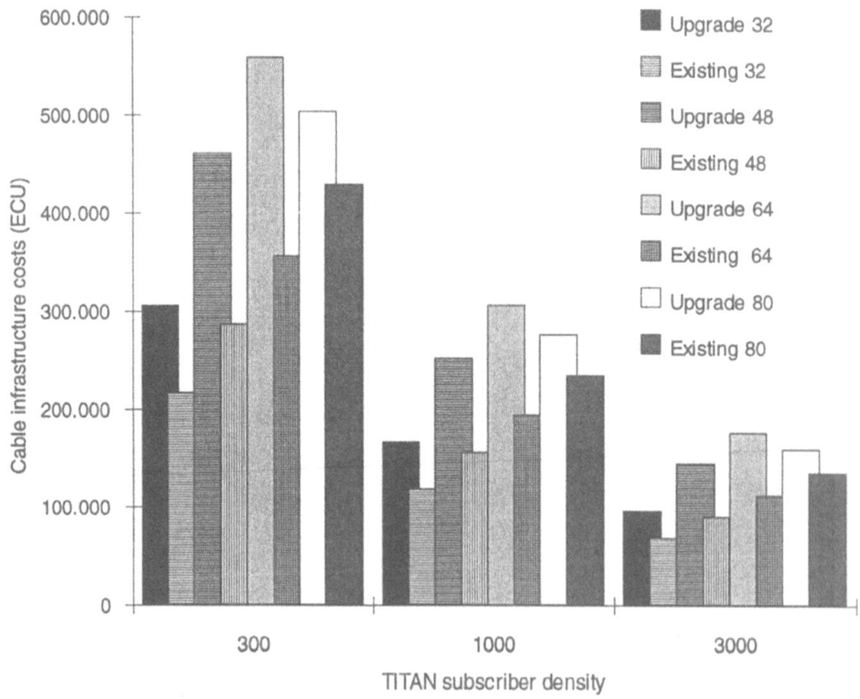

Fig 18.10 Upgrade costs and existing network costs as a function of number of nodes, with duct availability of 50 % for both projects.

Civil work costs of 25 ECU/m represent an average European minimum level as used in several TITAN studies [7]. In central city areas civil works costs as high as 50 ECU per metre are not unusual. However, in such areas there are often more ducts available than assumed in this study, which to some extent may compensate for even higher civil works costs per metre. The results show that for the business node densities studied the cost per node of installing a one way connection is only loosely dependant on the total number of nodes, as is also true for the cost of an upgrade to a two-way connection per node.

In Table 18.4 the costs for establishing a one way connection (existing) and upgrading to a two-way connection are shown for different total civil work costs. In brackets the pure civil work costs per node without fibre and installation are indicated. The costs of the upgrade project divided by the costs of the existing network for the same civil work costs are about 1.3.

In Fig 18.11 the cost of the upgrade to a two-way connection divided by the cost of the existing network is shown, with different sets of duct availability for the upgrade and the existing network. I.e. the dotted line in Fig 18.11 depicts the cost ratio for a situation in which the initial network was deployed with 0 % duct availability, whilst the duct availability for the upgrade is 50 %. This cost ratio is slightly higher for 48 and 64 big business sites compared to 32 and 80 nodes.

Table 18.4 Costs of existing and upgrade network projects as a function of total civil work costs per metre. The civil work costs without fibre and installation are shown in brackets

TCW/m (ECU)	*Existing one way cost per node (kECU)*	*Upgrade to two way cost per node (kECU)*
0	0.73 (0)	1.5 (0)
12.5	3.3 (2.6)	4.7 (3.2)
25	5.9 (5.2)	7.9 (6.4)
50	11 (10)	14 (12.8)

Fig 18.11 and Table 18.4 illustrate that the cable infrastructure costs of upgrading a one-way connection to a two-way full protection network are at the same cost level as the initial one-way infrastructure costs, or even higher, depending on the condition of the duct layout in the area.

The cost of upgrading a network with 48 nodes was studied in detail. Fig 18.12 shows the cabling structure with 48 nodes. This node density has been modelled with 32 of the large business customers located in FP1 points and the remaining located in 16 selected FP2 points. The solid lines represent the cable and duct layout.

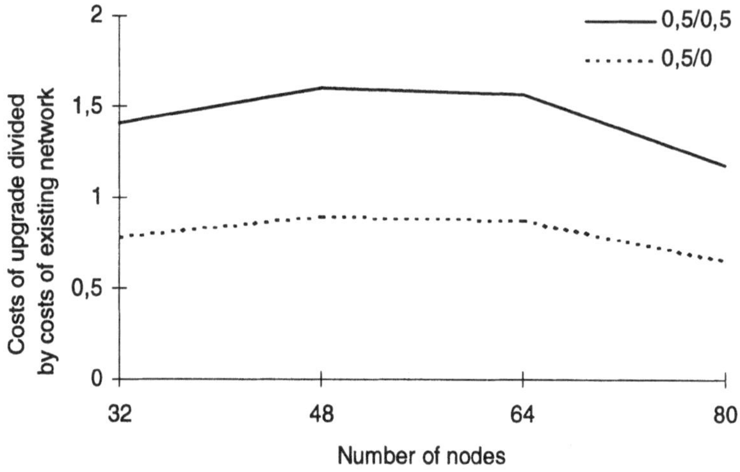

Fig 18.11 Costs of upgrade divided by costs of existing network for different number of optical nodes in the area.

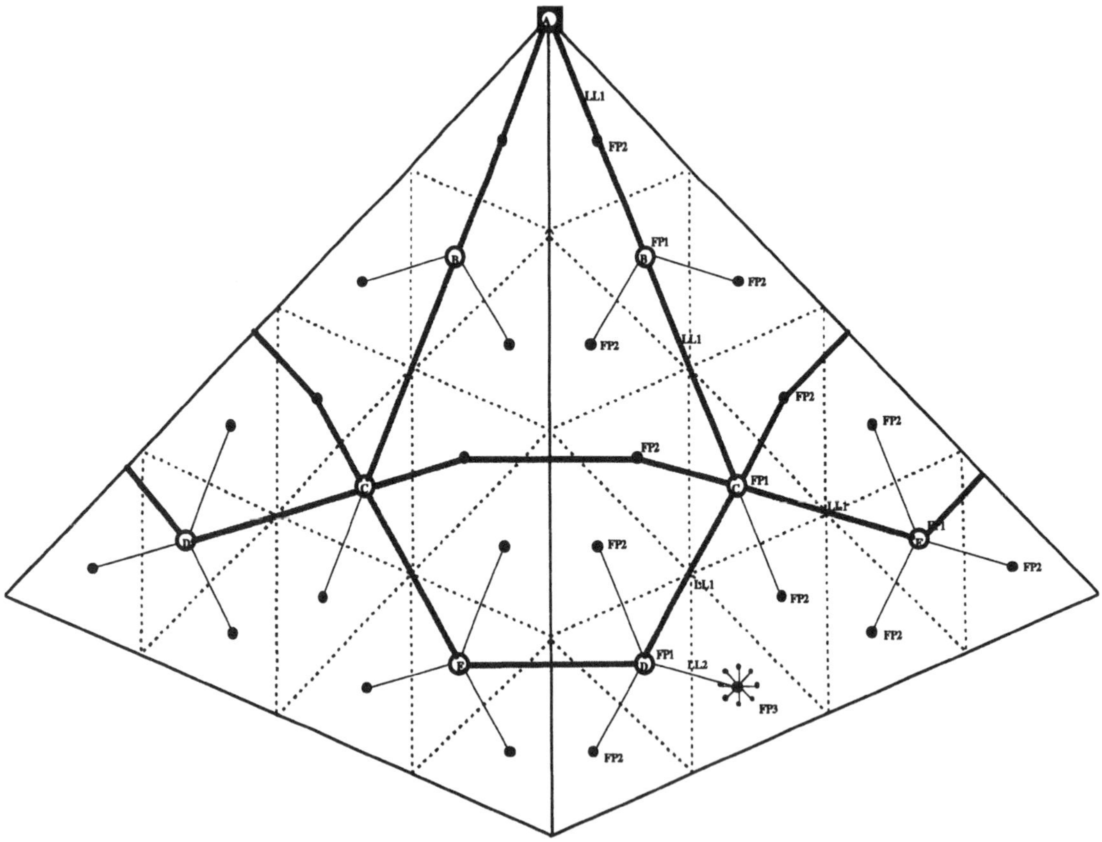

Fig 18.12 Cabling structure with 48 nodes.

Fig 18.13 shows the cost of the existing one-way connection to the buildings for three different subscriber densities (average loop lengths) as a function of duct availability. Similarly, Fig 18.14 shows the cost of the upgrade to two-way connections for the three same subscriber densities as a function of duct availability.

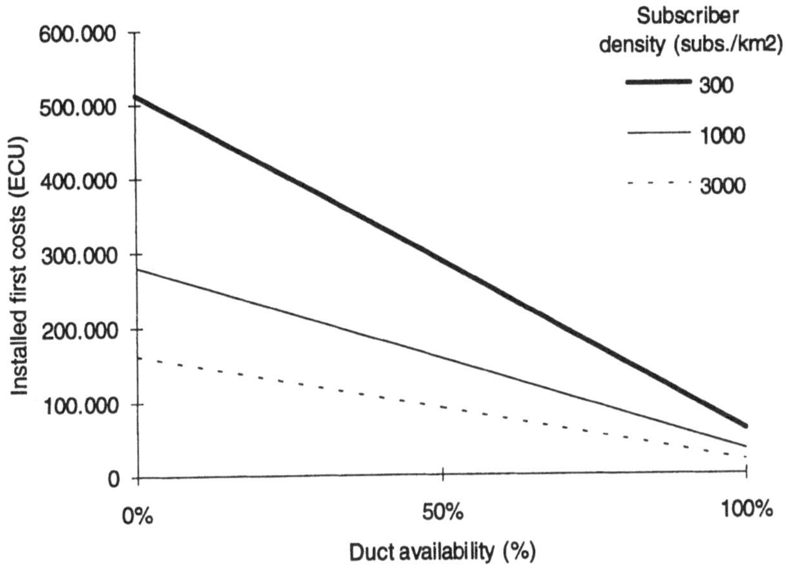

Fig 18.13 Cost of the existing one-way connection to the buildings for three different subscriber densities as a function of duct availability.

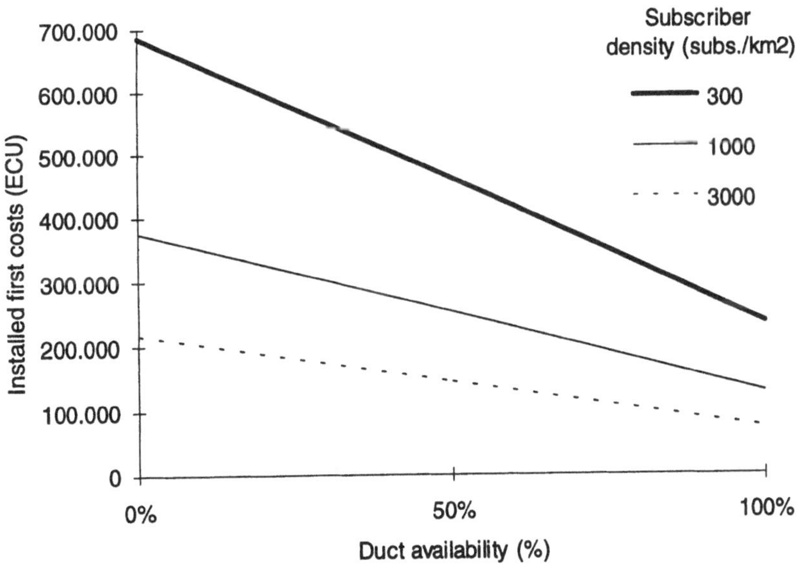

Fig 18.14 Cost of the upgrade to two-way connections for three different subscriber densities as a function of duct availability.

A comparison of the corresponding results in Fig 18.13 and Fig 18.14, shows that for all subscriber densities at 0 % duct availability, the cost of the upgrade (Fig 18.14) is 1.3 times the cost of the existing network (Fig 18.13). Similarly, the cost of the upgrade is 1.6 and 3.8 times the cost of the existing network for duct availabilities of 50 % and 100 % respectively.

18.7 UPGRADE ALTERNATIVES FOR 34 MBIT/S CAPACITY

Optical access network upgrade technologies for subscribers with high capacity demands (in excess of 2 Mbit/s) have been analysed for a ten year upgrade period starting in 1996. The four technical alternatives examined for the large business customers are SDH point-to-point architecture (Fig 18.15), SDH ring architecture (Fig 18.16), ATM point-to-point architecture (Fig 18.17) and ATM ring architecture (Fig 18.18). For all upgrades it is assumed that there already exists an SDH ring in the primary network between the LEX and the SAP. The SAP is equipped with an SDH ADM and connected to the LEX in a SDH ring structure. The upgrade refers to the secondary network between the SAP and the customer premises. The existing network in the secondary network consists of a one-way fibre connection between the SAP and the large business customers, which are provided with 4 Mbit/s, with no protection. Through the upgrade one new optical connection from the SAP to the customer is established for protection, in addition to the already existing one-way connection. This results in two separate optical connections from the customer to the SAP.

Initially, a fibre rich, system independent cable infrastructure is rolled-out, i.e. the cables contain enough fibre pairs to serve every of the large business nodes. This allows flexibility with respect to succeeding technology choices. In the cost calculations a faster price reduction within the ten year period of the new ATM technology has been assumed compared to the current SDH technology. The customer premises equipment is installed according to the take rate and the distribution network fibre infrastructure upgrade costs for the 34 Mbit/s connections are equally spread over the ten year period.

18.7.1 SDH point-to-point architecture

The SDH point-to-point architecture is shown in Fig 18.15. The figure also includes the enhanced copper architecture for the small and medium enterprises (SMEs) and the residential customers, based on asymmetric digital subscriber line (ADSL) and high speed digital subscriber line (HDSL). A 155 Mbit/s SDH terminal multiplexer (SDH-TM) is installed at the large business customer premises. The customer is connected through a distribution panel to a 34 Mbit/s tributary card with an optical interface in the SDH-TM, which in Fig 18.15 is depicted as A34M (34 Mbit/s ATM user network interface (UNI)).

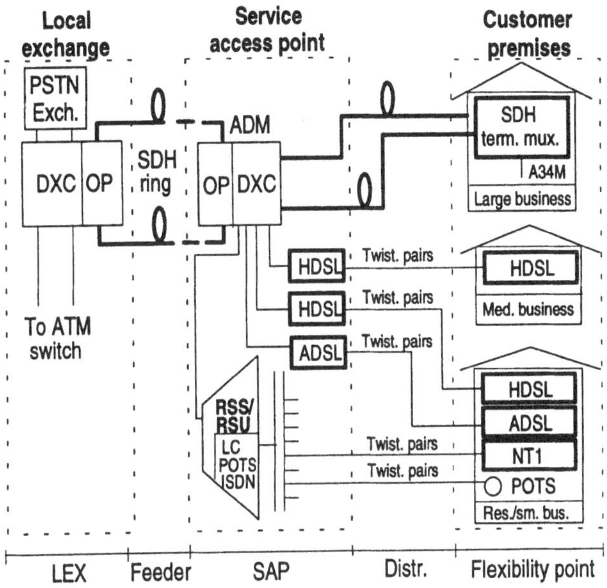

Fig 18.15 SDH point-to-point architecture with enhanced copper for the SMEs and the residential customers.

18.7.2 SDH ring architecture

Fig 18.16 depicts the SDH ring architecture. The enhanced copper architecture for the SMEs and the residential customers is included in the figure. A 155 Mbit/s SDH ADM is installed at the large business customer premises in a locked room. The customer is connected through a distribution panel to 34 Mbit/s tributary card with an optical interface in the SDH ADM. This is shown in the figure as A34M (34 Mbit/s ATM UNI). There are three nodes per ring connecting the customers to the SAP.

18.7.3 ATM point-to-point architecture

The ATM point-to-point architecture is illustrated in Fig 18.17. The illustration includes the ATM based passive optical network (PON) architecture for the SMEs and the residential customers, illustrating an upgrade strategy in which the synergy effect between the different customer groups is utilised through common ducting and sharing of civil works costs in the part of the secondary network closest to the SAP. The ATM PON is directly connected to the LEX via a separate feeder cable. The ATM PON ONU is installed in the basement of the apartment blocks. A 155 Mbit/s broadband optical network unit is installed at the large business customer premises. The customer is connected through a distribution panel to the 34 Mbit/s ATM UNI line card with an optical interface in the ONU.

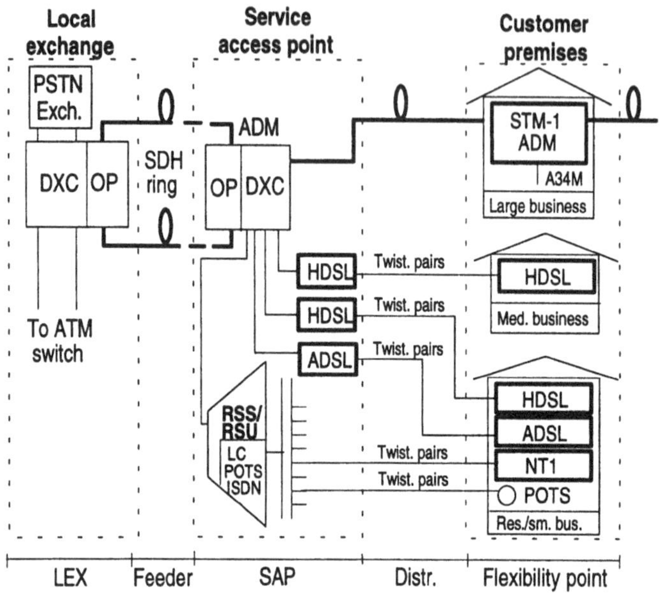

Fig 18.16 SDH ring architecture with enhanced copper for the SMEs and the residential customers.

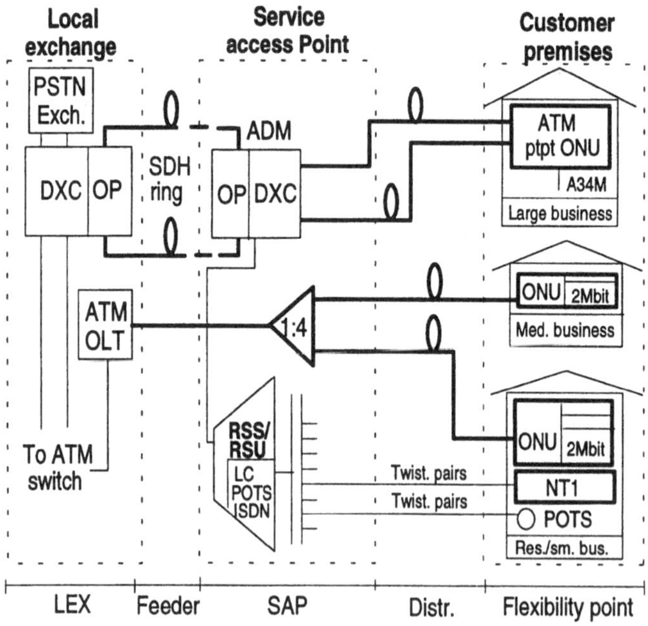

Fig 18.17 ATM point-to-point architecture with a parallel ATM PON for the SMEs and the residential customers.

18.7.4 ATM ring architecture

The ATM ring architecture is shown in Fig 18.18. In addition, the enhanced copper architecture for the SMEs and the residential customers is depicted. A 155 Mbit/s broadband optical network unit is installed at the large business customer premises in a locked room. The optical network unit is connected to the primary network SDH ring by a interworking unit (IWU). The customer is connected through a distribution panel to 34 Mbit/s ATM UNI tributary card with an optical interface in the ONU. There are three nodes per ring.

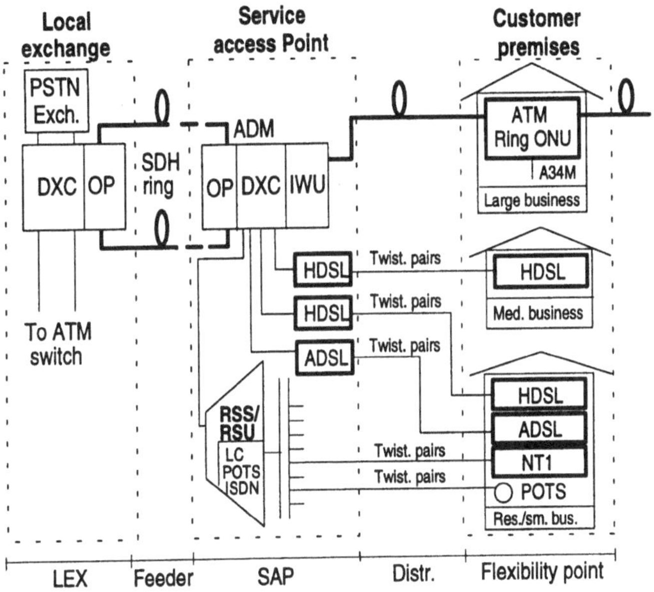

Fig 18.18 ATM ring architecture with enhanced copper for the SMEs and the residential customers.

18.8 ANALYSIS RESULTS, LARGE BUSINESS CUSTOMERS

The cost analysis shows that all the four upgrade alternatives have the same level of costs. The baseline equipment costs of any of the upgrade alternatives studied are between 6,000 ECU and 10,000 ECU per 34 Mbit/s connection. On average the 34 Mbit/s upgrade costs are equally split in three parts: customer premises costs, cable infrastructure costs and centrally located costs (LEX and SAP). In general, the electronics costs dominate in all solutions. The results show that even a fibre rich roll-out upgrade for two-way fibre connections in places where a significant number of the buildings have a fibre connection prior to the upgrade, may have small infrastructure costs compared to the overall upgrade costs.

Providing two-way connections for protection is twice as expensive as a pure capacity upgrade to 34 Mbit/s, given the assumption that there already exists a one-way fibre connection to the large business customer. In addition, the cable infrastructure costs may double the overall upgrade costs, depending on the duct availability and the civil works costs in the area.

18.8.1 Cost breakdown

Fig 18.19 shows the accumulated installed first costs (IFC) per 34 Mbit/s line as a function of network segment. The central network sections do not contribute substantially to the overall costs of broadband upgrade projects, except for the ATM point-to-point solution, which is discussed in greater detail later. The SDH based solutions and the ATM ring alternative have centrally located costs (LEX and SAP) on the same level. Nevertheless, the ATM ring alternative is approximately 20 % less expensive than the other upgrade options due to the significantly lower customer premises equipment costs for the ATM ring, 40-50 % of their SDH counterparts, combined with slightly cheaper SAP equipment.

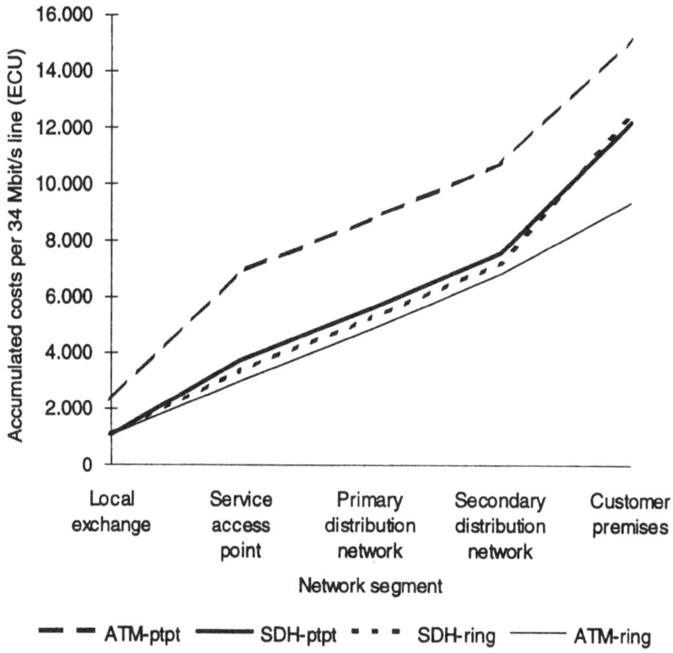

Fig 18.19 The accumulated IFC per 34 Mbit/s line as a function of network segment.

18.8.2 Costs of transport versus costs of switching

The high capacity requirements in the local transport network for a rapid evolution of 34 Mbit/s demand may be alleviated by providing access in the SAP to the statistical multiplexing properties inherent in ATM. Fig 18.19 illustrates the cost benefits of distributed ATM functionality, like in the ATM ring solution, relative to the higher transport capacity needs incurred by centrally located ATM functionality, like in the ATM point-to-point alternative.

In our study ATM point-to-point upgrades have higher costs than the SDH ring, SDH point-to-point and ATM ring alternatives. In the ATM point-to-point upgrade the customers are provided with a 34 Mbit/s ATM based user interface over a 155 Mbit/s transmission line rate. This leads to higher capacity requirements in the feeder network, due to the fact that this ATM based 155 Mbit/s bitstream is assumed to be transported all the way to the nearest ATM switch (in this case located in the LEX).

18.8.3 Number of nodes on the ring

Furthermore, the flexibility with respect to number of nodes on the ring is limited in SDH rings. There are three subscriber nodes on each ring for all the SDH and ATM ring solutions in the study. Increasing the number of nodes on the ring increases the costs of the customer premises equipment on the SDH ring by 25-30 % due to the need for upgrading the customer premises terminal equipment to synchronous transfer module (STM)-4. In contrast, the ATM ring costs are slightly reduced. This illustrates the benefits of the concentrating capability inherent in the ATM ring technology.

18.8.4 Flexibility points with or without active electronics

We have examined the difference in cost between upgrading an existing 1,024 homes passed flexibility point with no active electronics installed to a service access point (SAP) with an SDH node or alternatively, keeping this network point passive also after the upgrade. The results are shown in Fig 18.20.

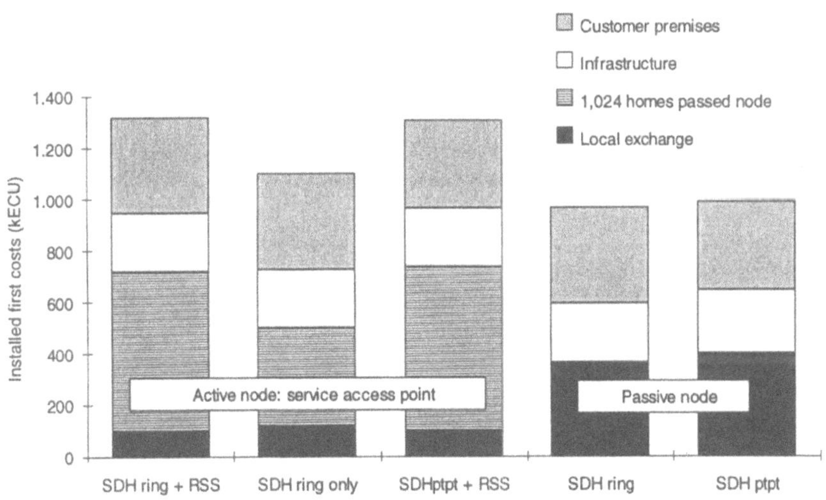

Fig 18.20 The IFC for the upgrade of an existing 1,024 homes passed flexibility point with no active electronics installed prior to the upgrade. Results are shown both for an upgrade to an active node (service access point) and for the case of keeping this network point passive also after the upgrade.

IFC are shown for three active node configurations and two passive node configurations after the upgrade. The cost breakdown shows customer premises costs, infrastructure costs and centrally located costs (1,024 homes passed node and local exchange). The upgrade to a SAP includes the establishment of a building, installation of the SAP located SDH ADMs for large business ring- or point-to-point configurations in the distribution part of the network, and an optional installation of a concentrator or a remote subscriber system (RSS) for POTS and N-ISDN lines. Two of the active node configurations include the introduction of such a concentrator for the narrowband subscribers in the area.

If the existing 1,024 homes passed flexibility point is upgraded to a SAP, an SDH ring structure is introduced in the feeder network between the LEX and the flexibility point, both for point-to-point and ring upgrades of the distribution part. However, if the flexibility point is retained as a passive network location, both the point-to-point connections and the ring structures are extended from the customer premises to the LEX.

For two-way protection upgrades there seems to be no significant cost difference between the ring and the point-to-point solutions, neither if the 1,024 homes passed point is upgraded to a SAP or not. The cost differences between the ring and the point-to-point solutions are marginal both in the LEX, SAP, the infrastructure and at the customer premises. In summary, the cost differences between point-to-point solutions and ring solutions are marginal. However, point-to-point solutions have a higher flexibility with respect to future customer specific/diverse capacity upgrades. For both ring and point-to-point alternatives, keeping the 1,024 homes passed point passive, i.e. simplifying the network structure by utilising optical fibre transmission technology, seems to give lower total investments than establishing a SAP.

18.8.5 The effect of civil works costs

Obviously the total civil works costs greatly depends on the duct availability and the civil works costs per metre. Fig 18.21 shows the IFC of four different large business architectures with enhanced copper solutions for SMEs. In addition, one architecture with fibre solutions for all customers, an ATM point-to-point architecture the large business segment with a parallel ATM PON for the residential, small- and medium business market. We have assumed a distribution network duct availability of 50 % and civil work costs of 25 ECU per metre. This yields a total civil work costs of 12.5 ECU per metre. Civil works, encompassing the digging, ducting and surface reinstatement, constitutes on average 20 % of the total upgrade costs.

Fig 18.21 shows that the total IFC of the upgrade ranges from one to three times the costs of equipment, depending on the penetration of fibre in the network, duct availability and the civil work cost. This is illustrated by the higher slope of the combined ATM point-to-point/ATM-PON solution.

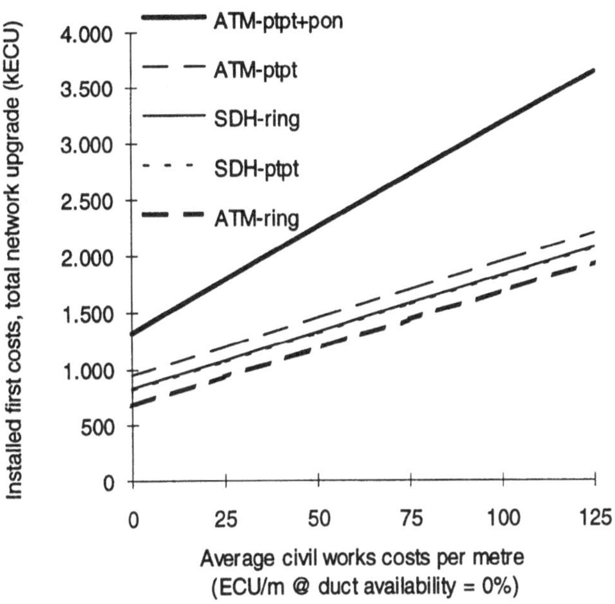

Fig 18.21 The IFC of four different large business architectures with enhanced copper solutions for SMEs, and one architecture with fibre solutions for all customers (ATM-ptpt + PON).

18.9 ANALYSIS RESULTS, RESIDENTIAL CUSTOMERS AND SMES

18.9.1 Optical network termination sharing

For the 2 Mbit/s demand in the residential, small and medium business market enhanced copper alternatives (HDSL and ADSL) were compared to a broadband upgrade with an ATM PON in a fibre to the building configuration (Fig 18.17). Fig 18.22 shows the IFC per 2 Mbit/s as a function of the average number of 2 Mbit/s subscribers per ONU. Results are shown for three TCW costs: 0 ECU/m, 25 ECU/m and 50 ECU/m.

In this case study the ATM PON is between two and three times more expensive than the enhanced copper upgrade. This is mainly caused by very few customers sharing each optical network termination, as pointed out above. On average there are four 2 Mbit/s subscribers per ATM ONU. The ATM PON solution benefits from cost reductions both due to reduced civil work costs and higher customer sharing of the ONUs if the ONUs are more centrally located. E.g. a four-fold increase in number of subscribers per ONU could probably reduce the costs per 2 Mbit/s line by at least 50 %. In addition, the ATM PON solution is very sensitive to civil works cost. It is evident that around 50 subscribers per optical node is the critical size for which ATM PON can be justified in pure economic terms, given the assumptions in this study.

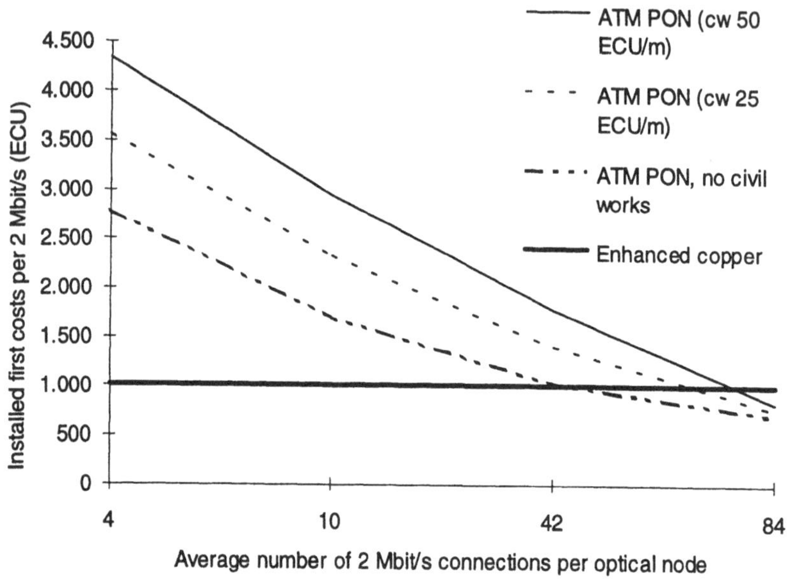

Fig 18.22 The cost per 2 Mbit/s as a function of number of subscribers per ONU.

18.10 SERVICE EVOLUTION SCENARIO ASSESSMENT

An optimistic, a pessimistic and a nominal service scenario have been studied in order to analyse the business opportunities of the upgrade options. In the nominal scenario the large and medium business customer segments convert their connections (2 Mbit/s and 64 kbit/s respectively) to higher capacities over the ten year period, as described in section 18.3. The optimistic scenario and pessimistic scenarios reflect faster (five years) and slower (20 years) transition rates than the nominal large and medium business conversion rates. Likewise the residential and small business market take rates in the optimistic and pessimistic scenarios are 50 % higher and lower than the nominal (Delphi study) take rate. During the period all the new broadband service tariffs are reduced annually by 10 %, the N-ISDN and the 64 kbit/s leased line tariffs by 5 % and the POTS tariffs are kept constant.

18.10.1 Revenue distribution

In the nominal and the optimistic scenario the total project revenue is almost constant during the study period (10 % reduction), as shown in Fig 18.23. In the pessimistic scenario the total revenue decreases with approximately 25 %.

In general, the medium business segment dominates, and make up almost two third of the initial total revenue volumes. During the upgrade the medium business share is reduced to one half of the final total revenue, while the revenues from the large business market slightly increase during the project period. The total residential/small business market revenues increase by more than 50 % during the period for the optimistic scenario, and decreases slightly in the pessimistic scenario. The residential and small business market dominates the total number of lines count, but initially only represent approximately 15 % of the total revenue volumes. The share of the revenues from this market segment increase slightly to about 20 % of the total market during the period.

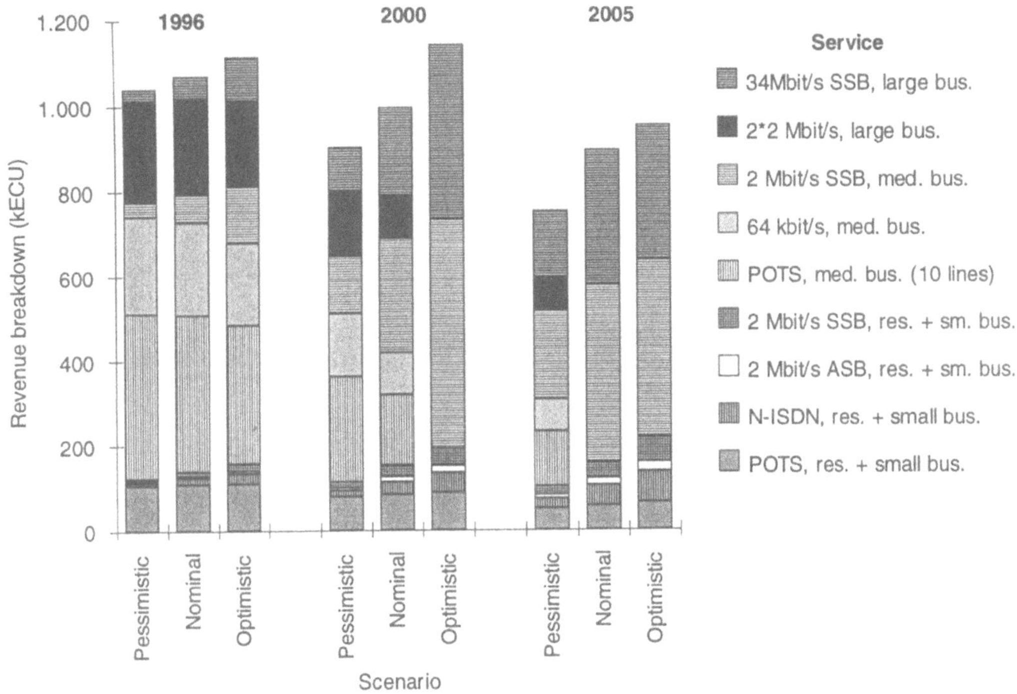

Fig 18.23 The development of the total revenues for the three service scenarios.

18.10.2 Line costs

The line costs for the 34 Mbit/s upgrade alternatives and the 2 Mbit/s upgrades are shown in Fig 18.24 and Fig 18.25 respectively.

The average costs per 34 Mbit/s line for the ten year upgrade period increase significantly as a function of the roll-out rate (high investments early in the project period), as illustrated by the higher optimistic scenario costs in Fig 18.24.

For the enhanced copper upgrade the 2 Mbit/s line costs are not significantly influenced by the different evolutions of the three scenarios, as seen from Fig 18.25.

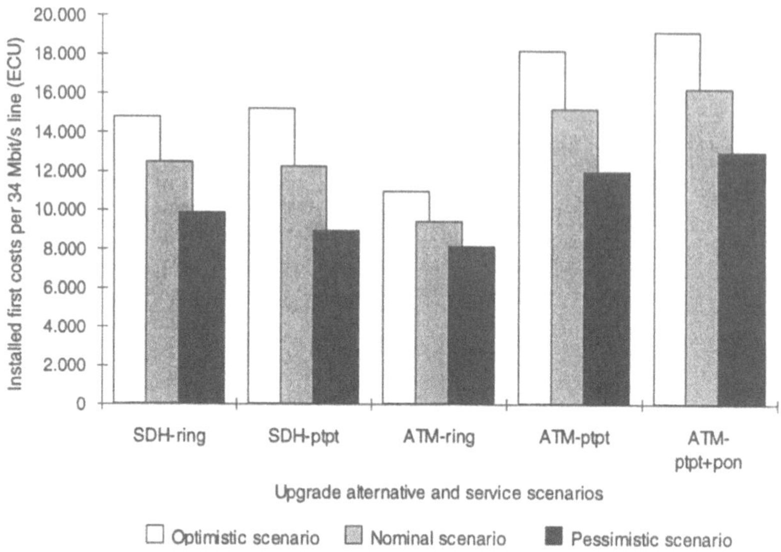

Fig 18.24 The average costs per 34 Mbit/s line for the ten year upgrade period.

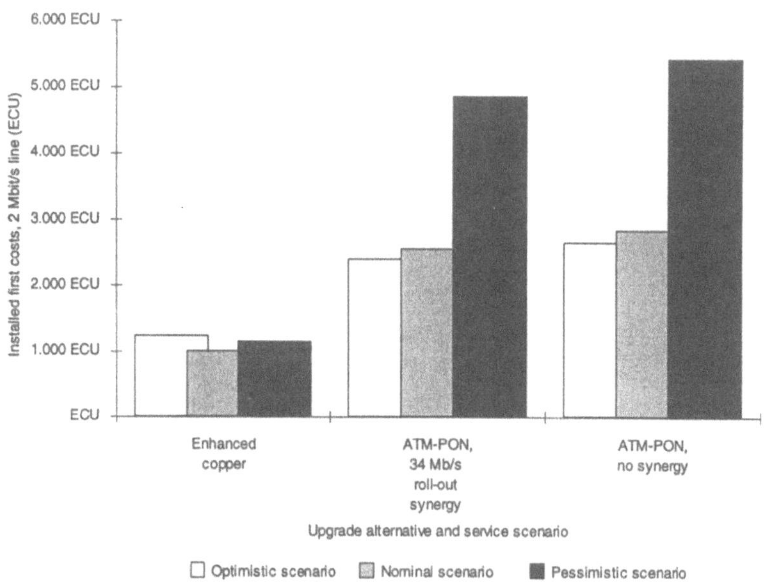

Fig 18.25 The average costs per 2 Mbit/s line for the ten year upgrade period. ATM PON upgrade costs are shown with and without the synergy effect of a simultaneous or preceding fibre network roll-out for the 34 Mbit/s subscribers in the area.

The 34 Mbit/s line costs in Fig 18.24 are higher for the upgrade alternative which combines an ATM point-to-point architecture for 34 Mbit/s customers with a parallel ATM PON for 2 Mbit/s customers than for the ATM point-to-point architecture with a parallel enhanced copper architecture. The former upgrade option illustrates a simultaneous roll-out of fibre infrastructure for both 34 Mbit/s customers and 2 Mbit/s customers in the initial year of the study period. However, in the latter case, and in all other 34 Mbit/s architectures examined, fibre is deployed to the 34 Mbit/s subscribers only at the time when the subscriber demands 34 Mbit/s capacity. Thus, in this case the cable infrastructure costs are spread throughout the study period, and there is no synergy between the network roll-out for the two market segments. This results in lower investments per 34 Mbit/s line due to the combined effect of price reduction and discounting.

Fig 18.25 shows the line costs for the 2 Mbit/s enhanced copper upgrade and for the ATM PON upgrade. For the latter costs are shown with and without the beneficial synergy effect of a simultaneous or preceding fibre network roll-out for the 34 Mbit/s subscribers in the area. In case of synergy, the civil works costs in the upper part of the distribution network are allocated to the 34 Mbit/s subscribers, as described previously. The ATM PON upgrades are more sensitive to the take rate, illustrated by a dramatic cost increase in the pessimistic scenario. The higher investments in the optimistic and nominal scenarios (rapid roll-out) is counterbalanced by the increased number of 2 Mbit/s connections per ONU in these scenarios as compared to the pessimistic scenario.

18.10.3 Net present value

Fig 18.26. depicts the net present value (NPV) for the 34 Mbit/s upgrades. The NPV for the 34 Mbit/s upgrades and the 2 Mbit/s upgrade are very sensitive to the service evolutions, as depicted by the high variations in NPV between the three scenarios in Fig 18.26. The figure shows the NPV of different upgrade alternatives for an existing 1,024 homes passed ring connected SDH node. Results are shown for the 34 Mbit/s upgrade only. Neither the costs nor the revenues from the residential and SME market are included. For the pessimistic scenario the NPV of the 34 Mbit/s upgrade project is on average one third of the upgrade investment, and as such the projects are not very efficient. This is illustrated in Fig 18.23 by payback periods in the range of ten years for the pessimistic scenario. For the nominal scenario the NPV is at the same level as the upgrade investment, on average 90 % of the investment level. Only for the optimistic scenario is the NPV larger than the investments, namely on average 1.35 times the upgrade costs. The 2 Mbit/s upgrade projects are relatively more heavy, except for the copper based upgrades, which yield NPV around three times the investment, independent of scenario. The ATM PON solutions are more prone to variations in take rates, with NPV close to zero for the pessimistic service evolution. Zero NPV represents return on investments within the project period.

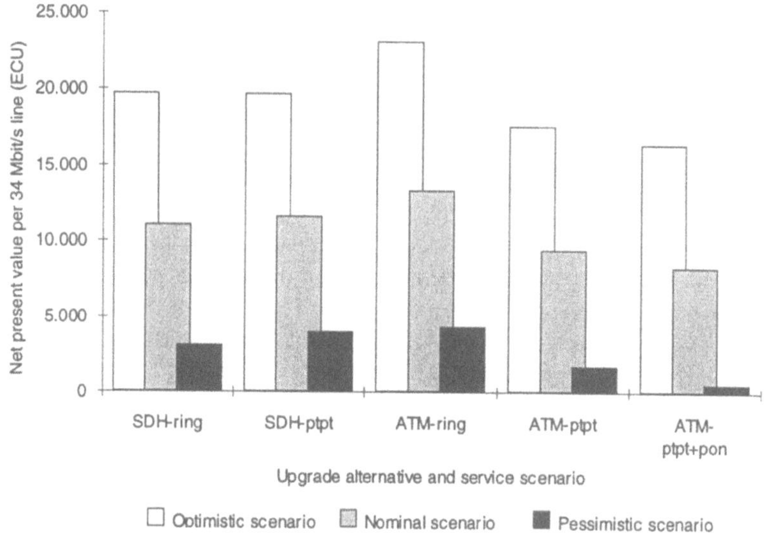

Fig 18.26 NPV of different upgrade alternatives for an existing 1,024 homes passed ring connected SDH node. Results are shown for the 34 Mbit/s upgrade only for three scenarios studied.

18.10.4 Payback period

Both upgrades have payback periods that are very sensitive to the service evolution, illustrated by a typical two fold increase in payback period from the optimistic to the pessimistic service scenarios. Fig 18.27 shows the payback periods of different upgrade alternatives for an existing 1,024 homes passed ring connected SDH node. Results are shown for the 34 Mbit/s upgrade only. Neither the costs nor the revenues from the residential and SME market are included. In the nominal scenario, the payback periods for the large business segment upgrade are typically around five years, independent of technology choice.

In addition, the analysis shows that the upgrade for the 2 Mbit/s residential and SME market with the ATM PON technology has payback periods very similar to large business segment upgrades. The enhanced copper upgrade alternative for the residential and SME market shows payback periods shorter than two years.

The significance of the 2 Mbit/s market (residential, small- and medium businesses) confronts the access network operator with the strategic choice of offering this customer group a future proof, full infrastructure fibre upgrade with ATM PON or maintaining the existing copper infrastructure for the provisioning of broadband services. Hence, the expected broadband market and operator market position must be among the most crucial single factors in the strategic decision on broadband upgrade technology choices for the 2 Mbit/s segment.

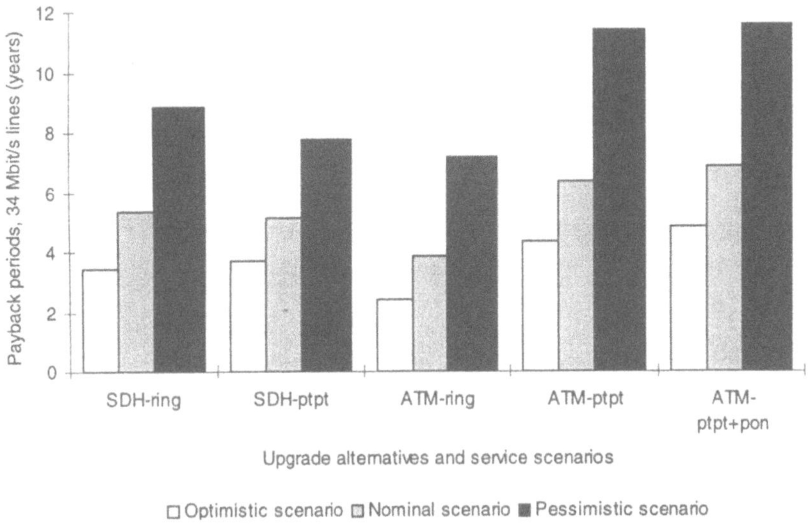

Fig 18.27 Payback periods of different large business upgrade alternatives for an existing 1,024 homes passed ring connected SDH node. Results are shown for three scenarios studied for the 34 Mbit/s upgrade only.

18.11 SUMMARY

The cable infrastructure costs of upgrading a one-way connection to a two-way full protection network are at the same cost level as the initial one-way infrastructure costs.

The baseline equipment costs of any of the large business upgrade alternatives studied are between 6,000 ECU and 10,000 ECU per large business connection (34 Mbit/s). The cable infrastructure costs may double the overall upgrade costs, depending on the duct availability and the civil works costs in the area.

With reasonable tariff assumptions, calculations show payback periods of large business upgrades between two and eight years for the above upgrade investments, depending on the expected demand evolution.

The ATM PON upgrades for the residential and SME market has payback periods very similar to large business segment upgrades. In fact, in this study the enhanced copper upgrade alternative for the residential and SME market shows payback periods shorter than two years. The significance of this market thus confronts the access network operator with the strategic choice of offering this customer group a future proof, fibre infrastructure or maintaining the existing copper infrastructure for the provisioning of broadband services.

This document is based on results achieved in a EURESCOM project; this does not imply that it reflects the common technical position of all the EURESCOM shareholders/parties.

REFERENCES

[1] Olsen, B T, Myhre, D, Ims, L A. Broadband access Aaternatives in the business dominated area, a techno-economic study. *In: Proc. Eurescom Workshop on Optical Multiservice Access Networks (WOMAN), Turin, Italy, September 23-24, 1996.*

[2] Ims, L A, Olsen B T, Myhre, D. Cost benefits and business opportunities of high capacity optical broadband access network upgrade technologies. *22nd European Conference on Optical Communication (ECOC '96), Oslo, September 15-19, 1996.*

[3] Ims, L A, Myhre, D, Olsen, B T, Stordahl, K. Paths towards a versatile broadband Access Network. *In Proc. International Conference on Telecommunications 95 (ICT '95),* Nusa Dua, Indonesia, April 3-5, 1995.

[4] Olsen, B T, Ims, L A, Myhre, D, Stordahl, K, Theivindrein, R. Technoeconomic Evaluation of Optical Broadband Access Network Scenarios for the residential and business market. *In: Proc. 21st European Conference on Optical Communications (ECOC '95), Brussels, Belgium September 14-17, 1995.*

[5] Myhre, D, Ims, L A, Olsen, B T, Lähteenoja, M, Mononen, J, Ferrero, U, Vangelista, L. Roll out Strategies and Upgrade Access Alternatives for the Mixed Residential and Business Market: A Techno-economic Evaluation. *In: Proc. Networks and Optical Communications '96 (NOC '96), Heidelberg, Germany, June 24-28, 1996.*

[6] Stordahl, K, Murphy, E. Methods for forecasting long term demand for wide and broadband services in the residential market. *IEEE Communications magazine*, 13, (2), 1995, 44-49.

[7] Olsen, B T, Zaganiaris, A, Stordahl, K, Ims, L A, Myhre, D, Øverli, T, Tahkokorpi, M, Welling, I, Drieskens, M, Kraushaar, J, Mononen, J, Lähteenoja, M, Markatos, S, De Bortoli, M, Ferrero, U, Ravera, M, Balzaretti, S, Fleuren, F, Gieschen, N, De Oliveira Duarte, M, de Castro, E. Techno-economic evaluation of narrowband and broadband access network alternatives and evolution scenario assessment. *IEEE Journal of Selected Areas in Communications,* 14 (6), 1996, 1184-1203.

Risk analysis of PNO and cable operator upgrades 19

Kjell Stordahl, Leif Aarthun Ims, Borgar Tørre Olsen

The chapter presents risk and techno-economic analyses of broadband access network upgrade strategies for public network operators and cable operators in the competitive residential and small business market. The effect of uncertainties in predictions of critical parameters such as demand forecasts and market shares are analysed. The assessed technology options include broadband twisted pair modems, hybrid fibre coax (HFC) networks and asynchronous transfer mode (ATM) based passive optical networks (PONs)[1].

19.1 INTRODUCTION

On the verge of full deregulation of the telecommunication services in Europe, public network operators (PNOs) are facing competition from cable operators in an almost saturated high-revenue telephony market. Most PNOs are, in turn, eager to provide distributive broadband services. Currently this cable operator-dominated market is characterised by a potentially high revenue, although the residential demand remains uncertain. The access network is expected to be one major battleground of network providers. The competitors will need quite different upgrade strategies, mainly due the difference in existing access networks. The decision to invest in new technology is subject to the risks and uncertainties inherent to the competitive marketplace. The economic viability of residential broadband upgrades and the associated risks need to be assessed, in order to identify minimum-risk introductory strategies [1-8].

This chapter addresses some of the different challenges and risks faced by the PNO and cable operators in adapting their present fixed network to competition. The analysis of broadband upgrade alternatives focuses on one of the most

[1] Based on "Risk analysis of residential broadband upgrade in a competitive and changing market", by L. A. Ims, K. Stordahl, B. T. Olsen. from *IEEE Communications Magazine,* vol. 35, no. 6, June. 1997.

competitive market segments: an urban, apartment block area with short outdoor average loop lengths, in which both the PNO and the cable operator have an established infrastructure for their own subscribers. Extensive risk assessments are performed based on forecasts of the residential and small business market, and market shares between PNO and cable operators for the services: plain old telephone service (POTS), narrowband integrated services digital network (N-ISDN), community antenna television (CATV), 2 Mbit/s asymmetric switched broadband (ASB), and 2 Mbit/s symmetric switched broadband (SSB) [9]. Other important economic variables like network component costs, civil work costs, operation, administration and maintenance (OAM) costs, tariffs etc. are predicted for each year, but are not varied according to a probability distribution. The analyses show the market evolution and how the related uncertainties influence the risks of PNO and cable operator projects.

19.2 THE BROADBAND UPGRADE CASE STUDY

A network evolution during a ten year period from 1997 to 2007 has been investigated. Upgrade boundary conditions for both the PNO and the CATV operator, like the demographic area, the existing networks, overall service take rate and market shares are included in the case study.

19.2.1 Network area characteristics

An urban, residential and small business area with customers living in apartment blocks with an average of 32 dwellings per block is analysed. The general assumptions are summarised in Table 19.1 (1 ECU ≈1.3 $US). The service access point (SAP) refers to the network localisation of the concentrator, as used in several European countries with modern access network infrastructures.

Table 19.1 General assumptions, network area characteristics.

General assumptions, area under study	
Number of households in area	1.024
Number of buildings in an area	32
Number of households per building	32
Duct availability, feeder	100 %
Duct availability, distribution, fibre	20 %
Duct availability, distr., coaxial cable	10 %
Civil works cost per metre	25 ECU
Global cable length, distribution network	7.46 km
Distance, SAP - building	400 m
Global cable length, indoor drop network	20.5 km
Average cable length, drop network	20 m

The average outdoor loop length from the network concentrator location to the buildings is 400 metres. Network architectures for PNO and cable operators have already been established in the area, and the availability of ducts for new cables is low (20 % for fibre and 10 % for coaxial cable). In the initial year the cable operator is assumed to have an existing 450 MHz distribution coaxial cable network, whilst the PNO has an existing twisted pair infrastructure for POTS and N-ISDN. For both operators the existing network includes optical access network nodes serving approximately 1,000 subscribers. All of the existing PNO and cable operator infrastructures have been fully amortised.

19.2.2 The market forecast

The scope of this paper is the study of the broadband upgrade of the above two networks. However, the new broadband residential and small business customer applications are still not clearly envisaged, in spite of recent field trials and market surveys. New services forecasting and demand projections have recently been reported, derived from current spending patterns of households or from market surveys [10]. In this study a common set of bearer services is assumed to be provided by both operators: POTS, N-ISDN, CATV, 2 Mbit/s ASB, and 2 Mbit/s SSB. The CATV service penetrations used here are European averages. The forecasts of the total market of the broadband bearer services are extracted from the Delphi survey on broadband demand, carried out by the project RACE (research for advanced communications in Europe) 2087/TITAN (tool for introduction strategies and techno-economic evaluation of access network) in ten European countries [11]. The experts were asked to indicate the percentage penetration of the residential market for ISDN, wideband (capacity ≤ 2 Mbit/s) and broadband (capacity > 2 Mbit/s) for the years 1995, 2000, 2005, 2010, and also the saturation penetration. The Delphi survey produced quantified information about the development of new services and applications which requires high capacity in the access network. Demand has been quantified in terms of the number of subscribers to one or more services.

19.2.3 Service take rate and market shares

In our study, competition is assumed from 1997 on all services, and hence the market will be shared between the PNO and the cable operator. The competition has been simulated by appropriate adjustments of the service penetrations. The PNO starts with the majority of the market share of the high-revenue narrowband switched services. The cable operator initially has the majority of the market share of the low-revenue CATV service. Fig 19.1 shows forecasts for the service penetration and Fig 19.2 shows the corresponding standard deviations in percentage of the small business and residential market for each service studied. The forecasts indicate the evolution from 1997 to 2006. In the simulations the total narrowband market (NB) which includes POTS and ISDN is assumed to remain stable at 100% penetration during the period. This may differ from the situation in USA and in some other countries, but it is in accordance with the experience from Europe. Since ISDN was introduced in the residential market in

Europe, new ISDN subscription has primarily substituted the existing POTS subscription, as verified by official statistics. There are several reasons for this:

1. Different terminals in the household are identified with a distinct ISDN number, hence two of the terminals can be used simultaneously.
2. A quasi-ISDN service has been introduced in Europe, giving the customers two independent analogue subscriptions and additional ISDN functionality for a price significantly lower than the price of two POTS subscriptions.
3. Most of the households had one POTS subscription prior to the introduction of ISDN in the residential market, typically with a mobile subscription as the second subscription. The average number of persons in a household in the Nordic countries is about two. The average is increasing up to about 3-3.5 when moving towards the Southern part of Europe. However, the POTS penetration is not as high in Southern Europe as in the Nordic countries.

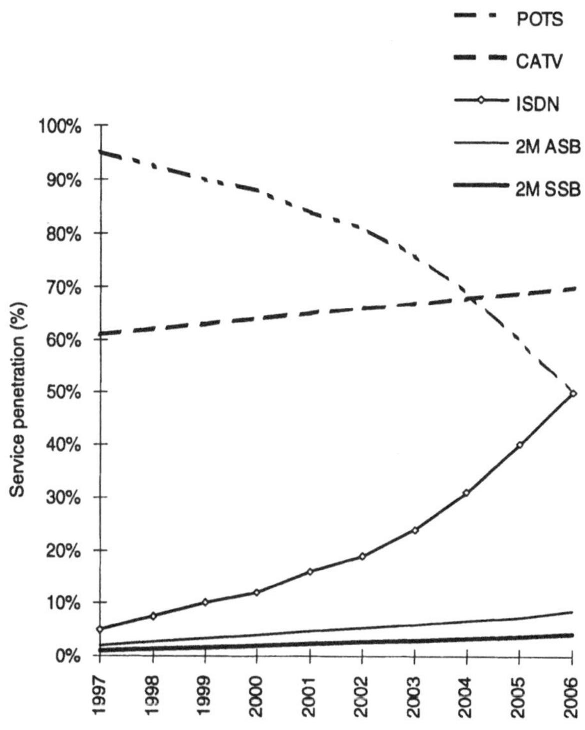

Fig 19.1 Forecasts for the years 1997 - 2006 of the service penetration (percentage of small business and residential market).

The correlation between the POTS and the ISDN markets has been taken into account. The simulated value for the POTS penetration is obtained by subtracting the simulated value of the ISDN penetration from the simulated value of the total narrowband market. Throughout the period it is assumed that the PNO operator will maintain 75 % of the POTS market and the N-ISDN market, the CATV operator will maintain 75 % of the CATV market, while the two operators equally share the broadband market: 50 % each. Section 19.5 describes how the penetrations and market shares are simulated around their expected valued according to a Normal distribution in order to examine how changes in penetrations and market shares influence the risks and the economic outcome of the projects.

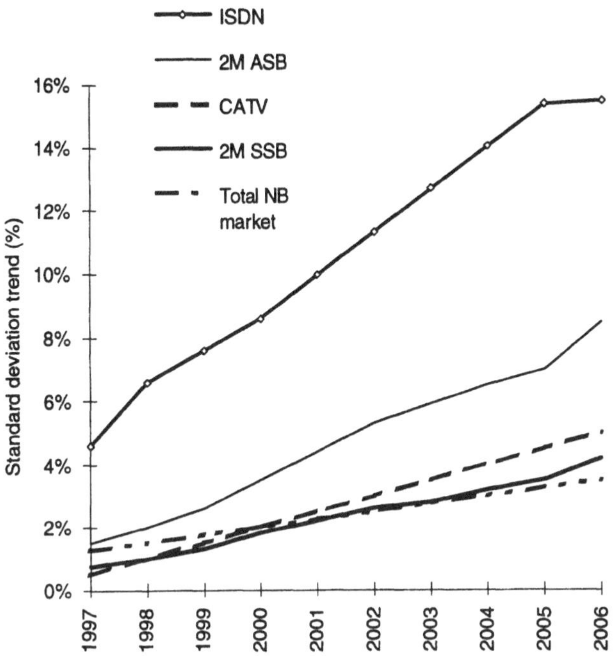

Fig 19.2 Estimated standard deviations of the forecasts for the services POTS, N-ISDN, CATV, 2 Mbit/s ASB, 2 Mbit/s SSB and the total narrowband (NB) market.

19.2.4 Tariff elasticity

Only the annual subscription tariffs for bearer services have been included in the analysis. Traffic income is assumed to be trunk network specific revenue, and as such is not accounted for in the analysis. The tariffs used represent European averages and are taken from the TITAN Delphi survey and other sources. The economic analysis incorporates a tariff structure with an evolution of the tariffs during the study period. The evolution of tariffs is strongly related to competition and penetration. Tariff elasticities for the broadband switched services are derived from the Delphi survey, as described in [12]. A similar approach has been taken in the modelling of the CATV tariff elasticity. The POTS and N-ISDN tariffs decrease annually by respectively 2 % and 4 %. The evolution of the annual tariffs are shown in Fig 19.3.

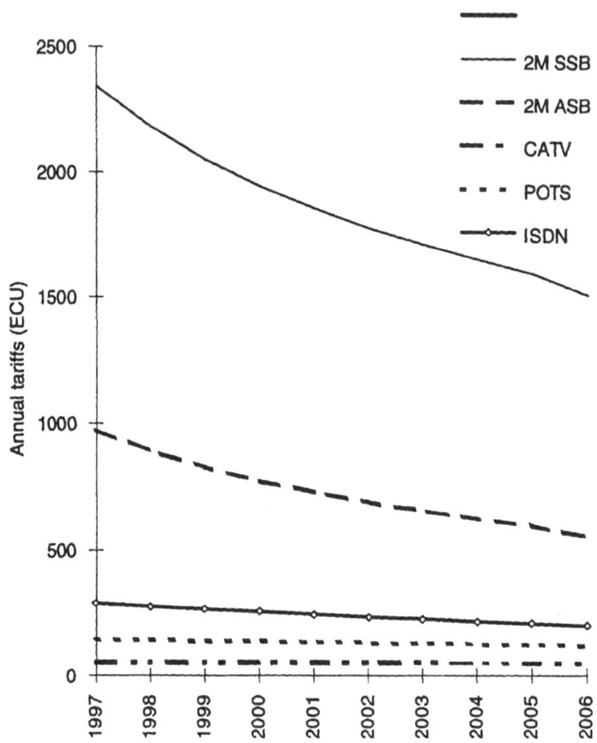

Fig 19.3 Tariff evolution, nominal penetrations.

19.3 BROADBAND UPGRADE ALTERNATIVES

Network operators face great strategic challenges in upgrading the physical layer of the very cost sensitive access network, both with respect to selection of transmission medium and system technology. Optical fibre cable, coaxial cable and twisted copper pairs are the relevant wireline transmission media, whilst there is a seemingly exhaustive choice of alternative system technologies [13]. These system technologies encompass enhanced copper options like asymmetric digital subscriber line (ADSL), very high speed digital subscriber line (VDSL) and high speed digital subscriber line (HDSL), the fibre option of asynchronous transfer mode based broadband passive optical network (ATM PON) and hybrid fibre coax (HFC) for coaxial cable networks. Two different access network upgrade architectures have been examined for each of the two operators:

1. PNO, alternative 1: Fibre to the node (FTTN) architecture with enhanced copper, 1.000 home passed per node.
2. PNO, alternative 2: Fibre to the building (FTTB) architecture based on ATM PON

3. Cable operator, alternative 1: FTTN architecture based on HFC with cable modems.
4. Cable operator, alternative 2: FTTB architecture based on HFC.

19.3.1 PNO upgrade alternatives

The alternative considered highlight the economic and technological implications for moderate operators which aim to utilise the existing twisted pair copper cables as the basic transmission medium, and aggressive operators which extensively upgrade the network with broadband fibre technology. The former represents a limited degree of service integration, whilst the aggressive upgrade enables an introduction of a full service fibre network. Detailed sketches of the architectures for the PNO alternatives are shown in Fig 19.4 (FTTN) and 19.5 (FTTB).

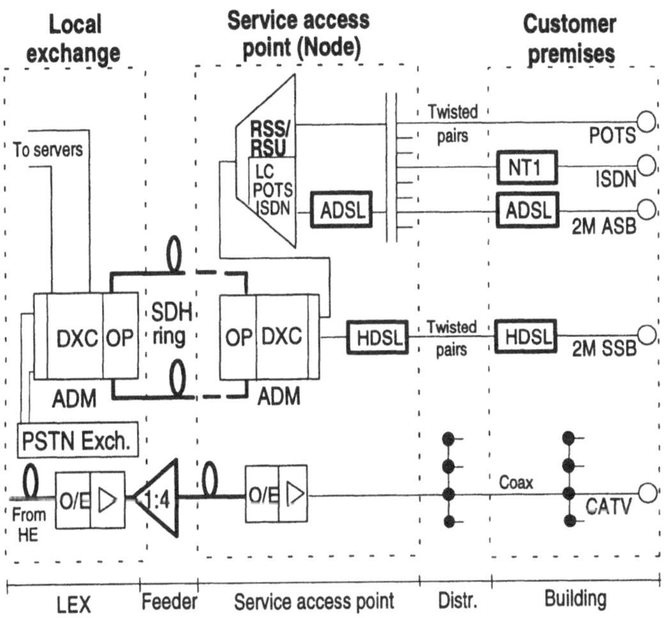

Fig 19.4 The PNO FTTN upgrade architecture with enhanced copper and a parallel distribution network for CATV.

PNO, fibre to the node

The plesiochronous digital hierarchy (PDH) ring between the local exchange (LEX) and the SAP is replaced by an synchronous digital hierarchy (SDH) ring in 1997. In the distribution network ADSL and HDSL equipment are installed to provide new services such as 2 Mbit/s ASB, i.e. service on demand (SOD) or fast Internet Access, and 2 Mbit/s SSB. In addition, a coaxial cable network is installed for CATV distribution.

With an average copper loop length of 400 metres as in this particular study, 6 Mbit/s ASB may be provided with the commercially available ADSL modems. For typical statistical copper loop length distributions within exchange areas in some European countries, a 90 % coverage even with 26 Mbit/s may be possible with the use of VDSL for this average copper loop length of 400 metres. The standards for VDSL modems are expected during next year, with the first products due to reach the market closely after that. The strategic aspects of how far the fibre should be deployed towards the customer (including the corresponding sizes of the optical nodes), in order to decrease the copper loop length, and hence increase the capacity available per customer, requires further sensitivity analysis. However, the objective of this work has primarily been an in-depth risk analysis of the changes in the market situations due to competition.

PNO, fibre to the building

The PDH ring connecting the SAP to the LEX is replaced by an SDH based ring structure in 1997. The deployment of an ATM-PON in an FTTB configuration is started in 1997 in order to provide new services like ASB and 2 Mbit/s SSB. In addition, a combined fibre-coax network for CATV distribution is installed in 1997.

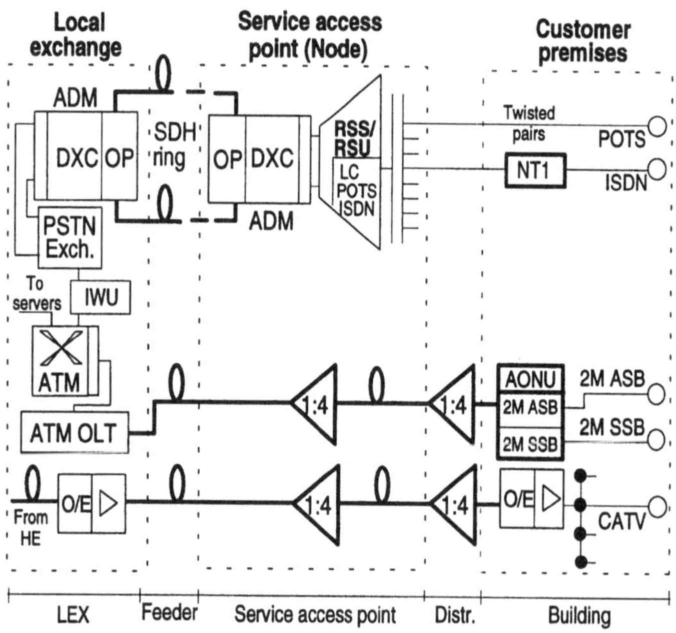

Fig 19.5 The PNO FTTB upgrade architecture with an ATM PON for broadband services and a parallel distribution network for CATV.

19.3.2 Cable operator upgrade alternatives

The cable operator upgrade alternatives represent moderate operators which only partially integrate their network by sharing duct layout, and aggressive operators which fully integrate the network by providing the service set over the same duct layout, transmission medium and network termination units as well. Detailed sketches of the architectures for the cable operator alternatives are shown in Fig 19. and Fig 19.7.

Cable operator, fibre to the node

The existing coaxial cable infrastructure between the access node serving 1.000 subscribers and the homes are retained during the upgrade period. In 1997 the CATV network is upgraded to a bi-directional network with return path. All services are then offered on the integrated coaxial cable network. Cable modems are installed at the customer premises. 10 Mbit/s shared access data modems are for used for 2 Mbit/s ASB.

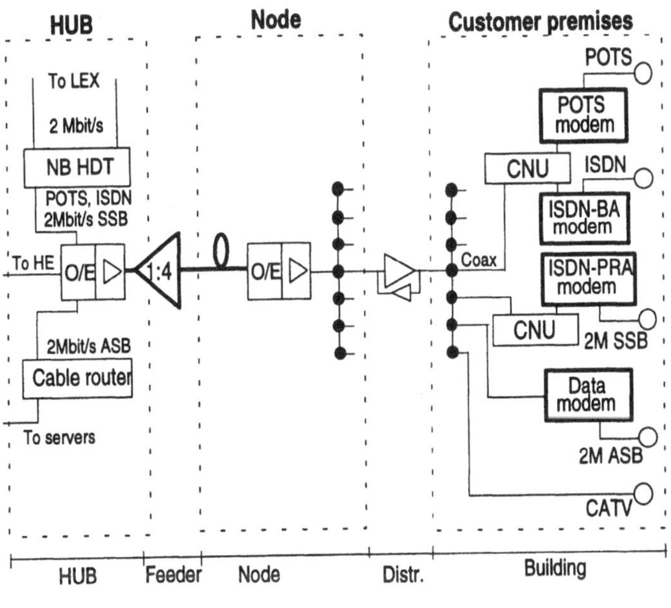

Fig 19.6 The cable operator FTTN upgrade architecture with a HFC network for all services.

Cable operator, fibre to the building

Fibre is deployed to the buildings in 1997. This yields a fibre rich network with separate fibres between the hub (serving approximately 4,000 homes) and the buildings. Cable modems are installed, with 2 Mbit/s dedicated channels access data modems for 2 Mbit/s ASB subscribers.

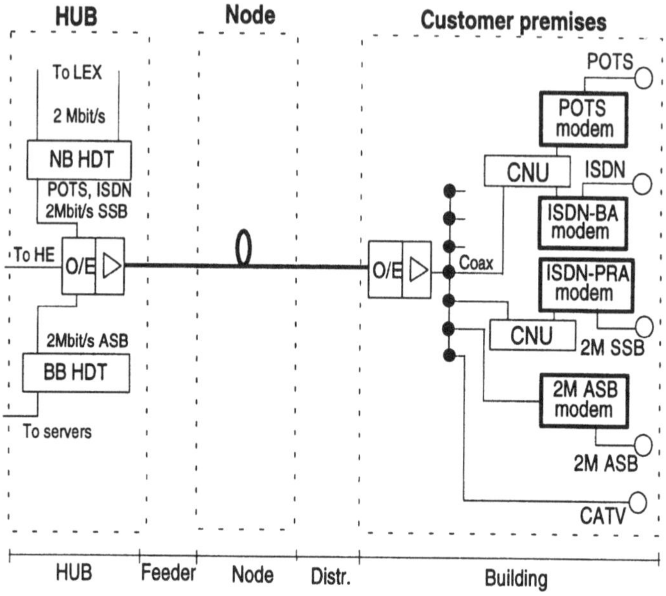

Fig 19.7 The cable operator FTTB upgrade architecture with a HFC network for all services.

19.4 METHODOLOGY

The methodology developed by the RACE 2087/TITAN project has been applied to evaluate the broadband upgrade case [4].The TITAN tool also contains a methodology for calculating economic risks. The main point in risk analysis is to introduce a probability distribution instead of the expected value for each critical parameter in the model. A simulation program called Crystal Ball®, based on Latin Hyper Cube simulations, is used in combination with the TITAN tool to calculate economic outputs based on various sets of the critical parameters. These includes a procedure for linking the parameter values in time based on a prediction function for the expected value and for the standard deviation. The methodology for risk assessment is used to analyse the effect of uncertainties in predictions of critical parameters. These include:

- the respective market shares (in percentage) of the PNO and the cable operators
- total market demand forecasts for POTS, N-ISDN, 2 Mbit/s ASB, 2 Mbit/s SSB and CATV.

19.5 RISK ASSESSMENT ASSUMPTIONS

The assumptions used in the risk assessments are shown in Fig 19.1, Fig 19.2 and Table 19.2. For each service penetration and each market share one expected value and one estimated standard deviation are given. For each variable the values uniquely defines a normal distribution. The set of normal distributions are used to simulate all service penetrations and market shares. Since all service penetrations and market shares are simulated simultaneously, it is necessary to perform a great number of repeated runs in order to achieve a representative output of the economic results of the different upgrade alternatives studied. In the risk analysis the simulated value is obtained by adding the product of the standard deviation values shown in Fig 19.2 and the Normal distribution (as shown in Fig 19.8) to the mean forecasted value of the market shares and penetrations.

Table 19.2 Expected values and estimated standard deviations of the market shares of the PNO and cable operator for each service studied.

Market parameters	*Value*	*Standard deviation*
PNO market share, POTS and ISDN	75 %	7.5 %
PNO market share, CATV	25 %	2.5 %
PNO market share, 2 Mbit/s ASB/SSB	50 %	5 %
Cable operator market share, POTS/ISDN	25 %	2.5 %
Cable operator market share, CATV	75 %	7.5 %
Cable operator market share, 2 Mbit/s ASB/SSB	50 %	5 %

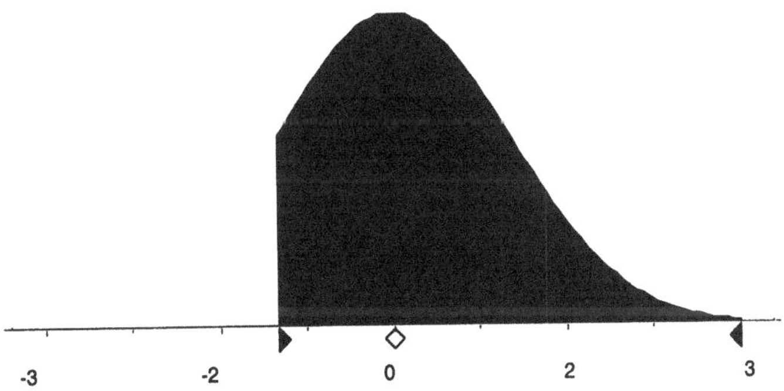

Fig 19.8 The Normal distribution applied to the total market forecast for ISDN.

The forecasted values used here are the net present value (NPV) of the project, the NPV divided by installed first costs (IFC) and the payback period. The NPV as used here is equal to the sum of all discounted retained cash flows (retained earnings) and the discounted rest value of the project. The straight line depreciation method is applied in the calculations. A tax rate of 30 % is used, and an annual discount factor of 7.5 %. The ratio NPV/IFC measures the efficiency of the project, and thus gives an indication of the level of the internal rate of return. The payback period is defined as the period from the start of the project to the time when the cash balance (cumulative sum of the retained cash flows) turns positive.

19.6 RESULTS AND DISCUSSION

The risk assessment has been applied to all four technical upgrade alternatives. The results illustrate the uncertainty in the different techno-economic outputs and the ranking of parameters having the most significant effects on the uncertainty. In general for all four upgrades, the analysis show that the uncertainty in the total market forecasts for the different services have the most significant influence on the variation of both the NPV, the ratio NPV/IFC and the expected payback period. Fig 19.9 depicts the ranking of the contribution of the four parameters having the most significant impact on the variances of the forecasted values. ISDN contributes significantly to the variation of the NPV and the ratio NPV/IFC. The forecasted values of the NPV and the payback period for all upgrade projects are shown in Fig 19.10 and Fig 19.11.

	Total NB market	POTS market share	ISDN		2M ASB		2M SSB		CATV	
	tm	ms	tm	ms	tm	ms	tm	ms	tm	ms
NPV										
PNO FTTN		3	1		4		2			
PNO FTTB		2	1	4			3			
Cable Op. FTTN		4	3		2		1			
Cable Op. FTTB			2		3		1			4
NPV / IFC										
PNO FTTN		2	1	4			3			
PNO FTTB		2	1	4	3					
Cable Op. FTTN		3	2		1		4			
Cable Op. FTTB			2		3		1			4
Payback period										
PNO FTTN		3			1	2	4			
PNO FTTB				2		4	1	3		
Cable Op. FTTN		3				2	1			4
Cable Op. FTTB		2				4	3	1		

: rank 1   : rank 2   : rank 3   : rank 4

Fig 19.9 The ranking of the contribution of the four parameters having the most significant impact on the variances of the forecasted values. Abbreviations in the figure: NB: narrowband market, tm: total market, ms: market share.

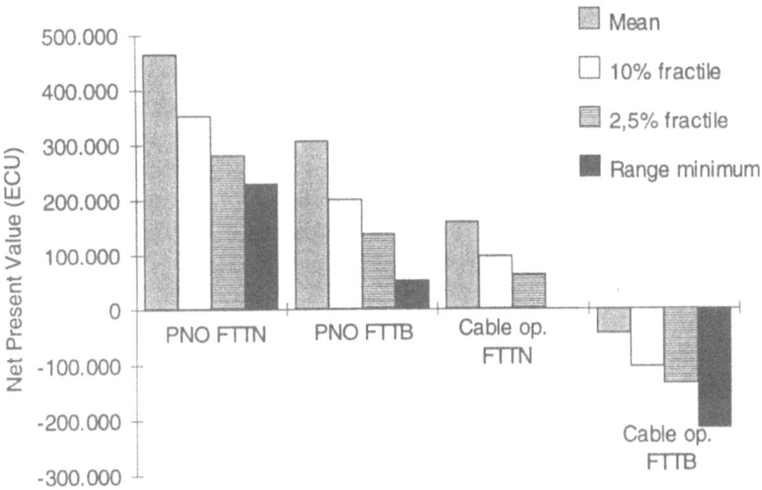

Fig 19.10 The forecasted (simulated) value of the NPV, with mean value, 10 % fractile, 2.5 % fractile and the minimum simulated values indicated.

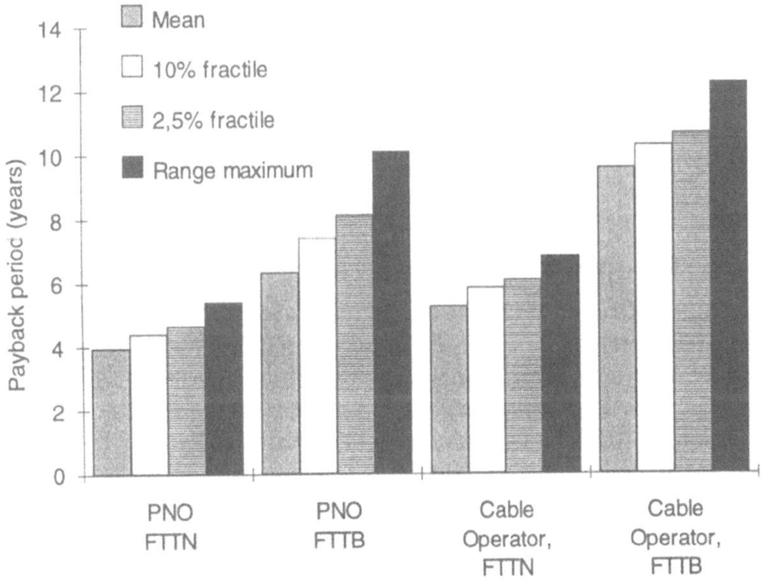

Fig 19.11 The forecasted (simulated) value of the payback period, with mean value, 10 % fractile, 2,5 % fractile and the maximum simulated values indicated.

19.6.1 PNO, fibre to the node

Figure 19.12 shows the frequency distribution of the NPV of the PNO FTTN project based on the variation of the market inputs. The expected NPV is about 465,000 ECU. There is a 10 % probability that the NPV will be less than 354,000 ECU. The most significant effects are contributed by the total market forecasts for ISDN and 2 Mbit/s SSB. Our study shows that the expected NPV divided by the IFC is about 1.04. There is a 10 % probability that the factor is less than 0.84. The most significant effects are contributed by the total market forecast for ISDN and the market share for POTS. The expected payback period in this case is 3.94 years. There is a 10 % probability that the payback period is greater than 4.41 years. In addition to the total market forecasts for the 2 Mbit/s services, the market shares of 2 Mbit/s SSB and POTS contribute significantly to the variation in the payback period.

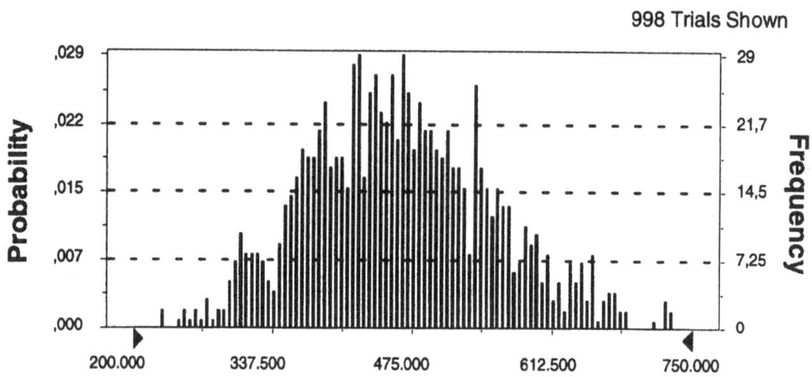

Fig 19.12 The frequency distribution of the NPV of the PNO FTTN upgrade.

19.6.2 PNO, fibre to the building

The expected NPV for the FTTB alternative is about 307,000 ECU. There is a 10 % probability that the NPV will be less than 201,000 ECU. The most significant effects are contributed by the total market for ISDN and the POTS market share. The expected NPV/IFC is about 0.48. There is a 10 % probability that the factor is less than 0.31. The most significant effect is contributed by total market for ISDN, 2 Mbit/s ASB and the market share for POTS. Figure 19.13 shows the cumulative frequency distribution of payback period based on the variation of the market inputs. The expected payback period is 6.34. There is a 10 % probability that the payback period is greater than 7.43 years. The total market for 2 Mbit/s SSB and the ISDN market share have the most significant influence on the variations in the payback period.

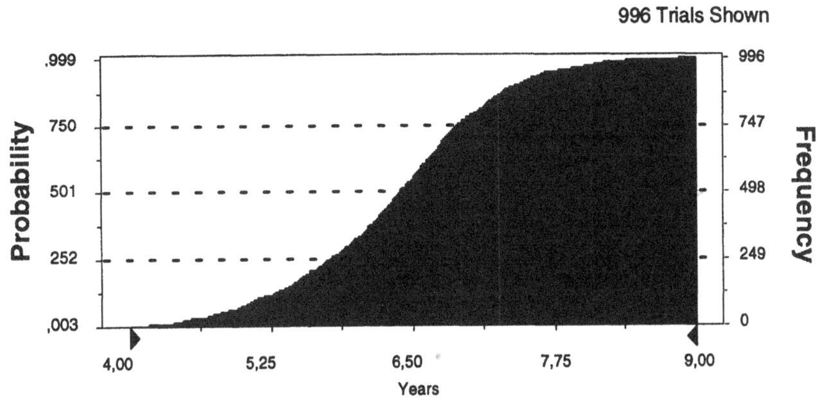

Fig 19.13 The cumulative frequency distribution of the payback period of the PNO FTTB upgrade.

19.6.3 Cable operator, fibre to the node

The expected NPV is about 160,000 ECU. There is a 10 % probability that the NPV will be less than 97,000 ECU. The uncertainty in the total broadband services market has the most significant influence on the variation of the NPV. The expected NPV/IFC is about 0.53. There is a 10 % probability that the factor is less than 0.37. The most significant effects are contributed by the total market for 2 Mbit/s ASB and ISDN. The expected payback period is 5.27 years. There is a 10 % probability that the payback period is greater than 5.83 years. In this case the uncertainty in the total market for 2 Mbit/s SSB and in the market shares of ISDN and 2 Mbit/s ASB have the most significant influence on the variation of the payback period.

19.6.4 Cable operator, fibre to the building

For the cable operator FTTB upgrade the expected NPV is negative, approximately -45,000 ECU. There is a 10 % probability that the NPV will be less than -105,000 ECU. The major contributions to the variance in the NPV are from the total market forecast of 2 Mbit/s SSB. The expected NPV/IFC is negative, about -0.06. There is a 10 % probability that the NPV/IFC ratio is less than -0.15. The total market forecasts for the 2 Mbit/s services and ISDN have the most significant influence on the variation in the NPV/IFC factor. The expected payback period is 9.59 years. There is a 10 % probability that the payback period is greater than 10.29 years. The uncertainty in the market shares for 2 Mbit/s SSB and POTS and the total market for 2 Mbit/s SSB significantly influence the variation of the payback period.

19.6.5 Evaluation of uncertainty factors

The results in sections 19.6.1-19.6.4 describe how variation in market variables influence economic outcome. For the PNO, the main uncertainty is caused by forecasts of the total market for 2 Mbit/s SSB and ISDN. The relatively strong influence is explained by high penetrations of 2 Mbit/s SSB and ISDN and corresponding tariffs which are significantly higher than for the CATV service. The market share for POTS has a significant influence on the NPV and the ratio NPV/IFC, while the market share for ISDN has the most significant influence on the payback period.

For the cable operator the forecasts of the total market for broadband services and ISDN are dominating the uncertainty in the variation of both the NPV and the ratio NPV/IFC. The penetration for 2 Mbit/s SSB and ASB are low compared to the penetration of CATV, POTS and ISDN, but their standard deviations are relatively high compared to the penetrations. In addition, the tariffs are considerably higher than for other services. For the two upgrade alternatives the total market forecast of 2 Mbit/s SSB and the market shares of POTS are the most important contributors to the variation in the payback period. The market shares of 2 Mbit/s ASB (FTTN) and 2 Mbit/s SSB has also a significant influence. The difference is due to the fact that the FTTB alternative has a long payback period (approximately ten years), compared to the FTTN alternative, for which the payback period is around five years. Hence, variables with the most significant contribution to the uncertainty during the first part of the ten year upgrade period have the highest weight.

19.6.6 Evaluation of upgrade alternatives

The risk analysis show that the NPV of the FTTN solution for the PNO may be reduced by 24 % from 465,000 ECU to 354,000 ECU with a 10 % probability, depending on the market evolution. The respective percentages for PNO FTTB, cable operator FTTN and cable operator FTTB projects are: 34 %, 39 % and 131 %. The analysis confirm a similar behaviour of the factor NPV/IFC for the different alternatives. The NPV/IFC factor is reduced by 20-35 %, except for the cable operator FTTB upgrade, for which the decrease is as high as 130 %. The cable operator FTTB alternative is definitely the most unprofitable upgrade project. This is underlined by a very long payback period. The expected values calculated for the alternative upgrade projects show that FTTN alternatives for both operators have the highest NPV. The ratio NPV/IFC range from 1.0 to 0.5 with a payback period from 3.9 to 5.3 years. In general, the PNO has the most profitable projects. Even the FTTB alternative for the PNO is comparable to the FTTN alternative for the cable operator, although slightly less profitable. With the given assumptions, the FTTB alternative for the cable operator is shown to be a very unpromising project

19.7 SUMMARY

The risk assessment and additional analysis show that the FTTN alternative of the PNO is the best economic alternative with the lowest risk. The FTTB project for the cable operator is the least economic alternative and also a high risk project. The service forecasts of the total market contribute more to the variation of the results than the service market shares of the operators. The ordinary POTS and ISDN services generates more uncertainty than the other services, mainly due to the high penetration. Depending on the technical alternative, type of operator and economic output, different set of market variables explain the main part of the variation.

In this paper only market variables (forecasts of total market penetration and market shares for a set of services) are simulated according to given probability distributions in order to evaluate their effect on the economic results for various upgrade alternatives. These variations are based on a set of assumptions mainly derived from a European Delphi survey. A supplementary analysis should include risk assessment of cost elements like civil works, network components and operation and maintenance, in addition to the tariffs.

REFERENCES

[1] Cook, A, Stern, J. Optical Fiber Access - Perspectives Towards the 21st Century. *IEEE Communications Magazine*, 32 (2), 1994, 78-86.

[2] LeBel, P, Oliver, R. Residential Broadband Architectures Strategies and Economics. *In: Proc. The 7th International IEEE workshop on Optical Access Networks (OAN '95), Nuremberg, Germany, September 24-27, 1995, 4.4-1 - 4.4-10.*

[3] Miki, T. Toward the Service-Rich Era. *IEEE Communications Magazine*, 32 (2), 1994, 34-39.

[4] Olsen, B T, Zaganiaris, A, Stordahl, K, Ims, L A, Myhre, D, Øverli, T, Tahkokorpi, M, Welling, I, Drieskens, M, Kraushaar, J, Mononen, J, Lähteenoja, M, Markatos, S, De Bortoli, M, Ferrero, U, Ravera, M, Balzaretti, S, Fleuren, F, Gieschen, N, De Oliveira Duarte, M, de Castro, E. Techno-economic evaluation of narrowband and broadband access network alternatives and evolution scenario assessment. *IEEE Journal on Selected Areas in Communications,* 14 (6), 1996, 1184-1203.

[5] Ims, L A, Olsen, B T, Myhre, D, Mononen, J, Lähteenoja, M, Ferrero, U, Zaganiaris, A. Multiservice Access Network Upgrading in Europe: a Techno-economic Analysis. *IEEE Communications Magazine,* 34, (12), 124-134, 1996.

[6] Olsen, B T, Ims, L A, Øverli, T, Stordahl, K, Myhre, D, Drieskens, M, Kraushaar, J, Tahkokorpi, M, Ravera, M, De Bortoli, M, Zaganiaris, A, Markatos, S, Lähteenoja, M, Mononen, J, Fleuren, F. PNO and Cable Operator broadband upgrade technology alternatives: a techno-economic analysis. *In: Proc. Optical Fiber Conference 1996 (OFC '96), San Jose, USA, Feb. 25 - March 1, 1996, 2, 140-141.*

[7] Olsen, B T, Ims, L A, Myhre, D, Stordahl, K, Theivindrein, R. Technoeconomic Evaluation of Optical Broadband Access Network Scenarios for the residential and business market. *In: Proc. 21st European Conference on Optical Communications (ECOC '95), Brussels, Belgium September 14-17, 1995.*

[8] Saijonmaa, J, Tahkokorpi, M, Welling, I. Cost of Investment and Revenue Modelling and Analysis of Various Networked Multimedia Services in PTO and CATV operator Environments. *In: Proc TELECOM '95, Technology Summit, Geneva, Switzerland, October 3-11, 1995, 2, 629-633.*

[9] Stordahl, K, Ims, L A, Olsen, B T. Risk assessment and techno-economic analysis of competition between PNO and CATV operators. *In: Proc. Networks '96, Sydney, Australia, November 25-29, 1996, 2, 405-410.*

[10] Luck, D P. Broadband To The Home: Evolution Scenarios for Australia. *In: Proc. 13th Annual Conference European Fibre Optic Communications and Networks (EFOC&N '95), Brighton, UK, June 27-30, 1995, 1, 173-176.*

[11] Stordahl, K, Murphy, E. Methods for forecasting long term demand for wide and broadband services in the residential market. *IEEE Communications magazine*, 13, (2), 1995, 44-49.

[12] Stordahl, K, Olsen, B T. Demand Elasticity and Forecasts of Wide and Broadband Services in the Residential Market Based on Results from an International Delphi Survey. *1995 International Communications Forecasting Conference, Toronto, Canada, June 13-16, 1995.*

[13] Pugh, W, Boyer, G. Broadband Access: Comparing Alternatives. *IEEE Communications Magazine*, 33 (8), 1995, 34-45.

[14] Siden, P. Modelling Tool for assessment of Business Opportunitites. *In: Proc. 11th International Symposium on Subscriber Loops and Services (ISSLS '96), Melbourne, Australia, February 4-9, 1996, 26-31.*

Summary and main conclusions 20

Leif Aarthun Ims, Borgar Tørre Olsen, Dagfinn Myhre

This chapter summarises some of the techno-economic assessment of broadband access network upgrade strategies presented in previous chapters. Main conclusions are drawn with respect to key issues faced by the infrastructure providers. The techno-economic studies from which the main findings are extracted, reflect the diversity in demographics, competition, market segmentation, demand for new services and corresponding network upgrade alternatives experienced by the network operators of today. The results emphasise the importance of factors like the quality of the existing infrastructure and demography, civil work costs, technology choice, roll-out year and capacity demand for the assessment of access network upgrade strategies [1,2][1].

20.1 INTRODUCTION

On the threshold of a wide-scale implementation of the "information superhighways" the access network infrastructure remain as one of the major obstacles for the deployment of a broadband (capacity per customer ≥ 2 Mbit/s) telecommunications network. The access network operators face several issues in preparing for a migration from narrowband to broadband connectivity [3,4]. The new broadband infrastructure most likely will be based on an upgrade of the existing access network, and hence the utilisation of the vast amount of capital invested in the current infrastructure. An access network infrastructure need to be established for delivery of a complex mix of broadband services like fast Internet access, telecommuting and video on demand in a market of great demographic diversity. The main challenge will probably be to find the appropriate combination of the relevant technologies for a versatile and scaleable multiservice solution for delivery of broadband services, and hence build a so-called full service

[1] Parts of this chapter are based on "Economics of Residential Broadband Acess Network Technologies and Strategies," by L. A. Ims, D. Myhre, B. T. Olsen. © 1997, IEEE. Reprinted with permission, from *IEEE Network*, vol. 11, no. 1, 58-64, Jan./Feb. 1997.

access network [5]. These facts advocate a techno-economic analysis of broadband access network upgrade projects, in order to gain an improved understanding of the business opportunities of broadband access delivery, and hence derive suitable minimum-risk introduction strategies. This chapter summarise the techno-economic assessment of access network upgrade strategies presented in previous chapters of this book. The work presented includes an in-depth sensitivity analysis of some of the key aspects and underlying strategic issues of upgrading the existing access network infrastructure to residential and mixed residential and business users.

20.2 FIVE STUDIES ON ACCESS NETWORK UPGRADE

The chapter mainly presents key findings from five different studies carried out over the past four years within the projects RACE (research and development in advanced communications technologies in Europe) 2087/TITAN (tool for introduction scenario and techno-economic evaluation of access network), EURESCOM (European institute for research and strategic studies in telecommunications) P306 (Access network evolution and preparation for implementation), EURESCOM P413 (Optical networking) and the Norwegian project HUGIN (Norwegian project for a national broadband infrastructure). The RACE 2087/TITAN project and EURESCOM P306 have had as one of their objectives the strategic evaluation of the access network alternatives for the residential and small business market in Europe, whilst the EURESCOM P413 and the HUGIN projects additionally have focused on the medium and large business customer segment. The studies include extensive techno-economic evaluations of upgrade technology options, ranging from enhanced copper alternatives, asymmetric digital subscriber line (ADSL) and high speed digital subscriber line (HDSL), to asynchronous transfer mode based passive optical networks (ATM PONs) and hybrid fibre coax (HFC) systems. Three of the studies have been on access network upgrades from the point of view of the public network operator (PNO). In one of the studies the technology alternatives and strategies for community antenna television (CATV) operators and PNOs where analysed with the emphasis on the effect of competition. The studies are briefly described as follows:

Study 1: PNO upgrade for the residential and small business market in urban and suburban areas

This study comprised five representative European cases, reflecting the wide diversity in Europe as regards existing network infrastructure, demography and regulatory conditions [6,7]. However, in this chapter, only results from one of these five cases are included: an urban area case. Study 1 is presented in more detail in chapter 13 of this book.

Study 2: PNO and CATV operator upgrades for the residential and small business market in an urban area

The study was focused on the effect of competition and the difference in existing networks and upgrade alternatives for the two kinds of operators [8,9]. Chapter 14 contains an in-depth documentation of the PNO and CATV operator study.

Study 3: PNO upgrade for the mixed residential and business market in downtown and suburban areas

Both the residential and small business market of the previous two studies as well as the medium and large business market were included in the analyses, which included two different cases: one downtown and one suburban area [10-12]. For more information on study 3, see chapter 16.

Study 4: PNO upgrade for the mixed residential and business market in a city area

The mixed residential and business market were analysed for an urban - «city» - area, focusing on different fibre optic transport technologies for a comparable, system independent fibre roll-out.

Study 5: PNO upgrade for the mixed residential and business market in urban areas

The analysis included both the residential and small business market as well as the medium and large business market, with emphasis on the cable infrastructure costs of different network roll-out strategies for to-way protection in a business dominated access area. Chapter 18 documents the details of this study.

For all five studies the percentage of the potential customers demanding 2 Mbit/s is depicted in Fig 20.1, illustrating the variety of new broadband service take-up covered by the analyses.

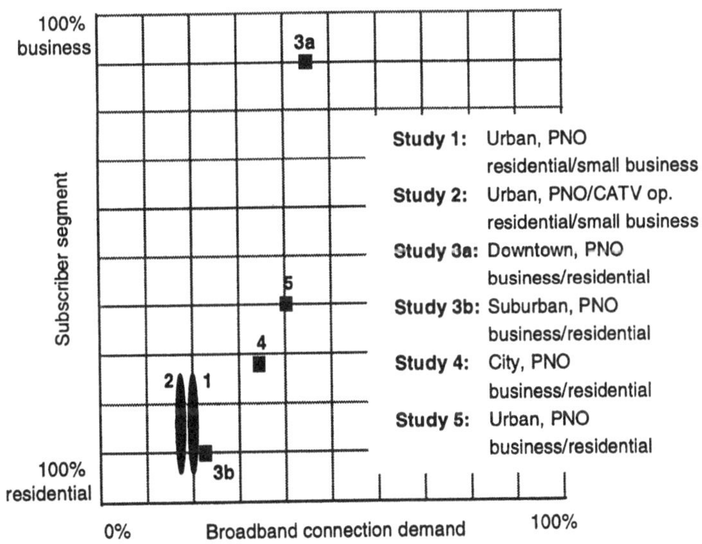

Fig 20.1 Each of the studies 1-5 mapped according to mix of customer segments and the percentage of the potential customers demanding 2 Mbit/s. Study 1: urban, PNO, residential/small business. Study 2: urban, PNO/CATV operator, residential and small business. Study 3a: downtown, PNO, business and residential. Study 3b: suburban, PNO, business and residential. Study 4: city, PNO, business and residential. Study 5: urban, PNO, business and residential.

The methodology and tool developed within TITAN has been applied in the techno-economic analyses in the above studies [13]. The demand forecasts for the selected bearer services for the residential and small business market as used in the described studies, are extracted from the RACE 2087/TITAN Delphi survey [14]. The business market services and penetrations are predicted based on available statistical material. The tariffs used are European averages from the Delphi survey and other sources. The relation between penetration and tariffs - tariff elasticity - has been incorporated in the economic analysis. The effect of competition is modelled through appropriate adjustments of market shares.

20.3 ANALYSIS OF KEY ASPECTS OF ACCESS NETWORK UPGRADE

In the following several conclusions on key aspects of access network broadband upgrade strategies are drawn in the context of what has been highlighted through the techno-economic analyses. The major topics under consideration are:

1. Upgrade costs compared to narrowband access network costs.
2. The effect of duct availability and civil works cost.
3. The influence of the costs of electronics.
4. Fibre roll-out as a strategic decision.
5. Fibre penetration and broadband capacity provided.
6. The effect on costs of the service take rate.
7. The effect on costs of optical network termination customer sharing.
8. The timing of the upgrade.
9. The timeframe of broadband access network upgrading.
10. The effect of revenue on the payback period.
11. The cash flow of the future broadband platform.

20.3.1 Upgrade costs compared to narrowband access network costs

Fig 20.2 shows the cost per new switched service connection for all five studies conducted. All new switched service connections are counted as one, independent of the capacity differences among the types of new service connections. Customers being connected prior to the upgrade are not accounted for. Neither customers demanding CATV connections are included in the overall number of new service connections, even if these connections contribute to the network costs.

In urban areas as the ones in studies 1 and 2, the installed first costs (IFC) per new switched service connection, plain old telephone service (POTS) and narrowband integrated services digital network (N-ISDN), in the existing access network typically is in the range of 400-500 ECUs. The costs of broadband upgrade varies from just below 400 ECU to 2.600 ECU per new connection, depending on factors like technology choice, dwelling distribution and civil work costs. This underlines the fact that a further upgrade of the access network probably will require huge investments for any technology selected, and that the operators will be faced with investment projects of similar or even higher financial burdens than the ones of establishing the narrowband access networks of today.

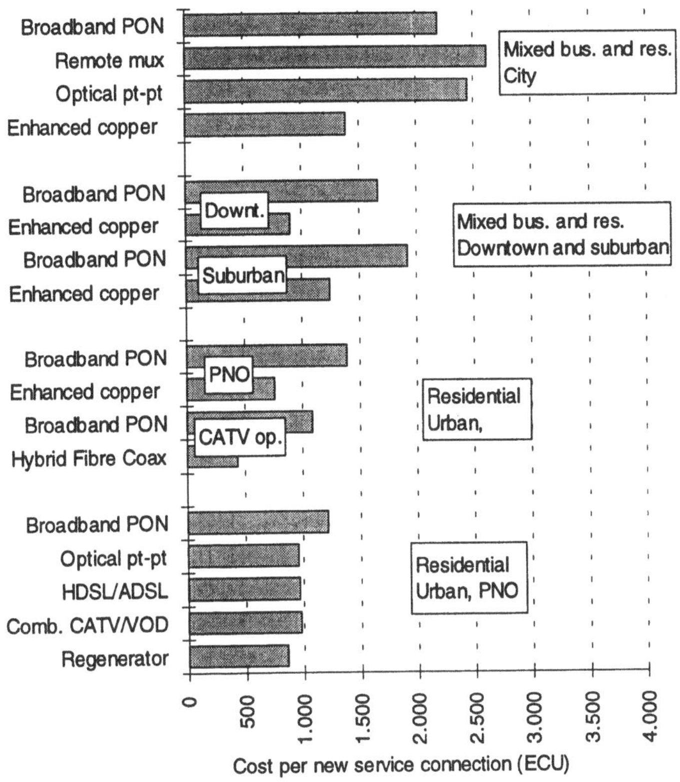

Fig 20.2 The cost per new switched service connection for studies 1-4.

In conclusion, the installed first costs of upgrading the existing narrowband access network to broadband seems to be comparable to and even in some cases up to several times higher than the overall costs of establishing the existing network.

20.3.2 The effect of duct availability and civil works cost

As mentioned in the previous section, the level of civil works cost in many situations will have a significant impact on the upgrade costs. In fact, the costs of civil works remain as one major obstacle for extensive introduction of fibre in areas with an established access network of good quality. This is illustrated in particular in Fig 20.3. The figure shows the IFC for the total network upgrade in study 5 as a function of average civil works costs per metre. In this study several different optical fibre upgrade alternatives for the large business customer market segment have been compared. For the small- and medium enterprise (SME) and residential market, enhanced copper have been taken as the default upgrade alternative irrespective of the large business technology choice. However, one ATM PON solution for the SME and residential market was analysed in combination with an ATM point-to-point upgrade alternative for the large business customers.

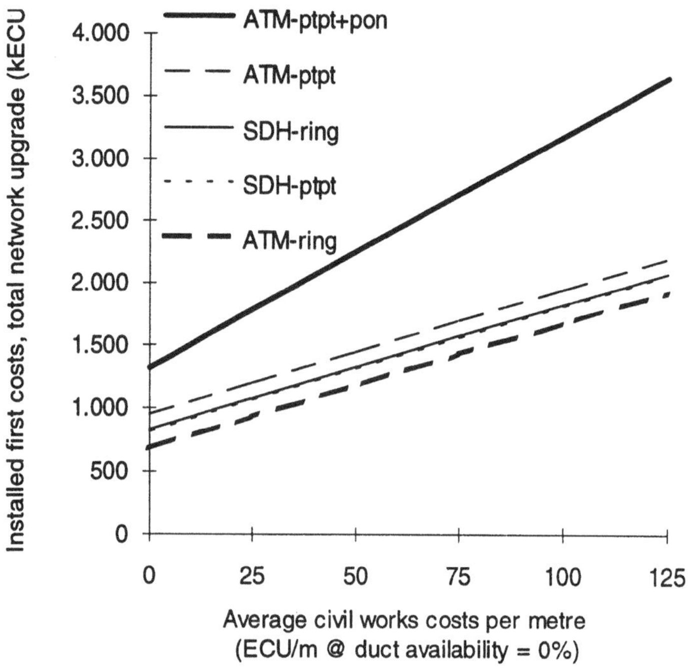

Fig 20.3 The IFC for the total network upgrade in study 5 as a function of average civil works costs per metre. Results are shown for four different large business architectures with enhanced copper solutions for SMEs, and one architecture with fibre solutions for all customer (ATM-ptpt + PON).

Fig 20.3 shows that the total IFC of the upgrade ranges from one to three times the costs of equipment, depending on the penetration of fibre in the network, duct availability and the civil works cost, illustrated by the higher slope of the combined ATM point-to-point/ATM PON solution.

Nevertheless, fibring the upper part of the access network is already likely to be cost effective in some cases, as illustrated in Fig 20.5 and Fig 20.6 by the lower costs of the fibre alternatives with a high number of potential broadband customers per ONU. However, the lower part of the network is very sensitive to civil works costs. The crossover point for fibre and copper depends on the amount of civil works required, as clearly seen by the effect of the duct availability on the upfront costs of the fibre alternatives. This illustrates that fibre deployment beyond the main flexibility point in the network increases the overall costs significantly, if ducts available are scarce. In conclusion, the costs of extensive fibre deployment is strongly related to civil works costs.

20.3.3 The influence of the costs of electronics

For all studies performed and all cases analysed the electronics costs are dominant, assuming moderate civil work costs [6,7]. The electronics part of the costs are in the range from 37 % to 68 % taking all cases into account and all technology options, from enhanced copper to advanced broadband fibre systems. This should be an obvious challenge for equipment manufacturers, maybe fibre system suppliers in particular, since even for such systems electronics seem to contribute at least as much as the optics to the overall costs.

Also, in absolute values the electronics costs of twisted pair, coaxial and fibre alternatives are comparable from case to case. The marginal costs of placing these investments in a broadband infrastructure, i.e. deploying fibre, depends on the availability of ducts. In our studies, the marginal costs of civil works related to establishing a fibre infrastructure varies between a few percent (90 % ducts available) and one third (20 % ducts available).

Hence, the studies confirm that an access network broadband upgrade probably will confront the network operator with the choice of placing significant electronics investments, which are on a very similar level independent of technology, in either a narrowband cable infrastructure (twisted pair copper) or a broadband cable infrastructure (fibre).

20.3.4 Fibre roll-out as a strategic decision

Given the intention of a fibre in the loop (FITL) upgrade, the establishment of the fibre infrastructure is in itself probably a strategically more important decision than the choice of system technology, provided a system independent infrastructure is rolled out. The reason behind this is two-fold: First the fibre infrastructure costs are likely to be significant and quite similar for all kinds of fibre systems, and that the system costs can be rather marginal between different technological options. Secondly, the expected technical lifetime of the fibre infrastructure is long compared to equipment.

Even if the access network operator should still postpone the decision on a fibre roll-out or not for some time, the relaxed importance of the fibre optic system choice could encourage the access network operator to consider substituting some of the vast amounts of pure copper cables that today are deployed for rehabilitation purposes with hybrid copper/fibre cables. This may enable beneficial savings on future civil works costs related to fibre upgrades, provided measures ensuring future-proof and fibre-rich infrastructures are taken whilst introducing such hybrid cables. In conclusion, rolling out fibre remains a key decision of great strategic importance in access network upgrading

20.3.5 Fibre penetration and broadband capacity provided

Given a migration towards fibre in the access network, how deep into the access network shall the fibre be deployed? This is one major question in the long term perspective for access network operators, and PNOs in particular. The recent developments in digital subscriber line (DSL) technology enables the PNO to provide broadband capacities on the existing copper network. However, in general for the DSL options, there is a trade off between copper loop distance and capacity available. The cable infrastructure costs varies significantly between network areas, as previously discussed, and hence quite different fibre penetration levels are likely to be seen.

This is reflected in particular in the results from a study on cable infrastructure costs, carried out in EURESCOM project P614 (Implementation strategies for advanced access networks) [15]. The analysis covers four network area types, a downtown area, an urban area, a suburban area and a rural area. The areas have been segmented and characterised according to average copper loop length in the existing access network (also reflects the density of living units in the area), availability of existing ducts and surface conditions with corresponding cable deployment type and civil works costs. Representative ranges of the characteristic parameters have been assigned to each network area segment.

Fig 20.4 shows the asymmetric capacity provided after the upgrade as a function of cable infrastructure investment per homes passed for the four network areas. Cable infrastructure investments encompass civil works costs (digging, ducting and surface reinstatement), costs of cable installation and cable costs. The upper and lower curves in each area represent minimum and maximum levels of civil works costs.

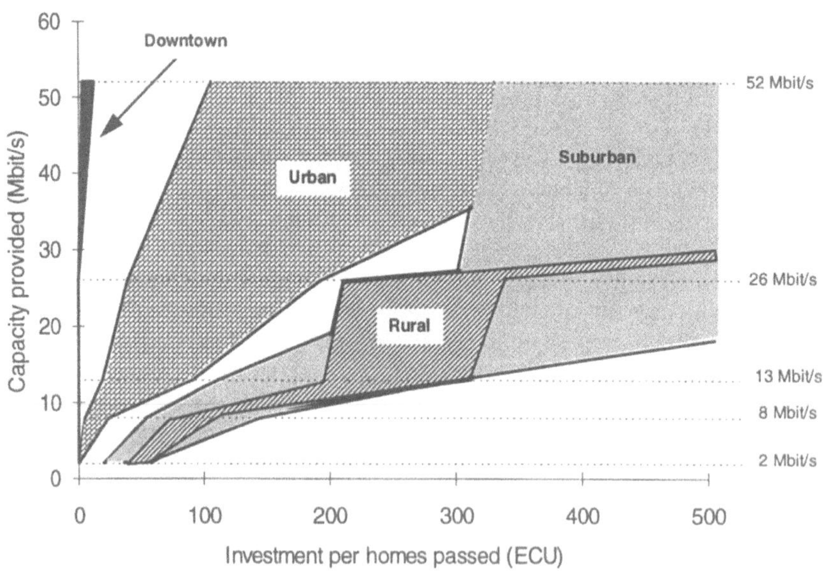

Fig 20.4 Asymmetric capacity provided after upgrade as a function of investment per homes passed for four network areas.

In general, for densely populated areas (downtown, urban and suburban areas) the investment levels required for upgrades to higher capacities have an increasingly strong dependency on the loop lengths for increasing asymmetric capacities.

The cable infrastructure cost levels presented here, indicate a need for differentiating between network areas with respect to service (capacity) offerings at asymmetric capacities beyond 2 Mbit/s.

Another aspect of this is the potential capacity increase gained with additional investments on top of the initial upgrade investments, i.e. a further upgrade of the network to the next higher asymmetric capacity level, e.g. from 2 Mbit/s to 8 Mbit/s or from 13 Mbit/s to 26 Mbit/s. The asymmetric capacities are indicated in the right side of the figure. The gain in available capacity per invested ECU decreases as a function of the population density, illustrated in Fig 20.4 by the decreasing slope of the investment level areas as we move from densely populated areas to more scarcely populated areas. Nevertheless, for all areas the additional investments for a further upgrade of the network to the next higher asymmetric capacity level are between twice and thrice the total initial investments.

20.3.6 The effect on costs of the service take rate

The different broadband upgrade alternatives have a different cost structure with respect to required upfront investment levels and service penetration dependant costs. Typically fibre upgrades incur significant upfront costs in terms of an initial deployment of fibre infrastructure, with associated risks of losses due to failing future service take up and corresponding revenues. However, for high service take rates, there is a good potential for reduction in line cost due to sharing of infrastructure. In opposite, ADSL upgrades in many cases only involve service penetration dependant investments, which the network operator do not need to carry out before the customer is connected. This alleviates the need for risky initial investments, at the expense though of a limited potential for line cost reductions for high take rates. Hence, the effect on costs of the service take rate need to be taken into account in cost comparisons of the different technology alternatives.

We have studied the effect on the overall upgrade costs for a network area, in order to determine for what level of average capacity demand a fibre deployment as compared to a ADSL/HDSL upgrade may be justified in pure economic terms by a PNO which has an installed copper base. In addition, the costs per line has been studied for broadband upgrades based on either fibre cable, coaxial cable or twisted pair copper cable.

Fig 20.5 dcpicts the costs of the two major types of solutions examined (enhanced copper and ATM based PONs) as a function of the final year (2005) average capacity demand within a network area of 1,024 potential customers. A ten year linear upgrade project has been studied. Results are shown for an urban area with a density of 3,000 subscribers per km^2. The IFC for ATM PON in fibre to the curb (FTTC) and fibre to the building (FTTB) configurations are included for three degrees of duct availability, 0 %, 50 % and 100 %. The FTTC and FTTB costs are plotted for eight and 32 potential customers sharing one optical network unit (ONU) respectively.

The costs of the two technology alternatives for each of the studies 1-4 are indicated by the black (enhanced copper) and white (ATM PON) squares, according to the study specific average subscriber capacity demands shown in Fig 20.1: study 1: 20 % broadband capacity demand, study 2 : 18 %, study 3a (downtown): 49 %, study 3b (suburban): 22 % and study 4: 36 %.

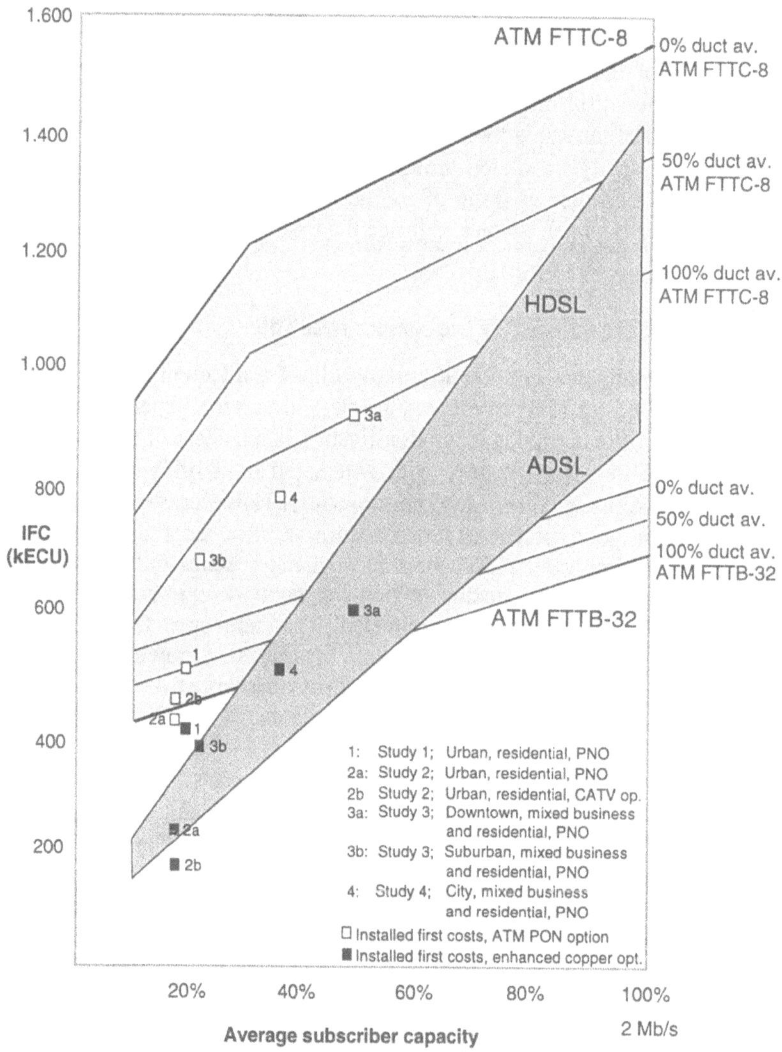

Fig 20.5 The IFC of the two major types of solutions examined (enhanced copper and ATM based PONs) as a function of the average capacity demand for 1,024 customers.

Fig 20.5 depicts the effect of the broadband service take up on the overall costs in an upgrade area. Accordingly, the broadband line costs is also influenced by the take rate. Fig 20.6 shows the IFC per 2 Mbit/s access line for a ten year linear upgrade project, as a function of broadband penetration in 2005.

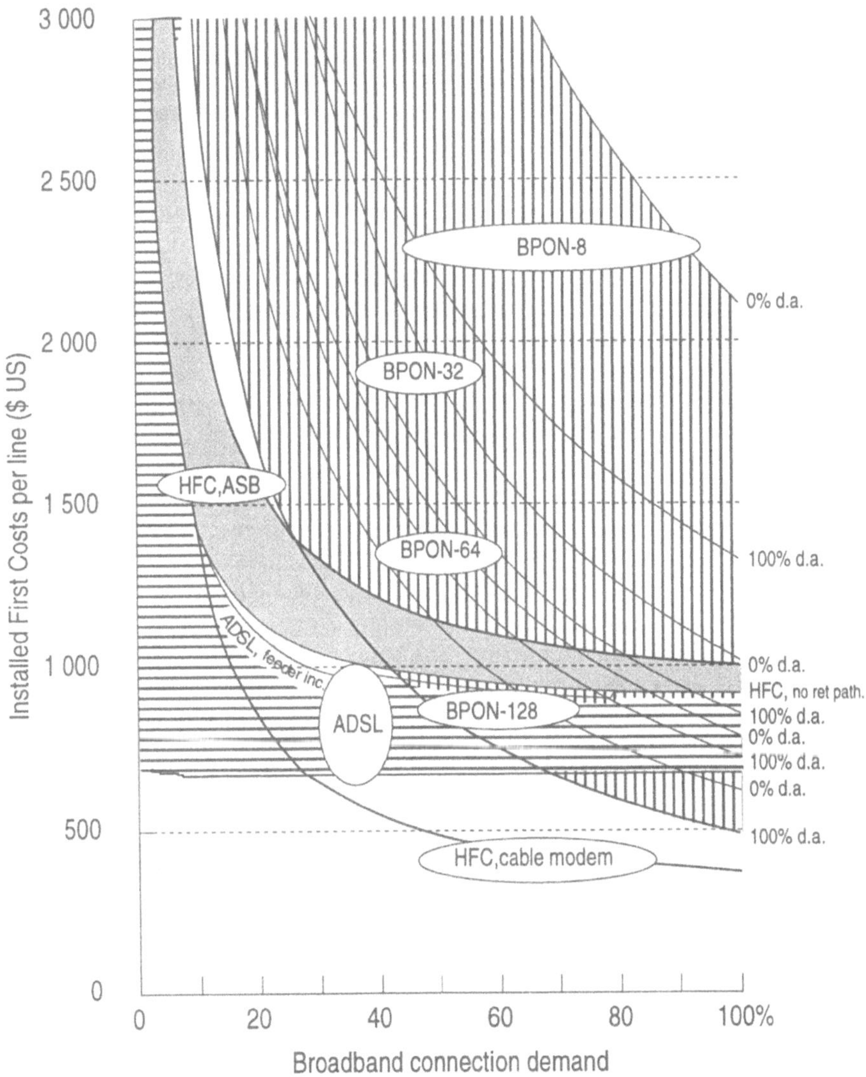

Fig 20.6. The cost per line in an urban area for the main wireline upgrade alternatives as a function of broadband connection demand (d.a.: duct availability).

Line costs are analysed for the main system alternatives for the three relevant wireline transmission media, namely fibre cable, coaxial cable and twisted pair copper cable. Results are shown for an urban, apartment block area with an average copper loop length of 400 metres between the optical node (service access point) and the buildings. The IFC for a broadband PON (BPON), or alternatively ATM PON, in the FTTB configuration are included for two degrees of duct availability, 0 % and 100 %. The FTTB costs are plotted for each ONU serving eight (BPON-8), 32 (BPON-32), 64 (BPON-64) or 128 (BPON-128) potential customers. 1 $US ≈ 1.25 ECU.

The results show that in general, for the fibre solutions and the dedicated channel HFC upgrade, i.e. HFC asymmetric switched broadband (ASB) or HFC symmetric switched broadband (SSB), the operator will have to rely on take rates of 50 % or higher in order to reach line cost levels below 1,000 $US. ADSL or cable modem upgrades are the only alternatives with prospects of line costs close to 500 $US for moderate take rates. The costs per line of upgrading existing twisted pair networks and coaxial cable networks with respectively DSL and cable modem technologies are comparable for take rates up to 30 %. For higher penetrations the cable modem technology seems to have a cost advantage, but a limitation in traffic capacity compared to the DSL technologies.

Fig 20.5 and Fig 20.6 illustrate that low demand for broadband services favours the copper based solutions whilst high capacity demand favours more aggressive fibre based access network solutions. The upfront costs are lower for the copper based solutions, whilst the additional costs of providing more capacity are smaller for the fibre based solutions. The break-even point in the figure indicates the capacity limit beyond which the demand can justify the deployment of optical solutions in pure economic terms. In general, if the demand is high enough the fibre based solutions will be the most cost-effective solution.

Optical fibre alternatives like BPONs will most likely have to rely on take rates in excess of 50 % and more than 50 subscriber per optical node in order to be justifiable in pure economic terms. The results illustrate that the costs of civil works and costs of ONUs remain as major obstacles for extensive introduction of fibre in areas with an established access network of good quality. However, optical fibre alternatives have the potential for providing higher capacities than ADSL. This is not fully reflected in Fig 20.6, which mainly relates to 2 Mbit/s broadband connections.

In summary, the studies have illustrated that the average capacity demand in the access network is a very important parameter to take into account in strategic considerations, since the overall investments associated with different upgrade technologies depend on the actual capacity provided. Moreover, the costs per line of upgrading existing twisted pair networks and coaxial cable networks with respectively DSL and cable modem technologies are comparable for take rates up to 30 %. For higher penetrations the cable modem technology seems to have a cost advantage, but a limitation in traffic capacity compared to the DSL technologies.

20.3.7 The effect on costs of optical network termination customer sharing

The influence on the overall costs of the sharing of network termination equipment is evident from Fig 20.5 and Fig 20.6. In Fig 20.5, the higher slope of the FTTC-8 curves for capacities below 30 % is due to the relatively high number of ONUs compared to broadband subscribers in this region. An FTTC alternative with 32 customers sharing each ONU would have slightly higher IFC than its FTTB counterpart, mainly due to the increased cost of outdoor housing. Similarly an FTTB configuration with eight customers per ONU would result in slightly lower cost than the FTTC-8 due to indoor housing savings.

The effect on costs of optical network termination customer sharing has been studied in detail in [16]. Chapter 18 in this book contains an in-depth documentation of the work, which analyses the IFC per 2 Mbit/s in the residential and SME market as a function of the average number of 2 Mbit/s subscribers per ONU. The analysis shows that the ATM PON solution benefits from cost reductions both due to reduced civil works costs and higher customer sharing of the ONUs if they are more centrally located. The results indicate that around 50 subscribers per optical node is the critical size for which ATM PON can be justified in pure economic terms. This illustrates the great benefit with respect to IFC of sharing the ONU costs between a few tens of subscribers, and confirms that a high splitting ratio is one crucial aspect of a cost effective broadband upgrade with PONs.

20.3.8 The timing of the upgrade

The appropriate timing of broadband upgrading is crucial, not at least in order to reduce the investments. The costs of upgrade strategies involving mass deployment of residential access fibre, is anticipated to become significantly lower if the system introduction is delayed enough to benefit from component cost reductions. Fig 20.7 shows the broadband line cost as a function of roll-out year and penetration for selected broadband PON (BPON), HFC and ADSL alternatives. It is assumed that the indicated penetration is obtained at time of roll-out.

Fig 20.7 illustrates the assumptions with respect to cost evolution embedded in the study results. The expected decrease in cost during the next years is clearly seen. However, in the competitive marketplace there is always a risk of loosing a significant share of the potential customer base if broadband networks and corresponding services are not offered in due time. The equipment cost of the three upgrade alternatives will most likely have quite similar cost reduction potential, as they all are emerging technologies in this market. The total ADSL line costs are expected to be reduced by two-thirds of the 1996 cost level during the ten-year period, whereas the resulting fibre and coaxial cable upgrades are expected to experience a 50 % cost reduction. The difference in cost reduction is attributed to the fact that the latter two upgrades involve civil works costs in addition to the equipment costs.

In conclusion, postponing the fibre roll-out may result in a cost advantage compared to the HFC and ADSL upgrades. In addition, prospects of future operation and maintenance savings might motivate for an extensive fibre deployment. So far, there is however no clear evidence of decreased operation and maintenance costs to offset the huge investments required.

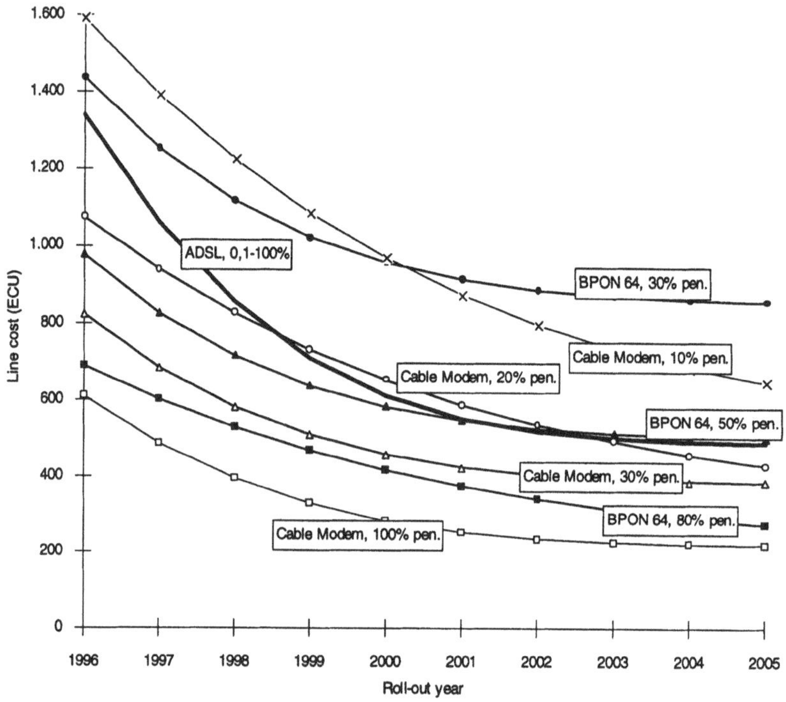

Fig 20.7 Line cost as a function of roll-out year and penetration. The indicated penetration is obtained at time of roll-out. The BPON alternatives are calculated with 50 % duct availability.

20.3.9 The timeframe of broadband access network upgrading

The net present value (NPV) divided by the IFC for studies 1, 2 and 3 is shown in Fig 20.8. The NPV used here is the cumulative sum of the cash flow from the start of the model year to the final year. The residual value of the network is not included. The ratio of NPV to IFC gives an indication of the level of the yearly internal rate of return (IRR) of the project. A ten year linear project with a yearly IRR of 10 % will have a NPV/IFC ratio of ~2,6.

The NPV/IFC ratio of the upgrade projects ranges from just above 0 to close to 5, suggesting that the project's payback periods are comparable to the establishment of the present day narrowband networks, and even lower in some instances. This illustrates that access network broadband upgrades are likely to turn out to be long term projects with payback periods in the range of five to ten years.

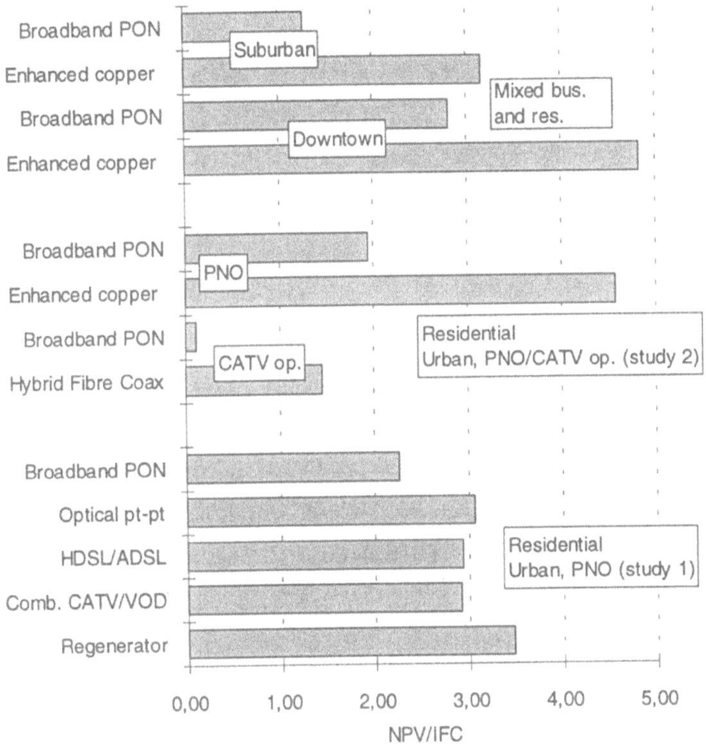

Fig 20.8 The NPV divided by IFC for studies 1-3.

20.3.10 The effect of revenue on the payback period

It is obvious that the timeframe and corresponding payback periods of the upgrades to a large extent relies on the revenue levels. Fig 20.9 shows the payback period as a function of average annual access network related revenue per line for a ten year linear upgrade project.

The payback period is defined as the period from the start of the project to the time when the cash balance (cumulative sum of the retained cash flows) turns positive. The indicated penetration is the saturation level of 2 Mbit/s asymmetric switched broadband service penetration in 2005. The BPON alternatives are calculated with 50 % duct availability.

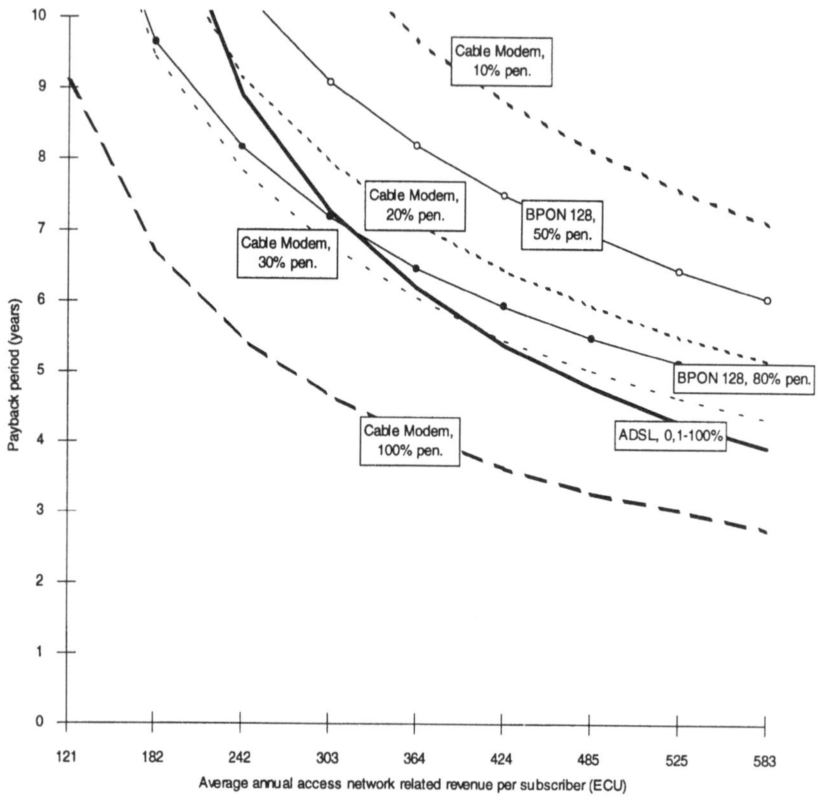

Fig 20.9 Payback period as a function of average annual revenue per line for a ten year linear upgrade project 1996-2005.

The results indicate that the cost level of broadband access upgrades is likely to be within the investment range in which payback periods in the order of five years may be expected with annual revenues per access line of 400 ECU. However, it must be emphasised that these calculations are based on an early (1996) and quite extensive roll-out. A more gradual deployment which in addition is postponed in time, will require a lower turnover per access line in order to achieve acceptable payback periods. Hence, the payback periods of broadband access is likely to be in the order of five to ten years with annual revenues per access line of 400 ECU.

20.3.11 The cash flow of the future broadband platform

The upgrade effects on the prospects looking beyond the upgrade period are not evident from the NPV and the payback period. The cash flows in the final year is useful as a predictor for the long term effect of the expanded revenue base. Fig 20.10 shows the cash flow before and after upgrade for studies 1-3.

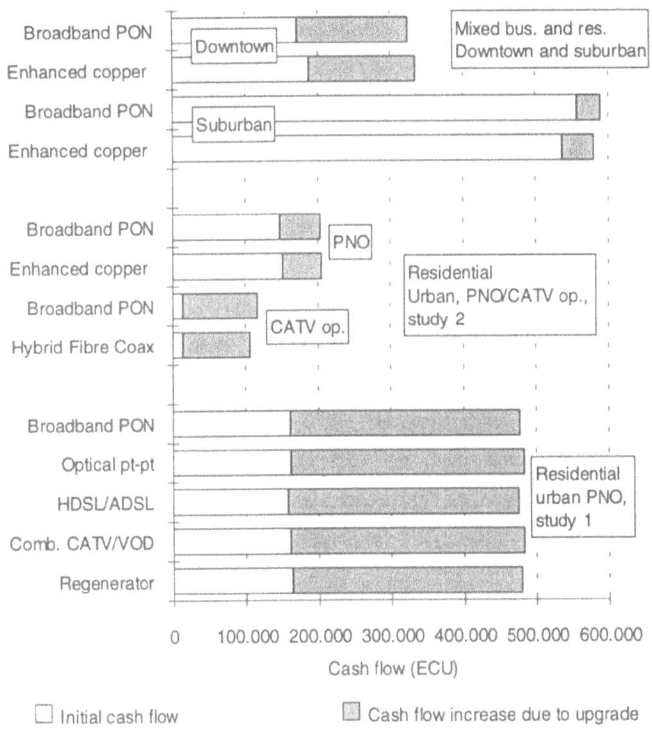

Fig 20.10 Cash flow before and after upgrade for studies 1-3.

With the given tariffs and penetrations all of the upgrade projects have gained strength through the broadband upgrade compared to the initial annual cash flow, i.e. the final year cash flow is higher than the pre-upgrade cash flow.

The results illustrate that access network broadband upgrades represent long term projects which are assumed to create future cash flows. As such, the establishment of a broadband access network platform represents a challenge comparable to the build-up of the present narrowband infrastructure: heavy investment projects with associated investment levels which in the short and medium term result in a weakened NPV. However, at the same time these investments are a requirement in order to maintain the revenue level in the long term. In other words, a necessity in order to establish an effective, broadband platform, which is the basis for the future "money machine".

20.4 CONCLUDING REMARKS

The issue of broadband access network upgrade remains a major challenge for operators due to the high cost sensitivity of this network segment and the high uncertainties of future service take up. This chapter has highlighted some of the key issues of broadband access upgrade and their possible impact:

- Upgrade costs are comparable to narrowband access network costs
- The costs of extensive fibre deployment is strongly related to civil works costs
- The costs of electronics are high for any technology alternative
- Fibre roll-out seems to be a key strategic decision
- For all areas the additional investments for a further upgrade of the network to the next higher capacity level are between twice and thrice the total initial investments
- The average capacity demand differentiates costs of copper and fibre upgrades
- Cost reduction relies heavily on optical network termination customer sharing
- Access network upgrading generally turns out to be long term projects
- The payback periods of broadband access upgrades is likely to be in the order of five years with annual revenues per access line of 400 ECU
- Access network broadband upgrades will create increased future cash flows, and constitute the basis for the future "money machine".

REFERENCES

[1] Olsen, B T, Ims, L A, Myhre, D. Techno-economic analysis of Broadband Access Network Upgrading. *In: Proc. Networks '96, Sydney,* Australia, *November 25-29, 1996, 1, 373-378.*

[2] Ims, L A, Olsen, B T, Myhre, D. Economics of residential broadband access network technologies and strategies. *IEEE Network*, 11, (1), 1997, 58-64.

[3] Cook, A, Stern, J. Optical Fiber Access - Perspectives Towards the 21st Century. *IEEE Communications Magazine*, 32 (2), 1994, 78-86.

[4] LeBel, P, Oliver, R. Residential Broadband Architectures Strategies and Economics. *In: Proc. The 7th International IEEE workshop on Optical Access Networks (OAN '95), Nuremberg, Germany, September 24-27, 1995, 4.4-1 - 4.4-10.*

[5] Barlet, W C. A copper-plated full service network. *Telephony, January 15,1996, 20-26.*

[6] Ims, L A, Olsen, B T, Myhre, D, Mononen, J, Lähteenoja, M, Ferrero, U, Zaganiaris, A. Multiservice Access Network Upgrading in Europe: a Techno-economic Analysis. *IEEE Communications Magazine,* 34, (12), 124-134, 1996.

[7] Ims, L A, Myhre, D, Olsen, B T, Mononen, J, Lähteenoja, Ferrero, U, Zaganiaris, A, Szomolányi, K, Orfanos, I. Advanced Multiservice Scenarios for Europe. *In: Proc. 13th Annual Conference European Fibre Optic Communications and Networks (EFOC&N '95), Brighton, UK, June 27-30, 1995, 1, 177-183.*

[8] Ims, L A, Olsen, B T, Stordahl, K, Myhre, D, Zaganiaris, A. Upgrading the PNO and Cable Operator access networks to advanced broadband services: technology alternatives and their economic implications in a competitive environment, *In: Proc. Telecom America 96, Technology Summit, Rio de Janeiro, Brasil, June 10-15, 1996, 243-249.*
[9] Olsen, B T, Ims, L A, Øverli, T, Stordahl, K, Myhre, D, Drieskens, M, Kraushaar, J, Zaganiaris, A. Techno-economic analysis of PNO and CATV operator strategies for broadband upgrade of the access network. *In: Proc. International Teletraffic Seminar 1995, Bangkok, Thailand, Nov. 28 - Dec. 1, 1995, 54-1 - 54-12.*
[10] Myhre, D, Ims, L A, Olsen, B T, Stordahl, K. Access Network alternatives for broadband services. *In: Proc. ConTEL '95, Zagreb, Croatia, June 7-9, 1995, 17-24.*
[11] Ims, L A, Olsen, B T, Myhre, D, Stordahl, K. Economic Implications of Broadband Access Network Upgrade. *In: Proc. Broadband Islands 1995 (BRIS '95), Dublin, Ireland, September 4-5, 1995, 33-43.*
[12] Olsen, B T, Ims, L A, Myhre, D, Stordahl, K, Theivindrein, R. Technoeconomic Evaluation of Optical Broadband Access Network Scenarios for the residential and business market. *In: Proc. 21st European Conference on Optical Communications (ECOC '95), Brussels, Belgium September 14-17, 1995.*
[13] Olsen, B T, Zaganiaris, A, Gieschen, N, Tahkokorpi, M, Drieskens, M, Markatos, S. RACE 2087/TITAN: Tool for Introduction scenarios and Techno-economic studies for the Access Network. *In: Proc. RACE Open Workshop on Broadband Access, Nijmegen, the Netherlands, June 7-8, 1993.*
[14] Stordahl, K, Murphy, E. Methods for forecasting long term demand for wide and broadband services in the residential market. *IEEE Communications magazine*, 13, (2), 1995, 44-49.
[15] Ims L A, Mononen, J, Lähteenoja, M, Budry L, Salerno, M, Ferrero, U, Myhre, D, Olsen, B T, Aas K C. Key factors influencing the overall costs and project values of broadband access network upgrades. *European Conference on Networks & Optical Communications (NOC '97), Antwerp, Belgium, June 17-20, 1997.*
[16] Olsen, B T, Myhre, D, Ims, L A. Broadband access Alternatives in the business dominated area, a techno-economic study. *In: Proc. EURESCOM Workshop on Optical Multiservice Access Networks (WOMAN), Turin, Italy, September 23-24, 1996.*

APPENDIX A

List of acronyms

ADM	add and drop multiplexer
ADPCM	adaptive differential pulse code modulation
ADSL	asymmetric digital subscriber line
AONU	broadband optical network unit
ASB	asymmetric switched broadband
ATM	asynchronous transfer mode
ATM PON	asynchronous transfer mode based passive optical network
BA or BRA	basic (rate) access
BB	broadband
BB HDT	broadband host digital terminal
BONU	broadband optical network unit
BS	base station
BT	British telecom
CAP	competitive access provider
CATV	community antenna television
CCU	central control unit
CEPT	conférence Europeenne des administrations postes et de télécommunications
CDMA	code division multiple access
CNU	coaxial network unit
CP	customer premises
CPE	customer premises equipment
CPNI	customer premises network interface
CSDN	circuit switched datacommunications network
CTTB	copper to the building
CT2	cordless telephone, second generation
CUG	closed user group
DB	database
DB	declining balance method
DCS	digital cellular system
DECT	digital European cordless telecommunication
DSL	digital subscriber line
DTH	direct to the home
DTMF	dual tone multi-frequency
DXC	digital cross connect
EDFA	erbium doped fibre amplifier

EDP	electronic data processing
EP	evolutionary path
EPF	evolution prospects and framework
ETS	ETSI technical standard
ETSI	European telecommunications standards institute
EU	European union
FDM	frequency division multiplexing
FDMA	frequency division multiple access
FP	flexibility point
FITL	fibre in the loop
FM	frequency modulation
FTTB	fibre to the building
FTTC	fibre to the curb
FTTCab	fibre to the cabinet
FTTH	fibre to the home
FTTLA	fibre to the last amplifier
FTTN	fibre to the distribution node
FTTR	fibre to the remote node
FTTSA	fibre to the serving area
GSM	global system for mobile communication
HDSL	high speed digital subscriber line
HDT	host digital terminal
HDTV	high definition television
HDWDM	high density wavelength division multiplexing
HE	headend
HP	homes passed
IDS	integrated delivery system
IFC	installed first costs
IRR	internal rate of return
ISDN	integrated services digital network
ITU-T	international telecommunications union-telecommunication sector
IWU	interworking unit
LAN	local area network
LC	line card
LCC	life cycle costs
LEO	low-earth orbit
LEX	local exchange
LIM	line interface module
LMDS	local multipoint distribution service
LT	line termination
MBS	European mobile broadband system
MDF	main distribution frame
MEO	mid-earth orbit
MMDS	multipoint multichannel distribution system
MPEG	motion picture expert group
MPMP	microwave point-to-multipoint
MS	market share
MSC	mobile switching centre

MTBR	mean time between repair
MTTR	mean time to repair
MVDS	multipoint video distribution system
NB	narrowband
NB HDT	narrowband host digital terminal
N-ISDN	narrowband ISDN
NMT	nordic mobile telephone
NP	network provider
NPV	net present value
NT	network termination
NT1	network termination, narrowband ISDN
NVOD	near video on demand
OAM	operation, administration and maintenance
O/E	optical/electrical converter
OFA	optical fibre amplifier
OFDM	optical frequency division multiplexing
OLT	optical line termination
ONP	open network provisioning
ONU	optical network unit
OP	optical port
PBX	private branch exchange
PC	personal computer
PCM	pulse code modulation
PCN	personal communications network
PDH	plesiochronous digital hierarchy
PH	portable handset
PNO	public network operator
PON	passive optical network
POTS	plain old telephone service
PPC	pay per channel
PPV	pay per view
PRA	primary rate access
PSDN	packet switched datacommunications network
PSTN	public switched telephone network
PVCF	present value of cash flow
QOS	quality of service
RACE	research for advanced communications in Europe
REX	radio exchange
RF	radio frequency
RIF	radio interface
RLL	radio in the local loop
RFP	radio fixed point
RPR	return path receiver
RPT	return path transmitter
RSS	remote subscriber system
RSU	remote subscriber unit
RX	receiver
SAP	service access point
SCM	subcarrier multiplexing

SCMA	subcarrier multiple access
SDH	synchronous digital hierarchy
SDSL	symmetric digital subscriber line
SLN	straight line depreciation
SME	small and medium enterprise
SP	service package
SSB	symmetric switched broadband
STB	set-top box
STM	synchronous transfer module
SYD	sum-of-the-years'-digits method
TACS	total access communication system
TCW	total civil works
TDM	time division multiplexing
TDMA	time division multiple access
TITAN	tool for introduction strategies and techno-economic evaluation of access network
TPON	telecommunication over passive optical networks
TO	telecommunications operator
TX	transmitter
UK	United Kingdom
UNI	user network interface
UMTS	universal mobile telephony system
UPT	universal personal telecommunications
VB	voice band
VCR	video cassette recorder
VDSL	very high speed digital subscriber line
VOD	video on demand
WDM	wavelength division multiplexing
WDM	wavelength division multiple access

APPENDIX B

List of publications

[1] Ims, L A, Stordahl, K, Olsen, B T. *Risk analysis of residential broadband upgrade in a competitive environment. IEEE Communications Magazine,* June 1997.

[2] Ims, L A, Olsen, B T, Myhre, D. Economics of residential broadband access network technologies and strategies. *IEEE Network*, 11, (1), 1997, 58-64.

[3] Ims, L A, Olsen, B T, Myhre, D, Mononen, J, Lähteenoja, M, Ferrero, U, Zaganiaris, A. Multiservice Access Network Upgrading in Europe: a Techno-economic Analysis. *IEEE Communications Magazine,* 34, (12), 1996, 124-134.

[4] Stordahl, K, Ims, L A, Olsen, B T. Risk assessment and techno-economic analysis of competition between PNO and CATV operators, *In: Proc. Networks '96, Sydney, Australia, November 25-29, 1996, 2, 405-410.*

[5] Olsen, B T, Ims, L A, Myhre, D. Techno-economic analysis of Broadband Access Network Upgrading. *In: Proc. Networks '96, Sydney,* Australia, *November 25-29, 1996, 1, 373-378.*

[6] Ferrero, U. Broadband optical access network: co-operative work among European PNOs. *In: Proc. EURESCOM Workshop on Optical Multiservice Access Networks (WOMAN), Turin, Italy, September 23-24, 1996.*

[7] Stordahl, K. Application and traffic demand, the key to network evolution. *In: Proc. EURESCOM Workshop on Optical Multiservice Access Networks (WOMAN), Turin, Italy, September 23-24, 1996.*

[8] Welling, I. Economics of VOD and Internet access service upgrade. *In: Proc. EURESCOM Workshop on Optical Multiservice Access Networks (WOMAN), Turin, Italy, September 23-24, 1996.*

[9] Tahkokorpi, M. TITAN methodology. *In: Proc. EURESCOM Workshop on Optical Multiservice Access Networks (WOMAN), Turin, Italy, September 23-24, 1996.*

[10] Olsen, B T, Myhre, D, Ims, L A. Broadband access Alternatives in the business dominated area, a techno-economic study. *In: Proc. EURESCOM Workshop on Optical Multiservice Access Networks (WOMAN), Turin, Italy, September 23-24, 1996.*

[11] Mononen, J, Lähteenoja, M. Broadband access alternatives in the residential dominated area, a techno-economic study. *In: Proc. EURESCOM Workshop on Optical Multiservice Access Networks (WOMAN), Turin, Italy, September 23-24, 1996.*

[12] Ferrero, U. Broadband optical access network: co-operative work among European PNOs. *In: Proc. 22nd European Conference on Optical Communication, ECOC '96, Oslo, 15-19 September 1996, 3, 3-3 - 3-10.*

[13] Ims, L A, Olsen B T, Myhre, D. Cost benefits and business opportunities of high capacity optical broadband access network upgrade technologies. *22nd European Conference on Optical Communication (ECOC '96), Oslo, September 15-19, 1996.*

[14] Stordahl, K, Olsen B T, Ims, L A. Do we need a Pan-European network and what is the need for new applications? *In: Proc. 22nd European Conference on Optical Communication (ECOC '96), Oslo, September 15-19, 1996, 2, 2-3 - 2-10.*

[15] Olsen, B T, Zaganiaris, A, Stordahl, K, Ims, L A, Myhre, D, Øverli, T, Tahkokorpi, M, Welling, I, Drieskens, M, Kraushaar, J, Mononen, J, Lähteenoja, M, Markatos, S, De Bortoli, M, Ferrero, U, Ravera, M, Balzaretti, S, Fleuren, F, Gieschen, N, De Oliveira Duarte, M, de Castro, E. Techno-economic evaluation of narrowband and broadband access network alternatives and evolution scenario assessment. *IEEE Journal on Selected Areas in Communications,* 14 (6), 1996, 1184-1203.

[16] Stordahl, K. Forecasting long-term demand for broadband services in the residential market: devicing a method for developing forecasts for new services. *In: Proc. Market forecasting in the telecoms industry, Institute of International Research, London, UK, 8-10 July, 1996.*

[17] Pousa, M, Verhofstadt, K, Corveleyn, F, Paksy, G, Kimer, O, Borges, I, Myhre, D, Ims, L A, Olsen, B T. Fiber Based Architectures for Large Business Users. *In: Proc. Networks and Optical Communications '96 (NOC '96), Heidelberg, Germany, June 24-28, 1996, 151-157.*

[18] Myhre, D, Ims, L A, Olsen, B T, Lähteenoja, M, Mononen, J, Ferrero, U, Vangelista, L. Roll out Strategies and Upgrade Access Alternatives for the Mixed Residential and Business Market: A Techno-economic Evaluation. *In: Proc. Networks and Optical Communications '96 (NOC '96), Heidelberg, Germany, June 24-28, 1996, 143-150.*

[19] Ims, L A, Olsen, B T, Stordahl, K, Myhre, D, Zaganiaris, A. Upgrading the PNO and Cable Operator access networks to advanced broadband services: technology alternatives and their economic implications in a competitive environment, *In: Proc. Telecom America 96, Technology Summit, Rio de Janeiro, Brasil, June 10-15, 1996, 243-249.*

[20] Ims, L A, Olsen, B T, Myhre, D, Stordahl, K. Migration paths to broadband access for the mixed residential and business market. *Accepted for publication in Telecommunications Systems.*

[21] Olsen, B T, Ims, L A, Øverli, T, Stordahl, K, Myhre, D, Drieskens, M, Kraushaar, J, Tahkokorpi, M, Ravera, M, De Bortoli, M, Zaganiaris, A, Markatos, S, Lähteenoja, M, Mononen, J, Fleuren, F. PNO and Cable Operator broadband upgrade technology alternatives: a techno-economic analysis. *In: Proc. Optical Fiber Conference 1996 (OFC '96), San Jose, USA, Feb. 25 - March 1, 1996, 2, 140-141.*

[22] Olsen, B T, Ims, L A, Øverli, T, Stordahl, K, Myhre, D, Drieskens, M, Kraushaar, J, Zaganiaris, A. Techno-economic analysis of PNO and CATV operator strategies for broadband upgrade of the access network. *In: Proc.*

International Teletraffic Seminar 1995, Bangkok, Thailand, Nov. 28 - Dec. 1, 1995, 54-1 - 54-12.

[23] Ims, L A, Olsen, B T, Myhre, D. The Economic Implications of Broadband Access Network Upgrading. *In: Proc. Broadband Networks, London, UK, November 8-9, 1995.*

[24] Saijonmaa, J, Tahkokorpi, M, Welling, I. Cost of Investment and Revenue Modelling and Analysis of Various Networked Multimedia Services in PTO and CATV operator Environments. *In: Proc TELECOM '95, Technology Summit, Geneva, Switzerland, October 3-11, 1995, 2, 629-633.*

[25] Ferrero, U, de Bortoli, M, Myhre, D, Ims, L A, Olsen, B T, Mononen, J, Lähteenoja, M, Zaganiaris, A, Pousa, M, Szomolányi, K, Orfanos, I. Access Network Upgrade Strategies: mapping the European Diversity. *In: Proc. 7th International IEEE workshop on Optical Access Networks, Nuremberg, Germany, September 24-28, 1995, 4.2-1 - 4.2-10.*

[26] Olsen, B T, Ims, L A, Myhre, D, Stordahl, K, Theivindrein, R. Techno-economic Evaluation of Optical Broadband Access Network Scenarios for the residential and business market. *In: Proc. 21st European Conference on Optical Communications (ECOC '95), Brussels, Belgium September 14-17, 1995.*

[27] Olsen, B T, Ims, L A, Myhre, D, Stordahl, K. Techno-economic Evaluation of Multiservice Broadband Access Network Alternatives. *In: Proc. Brazilian Telecommunications Symposium, Agua de Lindoia, Brasil, September 3-6, 1995, 2, 476-481.*

[28] Ims, L A, Olsen, B T, Myhre, D, Stordahl, K. Economic Implications of Broadband Access Network Upgrade. *In: Proc. Broadband Islands 1995 (BRIS '95), Dublin, Ireland, September 4-5, 1995, 33-43.*

[29] Van Hoecke, J. Access Network Issues and challenges from a manufacturing viewpont. *In: Proc. Techno-Economics of Access Networks, RACE 2087/TITAN Workshop, Aveiro, Portugal, July 4-6, 1995.*

[30] De Bortoli, M. Market Evolution of Optical fibre technologies. *In: Proc. Techno-Economics of Access Networks, RACE 2087/TITAN Workshop, Aveiro, Portugal, July 4-6, 1995.*

[31] Ferrero, U. Mapping the European Diversity: TITAN co-operation with EURESCOM. *In: Proc. Techno-Economics of Access Networks, RACE 2087/TITAN Workshop, Aveiro, Portugal, July 4-6, 1995.*

[32] de Oliveira Duarte, M. Presentation of Case Studies and Group Organisation. *In: Proc. Techno-Economics of Access Networks, RACE 2087/TITAN Workshop, Aveiro, Portugal, July 4-6, 1995.*

[33] Zaganiaris, A. Presentation of TITAN. *In: Proc. Techno-Economics of Access Networks, RACE 2087/TITAN Workshop, Aveiro, Portugal, July 4-6, 1995.*

[34] Stordahl, K. Market forecasts. *In: Proc. Techno-Economics of Access Networks, RACE 2087/TITAN Workshop, Aveiro, Portugal, July 4-6, 1995.*

[35] Olsen, B T. Forecasts of cost components. *In: Proc. Techno-Economics of Access Networks, RACE 2087/TITAN Workshop, Aveiro, Portugal, July 4-6, 1995.*

[36] Markatos, S. The TITAN cost database. *In: Proc. Techno-Economics of Access Networks, RACE 2087/TITAN Workshop, Aveiro, Portugal, July 4-6, 1995.*

[37] Drieskens, M, Tahkokorpi, M. TITAN methodology and Tool implementation. *In: Proc. Techno-Economics of Access Networks, RACE 2087/TITAN Workshop, Aveiro, Portugal, July 4-6, 1995.*
[38] Drieskens, M. Network structures and geometric modelling. *In: Proc. Techno-Economics of Access Networks, RACE 2087/TITAN Workshop, Aveiro, Portugal, July 4-6, 1995.*
[39] de Castro, E. Statistical survey of demographic characteristics in Europe. *In: Proc. Techno-Economics of Access Networks, RACE 2087/TITAN Workshop, Aveiro, Portugal, July 4-6, 1995.*
[40] Lähteenoja, M. Evaluation of ADSL and HDSL. *In: Proc. Techno-Economics of Access Networks, RACE 2087/TITAN Workshop, Aveiro, Portugal, July 4-6, 1995.*
[41] Øverli, T. Evaluation of Radio in The Local Loop solutions. *In: Proc. Techno-Economics of Access Networks, RACE 2087/TITAN Workshop, Aveiro, Portugal, July 4-6, 1995.*
[42] Olsen, B T, Ims, L A. Evaluation of access network architectures for PNO and CATV operator Multiservice Scenarios. *In: Proc. Techno-Economics of Access Networks, RACE 2087/TITAN Workshop, Aveiro, Portugal, July 4-6, 1995.*
[43] Olsen, B T, Ims, L A. Evaluation of five European multiservice access network cases. *In: Proc. Techno-Economics of Access Networks, RACE 2087/TITAN Workshop, Aveiro, Portugal, July 4-6, 1995.*
[44] Stordahl, K. Risk modelling and risk assessment. *In: Proc. Techno-Economics of Access Networks, RACE 2087/TITAN Workshop, Aveiro, Portugal, July 4-6, 1995.*
[45] Tahkokorpi, M. Applications of risk assessment in TITAN. *In: Proc. Techno-Economics of Access Networks, RACE 2087/TITAN Workshop, Aveiro, Portugal, July 4-6, 1995.*
[46] Ravera, M. Access Network opportunities and Scenarios. *In: Proc. Techno-Economics of Access Networks, RACE 2087/TITAN Workshop, Aveiro, Portugal, July 4-6, 1995.*
[47] de Oliveira Duarte, M, Gieschen, N. Validation of field trials in Europe - Portugal and Germany. *In: Proc. Techno-Economics of Access Networks, RACE 2087/TITAN Workshop, Aveiro, Portugal, July 4-6, 1995.*
[48] Ims, L A, Myhre, D, Olsen, B T, Mononen, J, Lähteenoja, Ferrero, U, Zaganiaris, A, Szomolányi, K, Orfanos, I. Advanced Multiservice Scenarios for Europe. *In: Proc. 13th Annual Conference European Fibre Optic Communications and Networks (EFOC&N '95), Brighton, UK, June 27-30, 1995, 1, 177-183.*
[49] Stordahl, K, Olsen, B T. Demand Elasticity and Forecasts of Wide and Broadband Services in the Residential Market Based on Results from an International Delphi Survey. *1995 International Communications Forecasting Conference, Toronto, Canada, June 13-16, 1995.*
[50] Drieskens, M, Stordahl, K, Olsen, B T, Ims, L A, Øverli, T, Zaganiaris, A, Ravera, M. Introducing Advanced Residential Services in a competitive environment. From Project 2087/TITAN. *1995 International Communications Forecasting Conference, Toronto, Canada, June 13-16, 1995.*

[51] Myhre, D, Ims, L A, Olsen, B T, Stordahl, K. Access Network alternatives for broadband services. *In: Proc. ConTEL '95, Zagreb, Croatia, June 7-9, 1995, 17-24.*

[52] Stordahl, K, Olsen, B T, Zaganiaris, A, Tahkokorpi, M, Kalervo, M, Naveau, P. Investment risks in telecommunications networks. *Annual meeting of the risk society, Stuttgart, Germany, May 21-24, 1995.*

[53] Ims, L A, Myhre, D, Olsen, B T, Stordahl, K. Paths towards a versatile broadband Access Network. *In: Proc. International Conference on Telecommunications 95 (ICT '95), Nusa Dua, Indonesia, April 3-5, 1995, 215-219.*

[54] Stordahl, K, Murphy, E. Methods for forecasting long term demand for wide and broadband services in the residential market. *IEEE Communications magazine*, 13, (2), 1995, 44-49.

[55] Zaganiaris, A, Olsen B T, Stordahl, K, De Bortoli, M, Tahkokorpi, M. Life-cycle cost and risk assessment of the access network. From RACE 2087/TITAN. *In: Proc. 6th International IEEE workshop on Optical Access Networks (OAN '94), Kyoto, Japan, Oct. 16-19, 1994, 1.4-1 - 1.4-10.*

[56] Stordahl, K, Olsen, B T, Murphy, E, Zaganiaris, A. de Castro, E, Ravera, M. Delphi Survey: Forecasting Demand for Wide- and broadband services. From RACE 2087/TITAN. *In: Proc 6th International IEEE workshop on Optical Access Networks (OAN '94), Kyoto, Japan, Oct. 16-19, 1994, 7.3-1 - 7.3-10.*

[57] Zaganiaris, A, Ferrero, U, Balzaretti, S, Tahkokorpi, M, Kalervo, M, Olsen, B T, Stordahl, K. Methodology for risk assessment and techno-economic evaluation of optical access networks. *In: Proc The 20th European Conference on Optical Communications (ECOC '94), Florence, Italy, September 25-29, 1994, 3, 83-90.*

[58] Warzanskyj, W, Ferrero, U. Access network evolution in Europe: a view from EURESCOM. *In: Proc The 20th European Conference on Optical Communications (ECOC '94), Florence, Italy, September 25-29, 1994.*

[59] Haugen, R B, Olsen, B T, Eskedal, B E. Radio in local loop for rural and suburban environments. *In: Proc. IEEE International Conference on Personal Wireless Communications, Bangalore, India, 1994.*

[60] Haugen, R B, Olsen, B T. Evolutionary Steps and Introduction Strategies for Mobile Systems. *ConTEL '94*, Zagreb, Croatia, June 1994. *In: Information, Telecommunications, Automata Journal*, 13, (1-3), 245-254.

[61] Olsen, B T, Zaganiaris, A, Markatos, S, Ferrero, U, Tahkokorpi, M, Schjetne, I, Øverli, T. Access network alternatives from RACE 2087/TITAN. *In: Proc. 12th Annual Conference European Fibre Optic Communications and Networks (EFOC '94), Heidelberg, Germany, June 21-24, 1994.*

[62] Stordahl, K. Application of expert surveys for making forecasts for new services in the residential market. *Telektronikk*, 90, (1), 53-64, 1994.

[63] Olsen, B T, Stordahl, K. Forecast of price development of network components, based on an extension of the learning curve model. *Telektronikk*, 90, (1), 166-172, 1994.

[64] Zaganiaris, A, Tahkokorpi, M, Drieskens, M, Van Hoecke, J, Olsen, B T, Markatos, S, De Bortoli, M. Life-cycle costs and economical budget of

optical and hybrid access networks. *In: Proc. 5th IEEE Conference on Optical/Hybrid Access Networks, Montreal, Canada September 7-9, 1993.*

[65] Zaganiaris, A, Gieschen, N, Olsen, B T, Stordahl, K, Tahkokorpi, M, Drieskens, M, Ajibulu, A. A methodology for achieving life-cycle costs of optical access networks-from RACE 2087/TITAN. *In: Proc. 11th Annual Conference European Fibre Optic Communications and Networks (EFOC '93), the Hague, the Netherlands, June 30-July 2, 1993, 2, 136-141.*

[66] Olsen, B T, Zaganiaris, A, Gieschen, N, Tahkokorpi, M, Drieskens, M, Markatos, S. RACE 2087/TITAN: Tool for Introduction scenarios and Techno-economic studies for the Access Network. *In: Proc. RACE Open Workshop on Broadband Access, Nijmegen, the Netherlands, June 7-8, 1993.*

APPENDIX C

TITAN 4.0 user manual

Contents

Introduction 1

1.1 TITAN METHODOLOGY

TITAN (tool for introduction strategies and techno-economic evaluation of access network) implements the methodology for assessment of access networks. The objective of TITAN is to assess any kind of access network architecture (e.g. star, bus, ring, or combinations), as well as, a variety of services with any estimated penetration within any desired study period. The flexibility in the length of the study period allows scenarios to be derived in which network and service evolution are taken into account.

In order to enable cost comparison of different architectures TITAN calculates the discounted system costs, operation, maintenance and powering costs, life cycle costs and finally the overall financial budget. The next sections present the methodology used in the TITAN tool.

1.1.1 Structure

Fig 1 shows the general structure and philosophy of the tool. TITAN consists of a *cost database* which contains information on network equipment and materials with their costs and cost evolution during time, a *geometric model* that calculates the amounts of materials and equipment needed in the specific access network and *general financial data* (e.g. taxes). A *shopping list* is created for the specified *study period* by choosing the types of materials and equipment needed according to the chosen network *architecture*. Revenue components are added to the list according to the *services* offered in the network, their estimated penetration and tariffs. The tool then finally calculates *economic results* by combining costs and revenues. Thus, the most economical solution can be derived by comparing cases with different kind of architectures and equipment

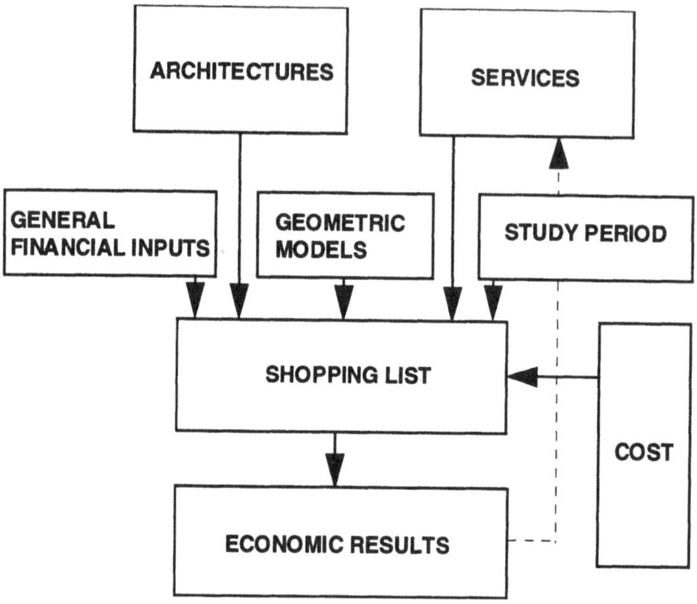

Fig 1 The logical structure of the TITAN model.

1.1.2 Architectures

A specific access network architecture must be chosen corresponding to an estimated subscriber density and a selected group of services in order to calculate the costs of access networks. The decision of architecture includes choosing specific equipment and materials used in the network. The architectures used in TITAN are based on a *generalised access network model.* This model is shown in Fig 2.

The model divides the access network into two parts, the feeder network and the distribution network. The model assumes in total four flexibility points (FPs) between the service access point (SAP) and the customer premises network interface (CPNI). The network components are assigned to these six points of location and the five link levels connecting them. If required, it is possible to assess only the distribution part of the access network or change the number of flexibility points in the model. The structure presented in Fig 2 is used mainly because it is a quite common structure in existing access networks.

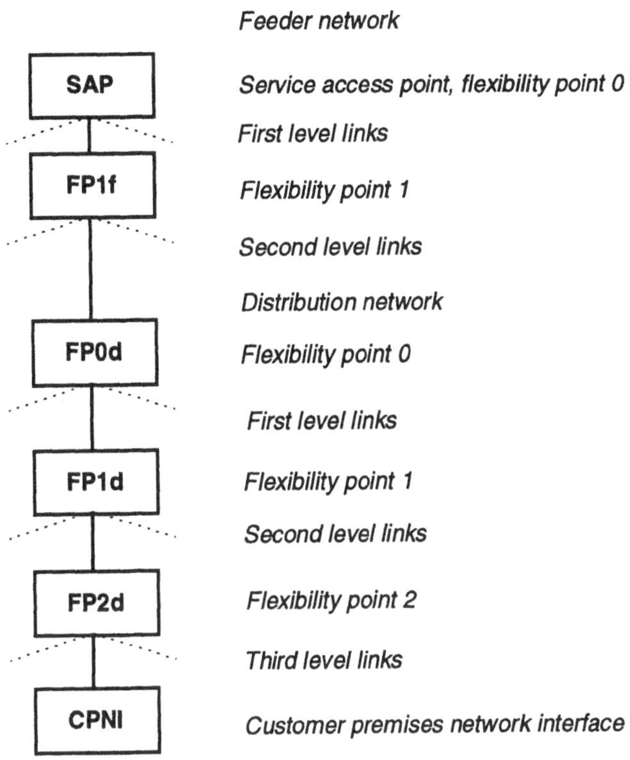

Fig 2 The generalised access network model used in TITAN.

1.1.3 Services

The network services must be specified in order to calculate the revenues generated by the network and the overall budget. The demand for each service within the potential market must be estimated in relation with the service tariffs. The evolution of each service can then be described with its estimated penetration and tariffs in each year of the study period.

1.1.4 General financial inputs

General financial inputs include the discount rates, tax rates and the depreciation methods used in the tool. Operation, administration and maintenance (OAM) inputs describe how the costs related to the OAM are derived.

1.1.5 Geometric model

Geometric models are used in the tool to calculate the straight distances (i.e. cable length) of different sections of the network. The models are based on a general access network model, and calculate the lengths depending on the subscriber density and chosen distribution ratios at each link level. The geometric models available in the current tool are:

- The SYNTHESYS model
- The modified SYNTHESYS model, with feeder part.

1.1.6 Costs

A cost database includes all necessary information on network equipment and materials needed for calculating the costs of the components. The component prices are referenced to a certain year. The following attributes are defined in order to calculate the price evolution of a component.

The *learning curve* describes how the production of a product will develop and therefore affect the component price.

The *volume class* describes the evolution of the component's production volume, which also affects the component price.

An *OAM class* is defined for every component in order to describe how OAM costs are related to that component.

The *write-off period* describes the time used to the financial write off of the component.

1.1.7 Study period

The *study period* specifies the starting year and the duration in time of the network study. The length of the study period can be chosen from two years up to 15 years.

1.1.8 Shopping list

The *shopping list* is a central part of the tool structure. The specific network materials and equipment (i.e. cost components) are added to the shopping list according to the chosen architecture. The discounted costs and OAM costs of these components are calculated from the cost database in order to produce life cycle costs for the specified study period. The amounts of equipment in each flexibility point level and cable lengths of each link level are calculated by the geometric model.

The revenue components are added to the list according to the specified services and their penetration.

1.1.9 Economic results

Finally the overall financial budget is calculated by integrating the demand and tariff data with the life cycle costs. The budget includes the following economic results:

- revenue
- investment
- depreciation
- running costs
- life cycle costs
- profit
- taxes
- interest payments
- retained profit
- cash flow
- retained cash flow
- cash balance
- payback period
- net present value of the project
- internal rate of return.

1.2 STARTING THE TOOL

1.2.1 TITAN Files

The TITAN tool runs under Microsoft Excel© 4.0. It consists of a TITAN.XLW workbook file, which contains all the functionality of the tool, a cost database file (.XLD) containing the cost information, main files (.XLS) containing the architecture and calculations and template files (.XLT) used by the macro file. Only one cost database file can be open at a time, but multiple main files can be open simultaneously, which allows comparison of different architectures. Fig 3 depicts the files in the tool and their inter-relation .

1.2.2 Getting started

The tool is started by opening the TITAN.XLW workbook file. The tool has two basic operating modes: *cost database mode* and *main mode*. Database mode is used for changing the databases if necessary. Main mode is used for actual calculations and reporting.

When started the tool is in the cost database mode. The user has to load a cost database file (.XLD) before he is able to change to the main mode This is done with the *file/open* menu command.

After loading the database, the user can either work with the database or switch to the Main Mode from the Mode Menu.

To learn how to use the database mode and cost databases read the Cost Database Mode/Using the Cost Database Mode section of this manual.

To learn how to use the main mode and main spreadsheets, please read the Main Mode/Using the Main Mode section of this manual.

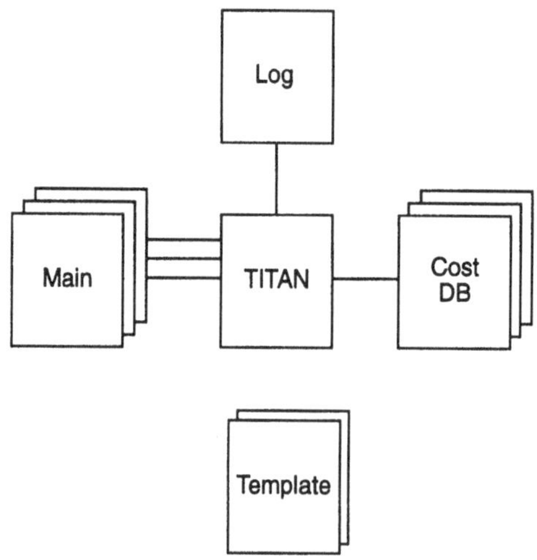

Fig 3 The file structure of the TITAN model.

Cost database mode

2

2.1 USING COST DATABASE MODE

The cost database mode is used for changing and operating the cost database. The mode of operation can be changed to cost database mode with the Mode/Cost Database menu command.

In the Cost Database mode the Cost Database file is open and can be worked with. Cost components and their information may be added, edited or removed. Specific components may be searched and the database can be sorted with a specified key.

Read the Using the cost database mode topics to get information on the procedures required when using the cost database mode. Alternatively, choose one of the reference topics to get specific help on a menu command or a section of the main sheet.

2.1.1 Opening and creating a cost database

The database file must be open in order to enable a change to main mode and main sheets to be opened. Only one cost database file may be open at a time.

To create a new cost database

1. Check that you are in cost database mode. If not, choose Cost Database from Mode menu.
2. From the File menu, choose New.

The program will open the cost database template file as a basis for the new cost database. Read the following topics to learn how to add cost components and work with cost database files.

To open a cost database

1. Check that you are in cost database mode. If not, choose Cost Database from the Mode menu.
2. From the File menu, choose Open.

3. Choose the cost database file (.XLD) according to standard procedure in Microsoft Excel (see Microsoft Excel Help for more information).

To close a cost database

1. Check that you are in cost database mode. If not, choose Cost Database from Mode menu.
2. From the File menu, choose Close.

2.1.2 Changing the view

In the same way, you can move around the cost database sheet with scroll bars or cursor keys as in Microsoft Excel. You may move the viewing area quickly to a specified section of the database with the commands of the View menu. The price evolution of an individual cost component may also be viewed graphically.

To change viewing area

1. From the View menu, choose one of the sections you want to move to from the list:
   a) Cost Components
   b) Learning Curve Classes
   c) Volume Classes
   d) OAM Classes
   e) Write-Off Classes
   f) Services

Changing the viewing area in this way in addition will change the database work area to the selected section, so that the Data menu commands are directed to it.

To view the price evolution of a cost component

1. From the View menu, choose Price evolution.
2. In the Paste cost component dialogue box choose the cost component to be displayed.
3. In the Start year dialogue box choose the starting year for the study.

The program will then display graphically the prices for that component for the time period of ten years. The chart can be printed with the PRINT button. You may exit the display and return to cost database mode by clicking the OK button.

To view the service evolution

1. From the View menu, choose Service evolution.
2. In the Select service name dialogue box choose the service component to be displayed.
3. In the Start year dialogue box choose the starting year for the study.

The program will then display graphically the tariff and penetration for that particular service for a time period of ten years. The chart can be printed with the

PRINT button. You may exit the display and return to cost database mode by clicking the OK button.

See also:
1. View menu commands
2. Sections of the Cost Database Sheet.

2.1.3 Inserting cost database items

There are two ways of inserting items to the database: with the Data/Form menu command or manually.

To insert cost database items

1. From the View menu, choose the section of the database that you want to insert items into.
2. From The Data menu, choose Form.
3. In the dialogue box, choose the New button.
4. Insert data of the item in the appropriate fields.
5. Choose the Close button.
6. Sort the database with the Data/Sort menu command.

To insert cost database items manually

1. Move to the row after the ZZZ_dummy row of the desired database section.
2. Insert the item data in the cells of the row or use the Formula Menu Commands to paste class names.
3. Sort the database with Data/Sort menu command.

See also:
1. Data menu commands
2. Formula Menu Commands.

2.1.4 Sorting the database

If you want to make a listing of the items in the database in a specific order or find components with some common attribute you can sort the database with a selected criterion.

To sort the database

1. From the View menu, choose the section of the database that you want to sort.
2. From Data menu, choose Sort.
3. In the Sort box, select the key to sort the database with.
4. Choose the OK button.

See also: Data menu commands.

2.1.5 Finding components

Finding a component manually may be difficult if the cost database is rather large. A component may be found in an easier way by the use of the Data/Form menu command.

To find a cost component

1. From the View menu, choose the cost components section.
2. From the Data menu, choose Form.
3. In the dialogue box, choose the Criteria button.
4. Insert appropriate criteria in the fields of your choice.
5. Choose Find next.
6. Use Find next/prev buttons to scan through the database items.
7. Choose the close button.

See also: Data menu commands.

2.1.6 Changing the data of the components

There are two ways to change the data of a cost component in the database: with the Data/Form menu command or manually.

To change the data of a cost component

1. From the View menu, choose the cost components section.
2. Select the component that you want to change with Find next and Find prev buttons or use appropriate criteria to find the component.
3. Change the data in the fields of your choice.
4. Choose the close button.

To change the data of a cost component manually

1. Move to the row of the component to be changed.
2. Edit the data of the cells to be changed or use the Formula Menu Commands to paste class names.

See also:

1. Data menu commands
2. Formula Menu Commands.

2.1.7 Changing to main mode

When you have loaded the database and/or edited the database and you want to start calculations, you first have to change the mode of operation to main mode. To get more information on operating in the main mode read the section Using the Main Mode.

To change to main mode

From the Mode menu, choose Main.

2.2 REFERENCE INFORMATION

2.2.1 Cost database mode menu commands

File Menu

The File Menu includes a restricted set of standard Excel commands (see Microsoft Excel Help for more information) and the commands listed below. With the Open command only valid database files [i.e. (costdatabase).XLD] can be opened. The New command opens a cost database template file as a basis of a new cost database (COSTDB.XLT).

Quit TITAN

Ends your TITAN tool session. Does not quit Microsoft Excel.

See also: Opening Cost Database.

Edit Menu

The Edit Menu provides a restricted set of standard Excel commands (see Microsoft Excel Help for more information).

Formula Menu

Note

Allows you to add, delete, edit, or view notes for specific cells in the active worksheet (see Microsoft Excel Help for more information).

Paste Learning Curve Class Name...

Inserts the selected Learning Curve Class Name into the formula bar.

Paste Volume Class Name...

Inserts the selected Volume Class Name into the formula bar.

Paste OAM Class Name...

Inserts the selected OAM Class Name into the formula bar.

Paste Write-Off Class Name...

Inserts the selected Write-Off Class Name into the formula bar.

See also:

1. Inserting cost components
2. Changing the data of cost components.

Mode Menu

Cost Database

Changes the mode of operation to the Cost Database mode.

Main

Changes the mode of operation to the Main mode. The Main mode can be selected only after a cost database has been loaded.

Data Menu

Form...

Displays a database form you use to view, change, add, and delete the cost database records currently viewed.

Sort

Rearranges the currently viewed records of the cost database in ascending order according to the column (key) specified by you.

See also:

1. Inserting Cost Components
2. Changing the Data of Cost Components
3. Sorting the Database
4. Finding Components.

Options Menu

The Options menu contains the standard Excel commands Set Print Area, Set Print Titles, Set Page Break (see Microsoft Excel Help for more information).

View Menu

From the View Menu the user can view the price evolution of a component or change the viewing area of the screen and the control to the selected section of the cost database. The available sections are:

- Cost Components
- Learning Curve Classes
- Volume Classes
- OAM Classes
- Write-Off Period Classes
- Services.

Price evolution...

Graphically displays the evolution of a selected cost component during a ten year period of time starting from a chosen year. The chart can be printed or copied to the Log file. In copying to the Log file only data for creating the chart are copied, not the chart itself.

Service evolution...

Graphically displays the evolution of the tariff and penetration of a selected service during a ten year period of time starting from a chosen year. The chart can be printed or copied to the Log file. In copying to the Log file only data for creating the chart are copied, not the chart itself.

See also: Changing the View.

Help Menu

Excel Help

Displays the Microsoft Excel Help table of contents.

TITAN Help

Displays the TITAN Help table of contents for Cost Database Mode.

About TITAN...

Displays the version number of the TITAN tool you are using.

2.2.2 Sections of the cost database sheet

Cost components section

This section contains all component specific information listed below. Cost components are added with the Data/Sort command or manually by adding the information above the "ZZZ_Dummy" rows. The current viewing area of the screen and part of the database to be controlled can be changed with the View menu command.

CostComponent

The name of the cost component. The name must be unique.

RefCostValue

Defines the reference cost value of the component.

CostUnit

The currency unit of the "RefCostValue".

RefYear

Defines the reference year. Refers to the year when "RefCostValue" is valid.

LearningCurveClass

The class of learning curve of the cost component. Refers to the Learning Curve Classes Section of the database, where the characteristics of the learning curves are defined.

VolumeClass

Defines the volume evolution class of the cost component. Refers to the Volume Classes Section of the database, where the characteristics of the volume classes are defined.

OA&MClass

Defines the operation, administration and maintenance class of the cost component. Refers to the OAM Classes Section of the database, where the characteristics of the OAM classes are defined.

WrOffClass

Defines the financial write-off time class of the cost component. Refers to the Write-Off Period Classes Section of the database, where the characteristics of the write-off period classes are defined.

Type ID

The type identification string of the cost component.

ConfClass

Defines the confidentiality class of the cost component. Refers to the Confidentiality Classes section of the database, where the characteristics of these classes are defined.

Importance

Defines the financial importance of the cost component.

Learning curve classes section

Defines a specific learning curve behaviour. The values of this section are used in the calculation of the life cycle costs of the network equipment and materials. See chapter 8 and the topic Component Price Versus Time in the user manual.

LearningCurveClass

Defines the learning curve class name. The name must be unique.

K Cons

Conservative K-value [%]

K Opt

Optimistic K-value [%]

Volume Classes Section

Defines a specific market volume evolution. The values of this section are used in calculation of the life cycle costs of network equipment and materials. See chapter 8 and the definitions section in the user manual.

VolumeClass

Defines the name of the volume class. The name must be unique.

n(0)

Relative volume at time t_0.

DELTAt

Growth parameter. Corresponds to the time for the volume to go from 10 % to 90 % of the saturation volume.

OA&M classes section

Defines specific operation, administration and maintenance cost behaviour. See chapter 5 and the definitions section in the user manual.

OA&MClass

Defines the name of the OAM class. The name must be unique.

OAMCost

Cost of OAM as percentage of the invested capital/year.

Write-off period classes section

Defines component lifetimes for the calculation of depreciation.

WriteOffClass

Defines the name of the financial write-off time class.

Write off period

Defines the financial write-off time in years.

Conf Classes Section

ConfClass

Class defining the uncertainty in the price.

Standard deviation

Relative standard deviation of the price.

Services Section

This section contains service specific information listed below. Services are added with the Data/Sort command, or by manually adding the information above the "ZZZ_Dummy" rows.

Revenue component

The name of the revenue component (service). The name must be unique.

Start tariff

Annual tariff at the time of the start year.

Tariff unit

The currency unit of the "Start tariff".

Start year

Defines the start year. Refers to the year when the "Start tariff" is valid.

K-value

Learning curve factor in percentage. Higher and lower than 100% respectively for increasing and decreasing tariff. See chapter 8 and the definitions section in the user manual.

Start penetration

Defines the penetration of the service in the start year.

End penetration

Defines the saturation or maximum penetration level of the service.

Delay time

Corresponds to the time for the penetration to increase from 10 % to 90 % of the saturation value.

Main mode 3

3.1 USING MAIN MODE

In TITAN the calculations are carried out in Main Mode. The mode of operation can be changed to main mode with the Mode/Main menu command.

In the Main Mode the main worksheet file is open and operated. The general inputs, geometric model, shopping list of cost components and revenue components need to be specified in order to calculate the economic results. Then, from the results reports with graphical charts are generated.

Read the *Using the main mode* topics in the user manual in order to obtain information on the necessary procedures in the use of the Main Mode, or choose one of the reference topics to get specific help on a menu command or a section of the main sheet.

3.1.1 Creating new main sheets

Main sheets can be created only after a cost database has been opened.

To create a new main sheet

1. Check that you are in main mode. If not, choose Main from the Mode menu.
2. From the File menu, choose New.

TITAN will then open a template main sheet to be used as a basis of your study. To learn how to use main sheets read the Using Main Mode topics.

3.1.2 Opening main sheets

Main sheets can be opened only after a cost database has been opened. Multiple main sheets may be open simultaneously, but they must use the same cost database.

To open a main sheet

1. Check that you are in Main Mode. If not, choose Main from the Mode menu.
2. From the File menu, choose Open.
3. Choose the main file (.XLS) from the file list (as in normal Microsoft Excel operation).
4. Click the OK button.

To close a main sheet

1. Check that you are in main mode. If not, choose Main from Mode menu.
2. From the File menu, choose Close.

3.1.3 Moving around the sheet

You can move around the main sheet with scroll bars or cursor keys as in Microsoft Excel. You may also move the viewing area quickly to a specified section of the main sheet with the commands in the View menu.

To change viewing area

1. From the View menu, choose one of the following sections respectively to which you want to move:

   a. Notes
   b. General inputs
   c. Geometric model
   c. Revenue components
   d. Cost components
   e. Economics.

2. If you choose the cost or revenue components section, the year that you want to set the view to must be chosen accordingly.

3.1.4 Input of general inputs

General inputs include financial, architecture and OAM inputs. These inputs are written directly in the cells (as in Excel) of the general inputs section of the main worksheet. The cells accepting inputs are marked with a shading. Read the Main Mode/Sections/Inputs section of the manual help in order to get information on each input parameter.

3.1.5 Input of geometric inputs

Geometric inputs include both inputs that are generic and inputs that depend on the chosen geometric model.

Geometric inputs are input directly to the main worksheet like general inputs. Read the Main Mode/Sections/Inputs section of this manual in order to get information on each input parameter.

3.1.6 Adding cost components

Cost components are added to the Cost Components section of the main sheet in order to specify the components and their costs used in the selected network architecture. The network architecture must be defined before cost components are selected.

To add cost components

1. From the Formula menu, choose Add Cost Component.
2. In the Add Cost Component box, select one of the possible cost component locations presented in the dialogue box.
   The program will automatically go to the specified place and start the Paste Cost Component Name command.
3. In the Paste Cost Component box, choose the component to be added.
4. Move to the cost components section, to the line of the new component. Insert the available volume of that component already in the network in the AvailVol column (shaded). Then insert values and/or formulas in the additional volumes (AddVol) columns (shaded) of each year of the study period. To use the cable lengths and number of flexibility points calculated by the geometric model use the names of the corresponding cells in the geometric model in your formulas. E.g. if you need two pieces of a certain equipment in each FP2d point, insert in the AddVol cell of that equipment the formula: = *2·NrOfFP2d*).
5. From the Options menu, choose Calculate document.

The program then calculates the correct values of the added cost component in the rest of the columns.

Note

The last section of cost components consists of OAM plane. You are supposed to use this section to include operation, administration and maintenance costs. Costs included here behave slightly differently from other cost components, since work is an intangible investment.

3.1.7 Adding revenue components

Revenue components are added to the Revenue Components section of the main sheet in order to specify the revenues related to the services provided by the network.

To add revenue components

1. From the Formula menu, choose Add Revenue Component.
2. The program will ask if you are willing to use a revenue component from the database. Choose OK if you are (go to step 3) or choose Cancel to add your own revenue component (go to step 4).
3. In the Select Service Name box, choose the service. The program will then add the revenue component to your main file with the tariff and penetration structure defined in the database. (go to step 6)
4. In the Input box, input the name of the Service/Revenue Component.
5. Move to the revenue components section, to the line of the new revenue component.

6. For each year of the study period insert the penetration of the service in the Penetration column (shaded) and insert the tariff of the service in the Tariff column (shaded).
7. From the Options menu, choose Calculate document.

The program then calculates the correct values of the added revenue component in the rest of the columns.

3.1.8 Setting the study period

The study period defines the starting year and the time span of the access network study, i.e. it controls from which year and how many years all inputs are inputted and calculations performed. The study period is restricted to between two and 15 years.

To set the study period

1. From the Options menu, choose Study Period.
2. In the Study Period box, input the starting year of the study and select the duration of the study (2 - 15 years).

Here the user should remember that the last two columns (provided that equation columns are demoted) in the range defining the study period are to be ignored. The last column is shown in red colour and it has the opposite value of the column next to the last column. These columns are only needed by the tool itself in the process of changing the study period. If you manually extract results from the tool, please ignore them.

3.1.9 Calculation

Calculation is the most central function of the main mode. It calculates the life cycle costs of the components, service revenues and the financial results. It uses the general inputs, the geometric model and the price formula. Because the Calculate Document command performs all the specified calculations, it may be a time consuming operation depending on the computer used. Therefore it may be wise to add all cost components before calculating the document.

To set the calculation options

1. From the Options menu, choose calculation.
2. In the Calculation options box, do any of the following:

a) Select Update DB-links/ Cost parameters if you want to update the links of cost components from main sheet to the database sheet before every calculation.
b) Select Update DB-links/Revenue parameters if you want to update the links of revenue components from the main sheet to the database sheet before every calculation.
c) Select Calculate Depreciation if you want calculation of the depreciation of the cost components to be performed.

d) Select Calculate IRR if you want the internal rate of return (IRR) to be automatically calculated when the document is recalculated.

Note

However, if you are performing risk analysis with Crystal Ball© you should not select this option. If you want IRR to be part of your risk analysis results, you must define CalcIRR macro in the Crystal Ball© user defined macros.

e) Choose the Optimistic/Conservative price evolution if you want that optimistic/conservative K-value respectively should be used in the learning curve calculations of the price evolution.

To calculate the document

1. From the Options menu, choose calculate document.

The program will then perform all calculations needed in the economic assessment of the network.

To calculate the internal rate of return

1. Choose the Calculate IRR option in the Calculation Options dialogue box and recalculate the model.

The program will then perform a goal seek operation and returns the IRR value.

3.1.10 Viewing results

Succeeding the Calculate Document command has been executed the complete budget can be seen in the Cost Components section, Revenue Components section and Economics section of the main sheet. You can change the viewing area of the screen to these sections from the View menu. Results can be displayed by the use of the log facility.

In the economics section the interest paid should be input to the shaded cells (for each year of the study period). The program will then calculate discounted values with the Calculate Document command.

3.1.11 Using log

Log is used to keep record of the use of the TITAN tool. Every time the tool is started or closed the exact time is saved to the Log file. With commands from the Log menu the user can add own self-defined texts, the general main input values, charts and selections to the log file.

The charts can be either time series charts or sensitivity charts. Time series charts can be created from various categories (e.g. revenues) and variables or results in those categories (e.g. penetration). The sensitivity analysis can be performed with either one or two variables and one result. The tool calculates the main sheet multiple times changing the variables according to the specification of the user and records the results for the creation of the chart.

The Log can be viewed, saved and edited.

To add text to the log

1. From the Log menu, choose Text.
2. In the Log Text box, type the text to be logged.

   The program will then write the text to the log file.

To add general main input values to the log

1. From the Log menu, choose Inputs.

The program will then read the general main inputs from the main sheet and write them to the log file.

To add chart and perform sensitivity analysis

1. From the Log menu, choose Chart. The dialogue box shown in Fig 4 will appear:

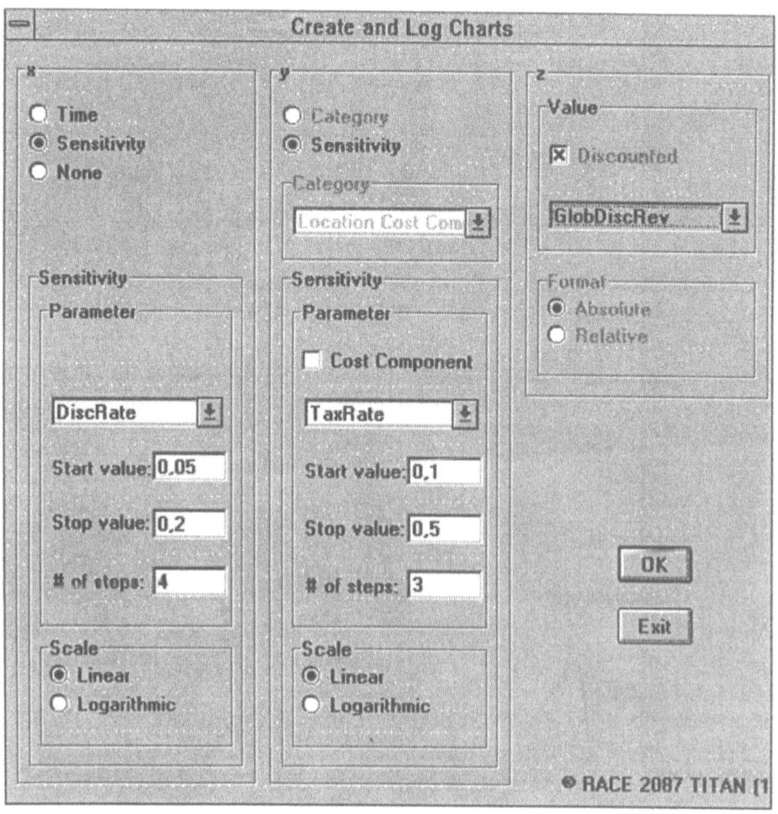

Fig 4 The screen window of the *Create and log charts* menu.

2. In the *Create and Log Charts* box choose time, sensitivity or none for the x-axis.

   If you choose sensitivity, choose the parameter to be studied, the start and stop values for the parameter, and number of steps between the start value and the stop value. Specify also linear or logarithmic scale for the x-axis.

   If you want to perform sensitivity analysis as function of only one variable, choose *none* for the x-axis. Define this variable on the y-axis section.
3. In the Create and Log Chart box, Choose category or sensitivity for the y-axis.
   a) If you choose category, pick the category type.
   b) If you choose sensitivity, choose the parameters as explained above for the x-axis. If you want to run sensitivity analysis based on variations in cost component price, mark the Cost Component box, choose the cost component of which the price is to be varied and choose other parameters as described before.
4. In the Create and Log Chart box, Choose variable or result in the selected category for the z-axis.

   Mark the box if you want to use discounted values.
5. Click the OK button.

   The program will then create a chart to the log according to your specifications. To view and edit the log see instructions below.
6. Exit Create and Log Chart dialogue box by clicking the Exit button.

Note that sensitivity analysis recalculates the document as many times as specified by the number of steps. This may be time consuming especially with large mainsheets (many cost components), as it may involve recalculations of the document several times.

To add selection to the log

1. First, select the cells to be logged in the main worksheet.
2. From the Log menu, choose Selection.

   The program will then write the cells to the log.

To view the log

1. From the File menu, choose Show Log.

   The program opens the log file. Then do any of the following:
   a) Move around the sheet as in Excel.
   b) Save log file with File/Save Log command.
   c) Clear log file with File/Clear Log command.
   d) Exit back to main mode with File/Exit Log command.

To edit the log file

1. From the File menu, choose Show Log.
2. From the File menu, choose Save Log. Choose the name of the file to which the log is to be saved.
3. From the File menu, choose Exit Log.
4. Exit TITAN normally with the File/Quit TITAN command.
5. Open the log file as a normal Excel file with Microsoft Excel.

Then you can print, edit and manipulate the log file as a standard document with all Excel options and tools.

3.2 REFERENCE INFORMATION

3.2.1 Main Mode Menu Commands

File Menu

The File menu includes a restricted set of standard Excel commands (see Microsoft Excel Help for more information) in addition to the following commands:

New...

Loads the TITAN template main file (MAIN.XLT).

Show Log

Displays the TITAN log file (TITAN.XLL).

Quit TITAN

Ends your TITAN tool session. Does not quit Microsoft Excel.

See also:
Starting up new Main Sheets
Opening Main Sheets

Edit Menu

The Edit Menu provides a restricted set of standard Excel commands (see Microsoft Excel Help for more information).

Formula Menu

Note

Provides insertion of comments into specific cells. Works as the standard Excel Note command (see Microsoft Excel Help for more information).

Add Revenue Component...

Adds a revenue component to the revenue component list.

Add Cost Component...

Adds the selected cost component to the selected level of the shopping list.

Paste Name...

Inserts the selected name into the formula bar. This function should be used only when you want to create a new architecture. It is for adding for example new link levels or SAPs, or for changing the links between the geometric model outputs and the number of cost component.

Paste Cost Component Name...

Inserts the selected cost component name into the formula bar.

See also:
Adding Cost Components
Adding Revenue Components.

Mode Menu

Cost Database

Changes the mode of operation to the Cost Database mode.

Main

Changes the mode of operation to the Main mode. The Main mode can be selected only after a cost database has been loaded.

Options Menu

The Options menu provides standard Set Print Area, Set Print Titles, Set Page Break Excel commands, and the following TITAN commands:

Study Period...

Selects the starting year and the duration of the study period of the model. The duration is limited to between two and 15 years.

Calculate Document

Calculates the costs and revenues according to the cost and revenue components added to the model. Operation may be time consuming and therefore should be performed after adding all of the components.

Calculation...

Controls how calculations are performed.

Open Crystal Ball

Opens the Crystal Ball© risk analysis and simulation add-in, and disables TITAN functionality. However, this requires a separately installed version of Crystal Ball©. To return back to TITAN choose Close Crystal Ball© from the Run menu. For more information about using Crystal Ball© read Crystal Ball© Help.

See also:
Setting study period

Calculation
Performing risk analysis with Crystal Ball©
Log Menu

Text...

A specified text may be saved to the log file (TITAN.XLL).

Inputs

Exports the current financial and architectural inputs to the log file.

Chart...

Creates a chart from specified financial outputs and exports it to the log file. Allows the user also to perform sensitivity analysis.

Selection

Exports the selected area of the spreadsheet to the log file.

See also:
Using Log.

View Menu

From View Menu you can select the viewing area of the screen from the following list:

Notes section
General inputs section
Geometric model section
Revenue components section
Cost components section
Economics.

See also:
Moving around the sheet.

Window Menu

Select...

Selects the main spreadsheet to be worked with from the previously loaded sheets.

Help Menu

Excel Help

Displays the Microsoft Excel Help table of contents.

TITAN Help

Displays the TITAN Help table of contents for the Main Mode.

About TITAN...

Displays the version number of the TITAN tool you are using.

3.2.2 The sections of the main sheet

The mainsheet is divided into three different sections. The first section consists of inputs which are independent of the study period. General economic inputs and geometric model inputs fall into this category.

The next section consists of sections that do depend on the study period. These are for example revenue components, cost components and economic results that are presented as time series over a set study period. The last two columns (when equation columns are demoted) of this block are to be ignored. The last column is shown in red colour and has the opposite value of the column before that. These columns are only needed by the tool itself in process of changing the study period. If you manually extract results from the tool, please ignore them.

The last section is the system area where internal calculations are performed. The user should never touch this area. The whole section is displayed with a grey background.

The following topics provide more detailed information about sections within the first two blocks.

Notes section

This section contains some general notes e.g. file version number.

Inputs section

This section is used for general financial inputs, general architecture inputs, general OAM inputs and geometric model inputs listed below. The input cells are marked with a shading. You should not input values to other cells since the tool will calculate them from your inputs.

- General financial inputs
- General architecture inputs
- General operation, administration and maintenance inputs
- Geometric model inputs, feeder part
- Geometric model inputs, distribution part

See also: Input of general inputs

Financial inputs

Discount rate

The discount rate to be used in the calculations of the model.

Tax rate

The tax rate on profit (country dependent).

Depreciation method

The depreciation method selector. The following depreciation methods are available:

- Straight line depreciation, *SLN*
- Sum-of-the-years-digits method, *SYD*
- Declining balance method, *DB*.

To select the depreciation method use the acronym listed after each name of the method.

Depreciation factor

Defines the depreciation factor for the declining balance method (DB). It has no impact on the other methods. Use the value 2 for double-declining balance method.

Architecture inputs

Copper pair occupation

Defines the occupation of the copper pair in percentages.

Operation, administration and maintenance inputs

Central management

Defines the use of centralised management in percentage. 100 % indicates that all equipment is managed from a central location. In contrast, 0 % indicates that no equipment is managed from a central location.

Operation and Administration savings

Defines the estimated savings on operation and administration (not maintenance) costs in percentage when centralised management is used.

Operation and administration part

Defines in percentages the operation and administration (OA) part of the operation, administration and maintenance (OAM) costs.

Percentage of operation, administration and maintenance savings (time dependent)

Defines in percentage how much of the calculated OAM savings is realised for each year separately. E.g. 100 % PercentageOfOAMsavings indicates that all of the e.g. 30 % savings is achieved. 90 % indicates that only 90 % of the 30 % savings is achieved. The percentages are inserted separately for each study year, into the shaded cells above the cost components section of the main sheet.

Note

There is also an alternative way to include OA&M costs. For more information see OA&M plane.

Geometric model inputs, feeder part

Feeder cable over length

Defines the over length of the feeder cables in percentage.

Distribution ratio at flexibility point Nf (N = 0..1)

Defines the cable distribution ratio at flexibility point Nf, in other words the number of cables leaving the FP Nf.

Duct availability in link level Nf

Defines the duct availability (the ducts already built and usable) in link level Nf in percentage of the total duct capacity needed in this level.

Medium in link level Nf

Defines the medium type used in this link level. The medium can be:

C Copper
F Fibre
R Radio

Use the letter in front of the medium type to define the medium used.

Distance of LLN

Defines the straight distance from FP(N-1) to FPN.

Geometric model inputs, distribution part

Potential subscriber density

Defines the potential subscriber density in the model area (number of subscribers/square kilometre).

Distribution cable over length

Defines the over length of the distribution cables in percentage.

Distribution ratio at flexibility point Nd (N = 0..2)

Defines the cable distribution ratio at flexibility point Nd, in other words the number of cables leaving the FP Nd.

Duct availability in link level Nd

Defines the duct availability (the ducts already built and usable) in link level Nd in percentage of the total duct capacity needed in this level.

Medium in link level Nd

Defines the medium type used in this link level. The medium can be:

C Copper
F Fibre
R Radio

Use the letter in front of the medium type to define the medium used.

Distribution cable over length LLNd

Defines the over length of the distribution cables in percentage for this link level (for bus architecture only).

Distance of LL3d drop

Defines the straight distance from FP2d to CPNI (for bus architecture only).

Geometric model section

This section identifies the geometric model used in the Main file. The section calculates the straight distances of different sections of the network (i.e., cable length) and number of flexibility points in each level (i.e. amount of required equipment). The outputs and possible inputs in this section depend on the chosen geometric model. Read the notes in the cells of this section to get information on used equations and defined cell names.

See also: Input of geometric inputs

Revenue components section

This section is used to describe the penetrations and tariffs of the services provided by the network and to calculate the revenues generated by those services for each year of the study period.

The revenue components are added to this section with the Formula/Add Revenue Component command. After adding a component the columns to be input manually are marked with shading in the sheet.

Columns of the revenue components

The following columns are repeated from year 0 to the end of the study period. All calculation columns will be hidden/shown by Excel demote/promote commands, so that only input columns for service penetrations are seen.

Penetration

Defines the penetration of the service in percentage of the total number of potential users.

NrOfUsers

Calculates the number of the service users from the total number of subscribers and the penetration percentage.

Tariff

Defines the tariff for the service for this year.

Revenue

Calculates the total revenue generated by the service for this year.

See also:
Adding revenue components
Calculation.

Cost components section

This section is the most important part of the main sheet because it defines the cost components used in the architecture. The section defines when each component is installed into the system. The section has been divided into subsections according to the generic network model defined in the geometric model section. In this section all the volumes are input or calculated from the geometric model for the whole system. All lengths (cables, ducts...) in this section are in kilometres and all prices in ECU or ECU/km.

The cost components are added to this section with the Formula/Add Cost Component command. The columns to be input manually are then marked with shading in the sheet.

Above cost components section is the time scale which shows the number of each study year section in the sheet. Below that is the percentage of OAM savings value which can be input separately for each year. See OAM inputs for more detail. Another way to include OAM costs is to use the OAM plane which is the last part of the cost components section.

Columns of the cost components section

The following columns except column "AvailVol" are repeated from year 0 to the end of the study period. All calculation columns will be hidden/shown by Excel demote/promote commands so that only input columns for additional investment are seen.

Note that the last year section is not the last year of the study period! It is not included in the calculations! It is used by the TITAN macro file for time period adjustments, i.e. to create new columns for calculation.

Cost component name

The name of the cost component.

AvailVol

Defines the volume of the cost component available in the system before the actual network construction begins. This could be for example copper pairs in a fibre to the curb (FTTC) configuration.

You should specify the "AvailVol" of a cost component after you have added the cost component to this section with the Formula/Add Cost Component command.

AddVol

Defines the number of components installed in this year. Regarding the basic components, which are service independent, the added volume is based on the penetration of the dominant service. For the service specific components, the added volume is based on the penetration of the appropriate service.

You should specify the "AddVol" of a cost component after you have added it to this section with the Formula/Add Cost Component command.

Price

The price of the cost component in this year.

The column contains the price formula that calculates the price of this component this year from the parameters in the database.

AddInvest

The total additional investment to this component this year. The column multiplies the column "Price" with the column "AddVol".

GlobVol

Calculates the total volume of this cost component in the system after the investment of this year. In year 0 this means addition of columns "AvailVol" and "AddVol". In later years this means addition of previous year's global volume and the current year's additional volume.

GlobInvest

Calculates the total value of the investments to this far for this cost component, using the current price of each year's investment.

OA&M Cost

Calculates the operation, administration and maintenance costs for cost component this year using the column "GlobInvest", the OAM inputs and the OAM information on the cost component from the cost database.

Depreciation

Calculates the accumulated depreciation on the installed base of this cost component.

RestVal

The residual value of the installed base of this cost component. It is calculated by subtracting "Depreciation" from "GlobInvest".

See also:
Adding cost components
Calculation.

Economics section

This section contains the final economic budget, i.e. the financial results of the investment and revenue calculations. The section is divided into subsections containing the following financial results:

- revenue
- investment
- depreciation
- running costs
- life cycle costs
- profit
- taxes
- interest payments
- retained profit
- cash flow
- retained cash flow
- cash balance
- payback period
- net present value of the project
- internal rate of return.

All results except interest payments are calculated by the program.
See also: Viewing results.

3.2.3 Adding link levels

You can add new link levels to the cost components section, if you need to assess networks with more than three link levels. Adding link levels to the main sheet requires modification and creation of names for worksheet cells and editing of formulas. The user should have some experience of using these functions in Microsoft Excel before adding link levels. If needed, look into Excel Help for more information.

To add link levels

1. Change TITAN to system mode by pressing CTRL + SHIFT + s.
2. In the cost components section select the rows of the last link level and flexibility point level (Click the row headings).
3. From the Edit menu, choose Copy.
4. Select the row after last flexibility point level (shaded).
5. From the Edit menu, choose Insert Paste.
6. If the last link level contains cost components already inserted, you may delete them from the new link level and flexibility point level.

LLN

7. Change the name of the link level to an appropriate one.
8. Select the name cell of the new link level.
9. From the Formula menu, choose Define Name.
10. In the Define Name box, insert into the Name field: EdgeLLN (N is the number of the level, e.g. EdgeLL4)
11. Choose the OK button.
12. Select the last row (empty!) of the new link level (click the row heading).
13. From the Formula menu, choose Define Name.
14. In the Define Name box, insert into the Name field: InsRowCompLLN (N is the number of the level, e.g. InsRowCompLL4).

FPN

15. Change the name of the flexibility point level to an appropriate one.
16. Select the name cell of the new flexibility point level.
17. From the Formula menu, choose Define Name.
18. In the Define Name box, insert into the Name field: EdgeFPN (N is the number of the level, e.g. EdgeFP4)
19. Choose the OK button.
20. Select the last row (empty!) of the new flexibility point level (click the row heading).
21. From the Formula menu, choose Define Name.
22. In the Define Name box, insert into the Name field: InsRowCompFPN (N is the number of the level, e.g. InsRowCompFP4).

Totals

23. Move to the TOTALS row of the cost components section.
24. To each cell containing a sum formula of the totals of each component level, add the totals of the new link level and flexibility point level.

AddInvest

25. From Formula menu, choose Define Name.
26. In the Define Name box, select name AddInvT0 from the Names in the Sheet box. Then, in the Refers to box, modify the area of the name (i.e. the AddInvest column) to include also the new link level and flexibility point level.
27. Choose the OK button.

CompAreaList

28. Move to the Work area for the program section of the worksheet.
29. Input the following to the rows after the last link level of the CompAreaList. For both new link levels input Link Level N (and N+1) and the name of the level (look at previous link levels for example).
30. From the Formula menu, choose Define Name.

31. In the Define Name box, select the name CompAreaList from the Names in Sheet box. Then in the Refers to box, modify the area of the name (i.e. the CompAreaList area) to include also the new link level and flexibility point level.
32. Choose the OK button.

The modification of the work sheet is now complete, and you should now save the work sheet and restart TITAN.

3.2.4 Performing risk analysis with Crystal Ball®

To perform risk analysis with TITAN, the Crystal Ball® add-in software is required. Crystal Ball® is a commercial general purpose risk analysis software package for Microsoft Excel manufactured by Decisioneering Inc.

With Crystal Ball® the user can assign probability distributed assumptions to the work sheet cells that are used as variables. Cells containing formulas referring to the assumption cells may be defined as forecasts. Crystal Ball® uses Monte Carlo or Latin Hypercube simulation for forecasting the resulting probability distributions.

To start Crystal Ball®

1. From the Options menu, choose Open Crystal Ball.

If Crystal Ball® has been installed, it is opened and the TITAN functionality will be closed during the analysis.

To initialise Crystal Ball®

If you are running Crystal Ball® for the first time on the computer, you must define TITAN macros that are run during each simulation cycle. In order to do this, Crystal Ball® has to be running.

1. Select *Run Preferences...* command in Run menu.
2. Click the *User Macros...* button.(if the User *Macros...* button is not visible then click the *More>>* button first).
   A dialogue box appears for definition of required macros as shown in Fig 5.

In the *Before Recalculations* box type:

'[titan.xlw]titan.xlm'!CalcDocument

(Note that in the example in Fig 5 you do not see the complete macro definition because the field size shown is too short.) If you are using internal rate of return as a value to be forecasted then type in the *After Recalculations* box:

'[titan.xlw]titan.xlm'!CalcIRR

3. Click the OK button in this and in previous dialogue box.

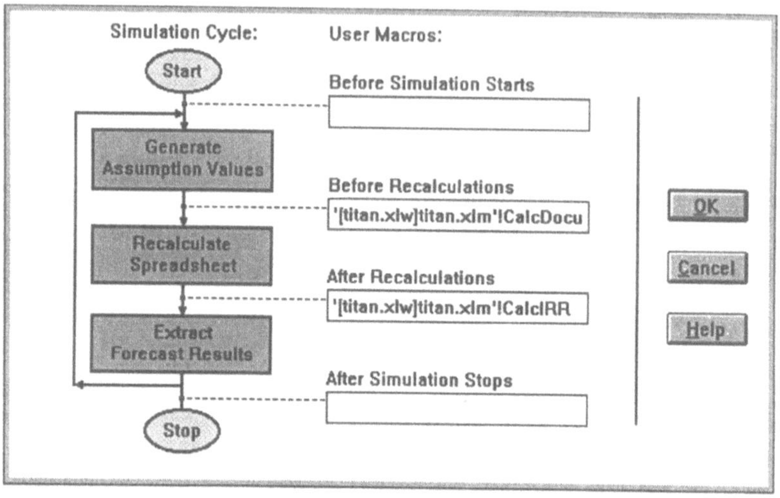

Fig 5 The Crystal Ball® window.

To perform analysis

1. Select the cell that represent uncertain variables in the model or the calculation .
2. From the Cell menu, choose Define Assumption.
3. Specify distribution type and corresponding attributes. For more information on defining assumptions, see Crystal Ball® help.
4. Repeat steps 1-3 for all assumption variable cells you require.
5. Select the cell to be forecasted.
6. From the Cell menu, choose Define Forecast.
7. Repeat the steps 4 and 5 for each cell required to be forecasted.
8. From the Run menu, choose Run.

The simulation calculates the TITAN main sheet multiple times and changes the values of the assumption cells before each calculation cycle randomly, according to the selected probability distributions. The results in the forecast cells are gathered for analysis, and reports can be generated. For more information on performing risk analysis, see Crystal Ball® help.

To close Crystal Ball®

From the Run menu, choose Close Crystal Ball. Crystal Ball® closes and TITAN functionality is re-established.

Definitions 4

A - C

Architecture. The architecture defines what kind of materials and equipment are used in the access network, and the overall geographical layout of the network, i.e. what kind of splitting ratios are used, etc. The architectures used in TITAN are based on generalised access network models.

Cash balance (cumulative cash flow). The Cash balance is equal to the cumulative sum of the retained cash flows from the start of the model run to the current year, plus the cash holdings at the beginning of the model run.

Cash flow (earnings). The cash flow is equal to the revenue minus the sum of running costs and investments.

Component price versus time. Usually, the cost trends are described by the *learning curve* relation according to Wrights' law:

$$P_{2n} = K \cdot P_n$$

where P_{2n} and P_n are the prices for the completion of the second and n^{th} units respectively. The clustering shown in Table 1 was established according to the K factor, and subsequently used in the tool.

Table 1 The cost component classes

Component class		*Learning curve (K)*	*Comments*
I.	Optical cables	85/75	
II.	Optical passive comp.	75/70	Other than cables
III.	Transmitters and receivers	80/73	
IV.	Electronics	80/73	
V.	Enclosures and cabinets	95/90	Including powering
VI.	Installation	105/95	
VII.	Civil works	105/95	
VIII.	Copper cables	103/98	Twisted pair and coax

Time dependence of the component price

The logistic formula for the accumulated component volume is:

$$N(t) = M \cdot \left[1 + e^{(a - bt)}\right]^{-1}$$

This yields:

$$a = \ln\left(\frac{M}{N(0)} - 1\right)$$

Let the normalised volume be: n(t)=N(t)/M; then,

$$a = \ln\left(\frac{1}{n(0)} - 1\right)$$

and the normalised formula becomes:

$$n(t) = \left\{1 + \exp\left[\ln\left(\frac{1}{n(0)} - 1\right) - bt\right]\right\}^{-1} \tag{A1}$$

Solving the above formula for $n(t_1) = 0.1$ and $n(t_2) = 0.9$ gives:

$$\Delta T = t_2 - t_1 = \frac{2 \cdot \ln(9)}{b}$$

or:

$$b = \frac{2 \cdot \ln(9)}{\Delta T} \approx \frac{4.4}{\Delta T}$$

Substituting this into (A1) gives the final form of the normalised accumulated volume forecast:

$$n(t) = \left\{1 + \exp\left[\ln\left(\frac{1}{n(0)} - 1\right) - \frac{2 \cdot \ln(9)}{\Delta T} \cdot t\right]\right\}^{-1} \tag{A2}$$

The above expression is now inserted into a learning curve of the form: $P_n = P_0 \cdot n^{-\alpha}$ for $n(0)$ and $n(t)$:

$$\frac{P_{n(t)}}{P_{n(0)}} = \left[\frac{n(t)}{n(0)}\right]^{-\alpha} \tag{A3}$$

Since $\alpha = -\log_2(K)$, we finally have:

$$P(t) = P(0) \cdot \left[\frac{n(t)}{n(0)} \right]^{\log_2(K)} \tag{A4}$$

$P(0)$ is the unit price at time t, whereas $P_{n(0)}$ is the price of the first component in a series.

The time equation

From the previous calculation, we conclude that every component can be classified by four parameters. These are the component price today, $P(0)$, the learning curve coefficient, K, the time it takes for the accumulated volume curve to increase from 10 % to 90 % of the saturation value, ΔT, and the value of the accumulated volume today, $n(0)$, normalised to the saturation value. The last parameter is the inverse of the assumed marked potential. ΔT and $n(0)$ are the only two parameters that need to be estimated by forecast methods. The forecast function for the evolution of the relative accumulated volume $n(t)$ is given by:

$$n(t) = \left\{ 1 + \exp\left[\ln\left(\frac{1}{n(0)} - 1 \right) - \frac{2 \cdot \ln(9)}{\Delta T} \cdot t \right] \right\}^{-1}$$

where the only inputs are $n(0)$ and ΔT. The previous relation is now inserted into a learning curve:

$$P(t) = P(0) \cdot \left[\frac{n(t)}{n(0)} \right]^{\log_2(K)}$$

This yields the final expression of the price of the component versus time:

$$P(t) = P(0) \cdot \left[\frac{1}{n(0)} \cdot \left\{ 1 + \exp\left[\ln\left(\frac{1}{n(0)} - 1 \right) - \frac{2 \cdot \ln(9)}{\Delta T} \cdot t \right] \right\}^{-1} \right]^{\log_2(K)}$$

which contains all the four parameters involved in the component price versus time evolution. The component price versus time is described in more detail in chapter 8.

Cost component. Network equipment, material or civil works that generates cost when added to the network.

Cost database file (.XLD). The cost database file contains information on the network equipment and materials costs (cost component name, cost value, OAM class) and the classes defining the price evolution of the components (learning curve class, volume class, write off period class). Only one cost database file may be open at a time, even though multiple databases may be created.

Cost database mode. The cost database mode is used for changing and operating with the cost database. The mode of operation can be changed to cost database mode with the Mode/Cost Database menu command.

In the Cost Database mode the cost database file is open and worked with. Cost Components and their information may be added, edited or removed. Specific components may be searched and the database can be sorted with specified keys.

Read the Using the Cost Database Mode topics to get information on the necessary procedures for the use of the cost database mode, or choose one of the reference topics to get specific help on a menu command or a section of the main sheet.

Crystal Ball©. Crystal Ball© is an Excel add-in program for performing forecasting and risk analysis. Crystal Ball© uses Monte Carlo or Latin Hypercube simulation for forecasting the resulting probability distributions.

Please note! Crystal Ball© is not automatically delivered with TITAN. Hence, if one wants to use it, it has to be separately purchased from Decisioneering Inc.

D

Decisioneering, Inc.. At the time of production of this manual the contact information of Decisioneering Inc. is:

1515 Arapahoe Street, Suite 1311
Denver, CO 80202
USA
tel. +1-303-534-1515
fax. +1-303-534-4818
http://www.decisioneering.com

Declining balance method (DB). A geometrically declining depreciation method in which the depreciation of each year is a fixed percentage of the rest value of the previous year. The depreciation factor describes the rate at which the balance is to be declined. E.g. if the factor is 1, then the annual depreciation is 10 % of the previous rest value. If the factor is 2 (double declined) then the percentage is 20.

Depreciation. Depreciation is a financial measure and represents the money put aside out of profits to pay for the replacement of worn-out equipment in the future. Its precise level depends upon the accounting method chosen. The available accounting methods in the tool are:

- Declining balance method
- Straight line depreciation
- Sum-of-the-years'-digits method

Discount formula. A discount formula is used in order to be enable comparison of cash flows of different years. First a discount rate, a, must be defined. Then the cash flows, C, of year t can be discounted to a cash flow at year 0, C_0, with a formula:

$$C_0 = \frac{C}{(1+a)^t}$$

Discounted costs. The discounted costs of the components are calculated by first calculating the costs of the components during time using the price formula and then discounting these investments with the discount formula.

I - L

Interest payments. The net interest payments to investors depend on the operator company profitability and are set by the operator itself.

Internal rate of return (IRR). Internal rate of return of the investment is the discount rate for which the net present value of the project is equal to zero.

Investment. The investment is the sum of expenditure on equipment to meet new demand and expenditure to replace worn out equipment.

Learning curve. The learning curve describes the price evolution of a product unit. First, the production starts with low volumes and no experience causing high costs and therefore high price. As more products are produced and learning has taken place, the production becomes more efficient resulting in lower costs and lower prices.

Usually, the cost trends are described using the learning curve relation according to Wrights' law:

$$P_{2n} = K \cdot P_n$$

where P_{2n} and P_n are the prices for the completion of the second and n[th] product units respectively. The K factor describes the speed of change. Combined with the volume equation the learning curve results in a price formula.

Life cycle costs. The life cycle costs describe the total cost of the project through it's life time. It equals the sum of accumulated discounted investment and accumulated discounted running costs.

Log. The log contains information on the tool sessions. The user may save input values, charts of the financial results or specific parts of the main sheets to the log. The log file may be saved and edited.

M - O

Main File (.XLS). The main worksheet file contains a description of a complete specific environment to be studied. It includes the architecture, cost calculations and financial results. Multiple main files may be open at the same time.

Main Mode. The main mode is used for TITAN calculations. The mode of operation can be changed to main mode with the Mode/Main menu command.

In the main mode the main worksheet file is open and worked with. The general inputs, geometric model, shopping list of cost components and revenue components need to be specified in order to calculate the economic results. Then, reports including graphics can be generated from the results.

Read the Using the Main Mode topics to get information on the necessary procedures for the use of the main mode or choose one of the reference topics to get specific help on a menu command or a section of the main sheet.

Modified SYNTHESYS model. The modified SYNTHESYS model is used for calculation of straight distances in the access network. The model has a feeder part of the network added to the original SYNTHESYS model. The feeder network includes two flexibility points and link levels. The lengths of the feeder cables are case dependent and therefore input separately.

Net present value of the project (NPV). The net present value equals the sum of all discounted retained cash flows and the discounted rest value of the investment.

OAM class. The operation, administration and maintenance (OAM) class defines the percentage of global investment that are the OAM costs of a cost item belonging to this class.

OAM costs. The operation, administration and maintenance costs include preventive maintenance, network management and use of installed plant and repair of defective equipment.

The analytical calculation methods of OAM costs are time consuming and are accurate only for a very specific network. Therefore TITAN has adopted a "proportional" approach to OAM costs that relates the costs to the annual capital investment. The justification of this approach is based on the fact that capital can usually be related in some way to complexity, functionality, size, power and construction of hardware.

The costs are calculated as a percentage of the global investment each year and then discounted with the discount formula. Alternatively, OAM costs may be included using the OAM plane.

OAM plane. The operation, administration and maintenance plane is an alternative way to include OAM costs.

If you use this plane to input OAM costs they should be fully depreciated in one year, hence leaving zero rest value. "Additional investment" of these components is added to the OAM costs instead of the investment costs. Other values in this plane have no impact on calculations.

P - R

Payback period. The payback period is the time from project start to the point in time when the revenues from the investment reach the investment and running costs, in other words when the cash balance turns positive.

If payback is not reached within the study period, TITAN calculates the slope of the cash balance curve during the last year and extrapolates payback period. If the slope is negative, 99 is returned as payback period.

Profit. The profit is equal to revenue minus the sum of running costs and depreciation.

Retained cash flow (retained earnings). The retained cash flow is equal to the revenue minus the sum of running costs, investment, net interest payments and taxes.

Retained profit. The retained profit is equal to the revenue minus the sum of running costs, depreciation, net interest payments and taxes.

Revenue. The revenue is the sum of the income or turnover earned by the operator on services.

Running costs. The running costs are the sum of operations, maintenance, administration and churn costs

S - T

Service. The network user services generate revenues for the access network operator. The network service might be telephone services like plain old telephone service (POTS) or integrated services digital network (ISDN).

Straight line depreciation (SLN). A depreciation method where the depreciation is evenly distributed over the years of the write-off period (i.e. life of the asset). E.g. if the write-off period is five years, then depreciation in each year is 1/5 of the total depreciation.

Sum-of-the-years'-digits method (SYD). A depreciation method where the part of the total depreciation belonging to each year of the write-off period (n years) can be calculated:

$$\frac{R}{n \cdot \frac{(n+1)}{2}} = \frac{2R}{n(n+1)}$$

where R is the number of years left in the write-off period including the year to be calculated. E.g., if the write-off period is five years then the depreciation of the first year is 5 / [5·(5+1)/2] = 5/15 of the total depreciation (second year 4/15, third year 3/15, fourth year 2/15, fifth year 1/15).

SYNTHESYS model. The SYNTHESYS model is used for calculation of straight distances in the access network. Fig 6 shows the model and illustrates how it relates to the generalised access network model described in the architectures section. For example first level links consist of the SYNTHESYS links A-B, B-C, C-D and C- E. The second level links are SYNTHESYS links D-F. The third level links are the links between the F points and the subscriber premises.

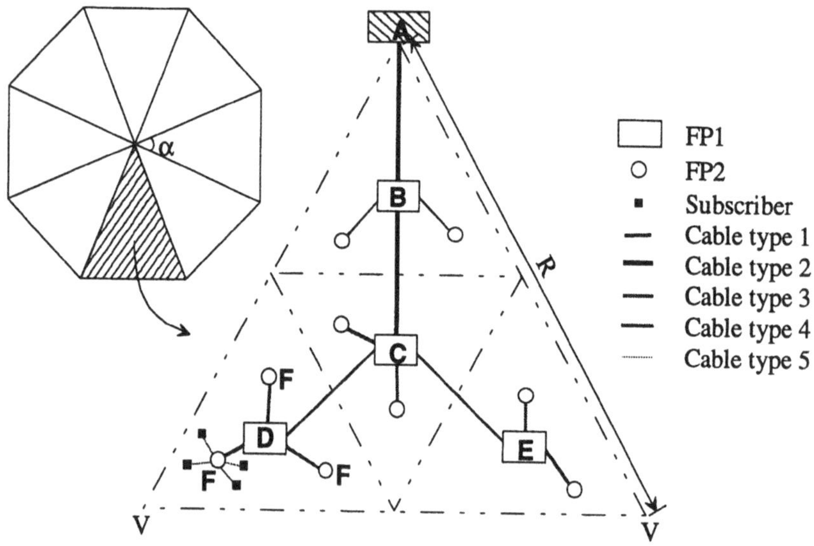

Fig 6 The SYNTHESYS model.

System mode. The third operation mode of TITAN. In system mode you are actually in standard Microsoft Excel environment with related TITAN files open. You can switch to system mode by pressing CTRL+SHIFT+s during a TITAN session. Back to main mode or database mode you get by pressing CTRL+SHIFT+n.

Warning! Use system mode with caution. Because all Microsoft Excel commands are available, it is easy to corrupt your TITAN files. You should use system mode only for adding link levels.

Taxes. The sum of taxes to be paid is equal to the tax rate percentage of the profit.

Template file (.XLT). A template file is used by the TITAN macro file as a base sheet for new main and cost database sheets. The macro file loads a template sheet when the user selects the *File/New* command. The user does not have to modify these files directly in order to use the tool.

TITAN. Tool for introduction strategies and techno-economic evaluation of access network.

U - Z

Utility file (.XLU). Utility files are used by the TITAN macro file as a workspace for internal calculations. The user does not have to modify these files in order to use the tool.

Volume equation. Describes how the production volume of a product develops during time. The logistic formula for the accumulated component volume *N(t)* is:

$$N(t) = \frac{M}{1 + e^{(a - bt)}}$$

where t is the time from the start of the production, M is the saturation value of the accumulated volume (i.e. the maximum world-wide production volume of the specified component) and a and b depend on the specific case. Combined with a learning curve the volume equation results in a price formula.

Workbook file (TITAN.XLW). The TITAN.XLW workbook file adds the functionality of the TITAN methodology to an Excel spreadsheet using macros. The workbook file contains a macro sheet and utility sheets. The macros customise the Excel environment to simplify the use of the tool. This is done by adding new menus and customised commands to the Excel environment. Macros also link the main files to the cost database file. The user does not have to modify this file in order to use the tool.

Write-off period. Defines the time in years used for the financial write-off of the component.

Index

Kolibri 360 GmbH
Ferdinand-Jühlke-Straße 7
99095 Erfurt, Deutschland
produktsicherheit@kolibri360.de